Antioxidant Biochemistry

Antioxidant Biochemistry

Edited by **Nick Gilmour**

SYRAWOOD
PUBLISHING HOUSE
New York

Published by Syrawood Publishing House,
750 Third Avenue, 9th Floor,
New York, NY 10017, USA
www.syrawoodpublishinghouse.com

Antioxidant Biochemistry
Edited by Nick Gilmour

International Standard Book Number: 978-1-68286-200-1(Hardback)

Printed in the United States of America.

Contents

Preface

This book has been an outcome of determined endeavour from a group of educationists in the field. The primary objective was to involve a broad spectrum of professionals from diverse cultural background involved in the field for developing new researches. The book not only targets students but also scholars pursuing higher research for further enhancement of the theoretical and practical applications of the subject.

There has been rapid progress in the field of antioxidant biochemistry and its applications are finding their way across multiple industries. This book is an assimilation of concepts and relevant topics such as antioxidant activity of different substances, oxidative stress, natural and synthetic antioxidants, dietary antioxidants, etc. It provides the information needed to efficiently translate new research findings and clinical experiences into novel applications. The book is an excellent source of reference for the students and researchers engaged in this field.

It was an honour to edit such a profound book and also a challenging task to compile and examine all the relevant data for accuracy and originality. I wish to acknowledge the efforts of the contributors for submitting such brilliant and diverse chapters in the field and for endlessly working for the completion of the book. Last, but not the least; I thank my family for being a constant source of support in all my research endeavours.

Editor

Avocado Seeds: Extraction Optimization and Possible Use as Antioxidant in Food

Francisco Segovia Gómez [1,2], Sara Peiró Sánchez [1], Maria Gabriela Gallego Iradi [1], Nurul Aini Mohd Azman [1] and María Pilar Almajano [1,]*

[1] Chemical Engineering Department, Technical University of Catalonia, Avda. Diagonal 647, 08028 Barcelona, Spain; E-Mails: segoviafj@gmail.com (F.S.G.); sapeisa@yahoo.es (S.P.S.); maria.gabriela.gallego@upc.edu (M.G.G.I.); aini.azman@gmail.com (N.A.M.A.)

[2] Chemical Engineering Department, Antonio José de Sucre National Experimental Polytechnic University, Avenida Corpahuaico, 3001 Barquisimeto, Venezuela

* Author to whom correspondence should be addressed; E-Mail: m.pilar.almajano@upc.edu

---

Abstract: Consumption of avocado (*Persea americana* Mill) has increased worldwide in recent years. Part of this food (skin and seed) is lost during processing. However, a high proportion of bioactive substances, such as polyphenols, remain in this residue. The primary objective of this study was to model the extraction of polyphenols from the avocado pits. In addition, a further objective was to use the extract obtained to evaluate the protective power against oxidation in food systems, as for instance oil in water emulsions and meat products. Moreover, the possible synergy between the extracts and egg albumin in the emulsions is discussed. In Response Surface Method (RSM), the variables used are: temperature, time and ethanol concentration. The results are the total polyphenols content (TPC) and the antiradical power measured by Oxygen Radical Antioxidant Capacity (ORAC). In emulsions, the primary oxidation, by Peroxide Value and in fat meat the secondary oxidation, by TBARS (Thiobarbituric acid reactive substances), were analyzed. The RSM model has an R^2 of 94.69 for TPC and 96.7 for ORAC. In emulsions, the inhibition of the oxidation is about 30% for pure extracts and 60% for the combination of extracts with egg albumin. In the meat burger oxidation, the formation of TBARS is avoided by 90%.

Keywords: RSM; avocado pit; ORAC; extraction; emulsion; oxidation; meat

1. Introduction

Vegetables and fruits are essential foods in our diet and also have many compounds that are beneficial for health due to minor components. These minor components include phenolic substances [1]. These are secondary metabolites of plants. They have an aromatic ring with one or more hydroxyl groups. Their complexity may be high, as for example quercetin, which is one flavone with several aromatic rings. The properties depend on the arrangement and/or structure of the molecule [2].

In recent times, many plants have been studied in order to characterize them depending on the amount of polyphenols they have and on their potential use [3].

The polyphenols are associated with the potential prevention of diseases which are due to the presence of free radicals, such as cardiovascular insufficiency, hypertension, inflammatory conditions, asthma, diabetes and Alzheimer's [4], thanks to their antiradical power. For this reason, they are very useful in food products, since they prevent lipid peroxidation due to the attack of free radicals [5]. They also protect against oxidation, direct or indirect, caused by metal cations [6]. These cations stimulate the creation of reactive oxygen species (ROS), which are harmful to the health. In some cases, polyphenols have been used as preservatives, protecting against microorganisms [7].

The process of food, especially for IV and V gamma products, produces many byproducts and waste. This type of waste has a significant environmental impact due to the organic charge. It also has associated handling, transport and storage costs, among others. Therefore, more and more alternative uses for these residues are sought, as for instance animal feed and fertilizers, among others. In the present case, it is interesting to obtain, through an optimized extraction process, harmless substances with high antioxidant power. Thus, what was a waste becomes a "high value-added" product [8,9]. Previous examples already studied [10–12] are the orange juice industry, where a large amount of skin and seeds are produced with a high content of polyphenols and the industry of processed apple, pear and peach, with a significant amount of skin byproduct. There is evidence that the skin may even have a greater amount of polyphenols than the flesh [13]. Also, the waste from wine and beer production includes phenolic compounds [9]. Other studies have focused on the shells of nuts, rice and wheat in which large amounts of polyphenols are found [14].

In the avocado industry the pulp is used, while the skin and the seeds are discarded as waste. These residues are rich in polyphenols with antioxidant and antimicrobial power [15]. Among the polyphenols the (+)-catechin and (−)-epicatechin [16] and chlorogenic and protocatechuic acid, are included [14]. Previous studies on this residue have been applied to pork burgers and have been shown to be effective in preventing oxidation and microbial growth [15].

Given the above, it can be concluded that polyphenols obtained from these industrial wastes can be potent antioxidants and, in some cases, they are better than synthetic antioxidants such as BHA or BHT which in high doses can become toxic [17].

In order to optimize the extraction process, response surface methodology (RSM) has been used. Phenolic compounds extraction optimization from strawberries [18], apple pulp [9] and residues of chestnuts [19], are examples of this. This method establishes a multivariable mathematic model to obtain the relationship between responses and independent variables [20,21] with the use of a minimal number of experiments.

This paper consists of two main objectives. First, a mathematical model was obtained to predict the best conditions of extraction of polyphenols from dried avocado seed. Second, an extract using these conditions was obtained and the effect of lyophilized powder in the delay oxidation in oil-in-water (O/W) emulsions and beef meat burgers analyzed.

2. Experimental Section

2.1. Materials

2,2′-Azo-bis(2-amidinopropane) dihydrochloride (AAPH), was used as peroxyl radical source. Trolox (6-hydroxy-2,5,8-tetramethylchroman-2-carboxylic acid), ethanol, fluorescein, AAPH, BHA, egg albumin, *p*-anisidine (4-amino-anisole; 4-methoxy-aniline), isooctane, potassium persulfate, acetic acid (glacial) and 2-thiobarbituric acid were purchased from Sigma-Aldrich Company Ltd. (Gillingham, UK). Folin-Ciocalteu reagent, sodium carbonate and 1,6-diaminohexane were supplied by Merck (Darmstadt, Germany). Trichloroacetic acid, hydrochloric acid and Tween® 20 (Panreac Química S.L.U, Barcelona, Spain) were acquired from Panreac Química S.L.U. (Barcelona, Spain). Refined sunflower oil, with no added antioxidants, was purchased from a local retail outlet. All compounds were of reagent grade.

2.2. Avocado Preparation

The avocado (*Persea americana*) was obtained in the local market; the seeds were separated from other edible parts. This waste was homogenized and frozen at −80 °C for lyophilization. Then the seeds were ground into a powder by using a Moulinex mill (A5052HF, Moulinex, Lyon, France). The particle size was standardized with a number 40 mesh sieve. Finally, the powder was stored in a dark bottle in a desiccator until use.

2.3. Extraction Procedure

Extraction was carried out in dark bottles: lyophilized sample powder (0.25 g) was blended with 15 mL of solvent of concentration specified by the experimental design (Table 1). This mixture was placed in a bath by stirring at the required temperature and time specified by the experimental design (Table 1). At the end, it was cooled in a refrigerator at 5 °C, centrifuged (Orto Alresa, Madrid, Spain) at 2500 rpm for 10 min, vacuum filtered and the lost solvent was replaced. The extract was stored at −20 °C until used for analysis.

2.4. Total Phenolic Content (TPC)

TPC was determined spectrophotometrically following the Folin-Ciocalteu colorimetric method [22]. A sample diluted 1:4 with milli-Q water was stirred in triplicate. The final concentration in the well (96 wells plate was used) was: 7.7% v/v sample, 4% v/v Folin-Ciocalteu's reagent, 4% saturated sodium carbonate solution and 84.3% of milli-Q water, all mixed. The solution was allowed to react for 1 h in the dark and the absorbance was measured at 765 nm using a Fluostar Omega (BMG Labtech, Ortenberg, Germany). The total phenolic content was expressed as mg Gallic Acid Equivalents (GAE)/g dry weight.

Table 1. Experimental design and responses for extraction.

Temperature (°C)	Ethanol Concentration (%)	Time (min)	TPC (mg GAE/g dw)	ORAC (mg TE/g dw)
60.00	60.00	25.00	41.00 ± 0.97	104.16 ± 2.13
60.00	93.63	25.00	35.10 ± 0.24	116.12 ± 1.03
80.00	80.00	5.00	46.78 ± 0.59	153.17 ± 3.84
26.36	60.00	25.00	40.78 ± 0.17	70.54 ± 0.97
60.00	60.00	25.00	41.10 ± 0.57	106.10 ± 2.40
40.00	40.00	45.00	43.24 ± 0.76	104.01 ± 2.35
80.00	80.00	45.00	45.43 ± 0.49	144.94 ± 2.84
80.00	40.00	45.00	45.37 ± 1.39	130.08 ± 2.65
80.00	40.00	5.00	43.70 ± 0.66	150.03 ± 1.73
60.00	60.00	25.00	40.90 ± 0.47	104.28 ± 1.03
60.00	60.00	55.22	42.87 ± 0.70	158.77 ± 1.33
40.00	40.00	5.00	41.19 ± 0.55	99.17 ± 1.81
60.00	60.00	2.77	42.92 ± 1.13	155.44 ± 2.71
93.64	60.00	25.00	46.95 ± 0.09	126.23 ± 3.35
60.00	26.36	25.00	42.33 ± 0.10	129.78 ± 3.84
40.00	80.00	45.00	38.98 ± 0.45	100.72 ± 3.27
40.00	80.00	5.00	35.48 ± 0.55	91.01 ± 3.51

GAE: Galic Acid Equivalents; TE: Trolox Equivalents; TPC: Total Phenolic Content; ORAC: Oxygen Radical Antioxidant Capacity.

2.5. ORAC Assay

Antioxidant activities of avocado seeds extracts were determined by the ORAC assay, as reported by Casettari *et al.* [23]. The assay was carried out using a Fluostar Omega equipped with a temperature-controlled incubation chamber. The incubator temperature was set to 37 °C. The extract samples were diluted 1:20 with milli-Q water. The assay was performed as follows: 20% of sample was mixed with Fluorescein 0.01 mM, and an initial reading was taken with excitation wavelength, 485 nm and emission wavelength, 520 nm. Then, AAPH (0.3 M) was added, measurements were continued for 2 h every 5 min. This method includes the time and decrease of fluorescence. The area under the curve (AUC) was calculated. A calibration curve was made each time with the standard Trolox (500, 400, 250, 200, 100, 50 mM). The blank was 0.01 M phosphate buffered saline (pH 7.4). ORAC values were expressed as mg Trolox Equivalents (TE)/g of dry weight.

2.6. Statistical Analysis

RSM was used to determine the optimal conditions of polyphenol extraction. A central composite design (CCD) was used to investigate the effects of three independent variables with two levels (solvent concentration, extraction temperature, and extraction time) with the dependent variables (TPC, ORAC activity). CCD uses the method of least-squares regression to fit the data to a quadratic model.

The adequacy of the model was determined by evaluating the lack of fit, coefficient of determination (R^2) obtained from the analysis of variance (ANOVA) that was generated by the software. Statistical significance of the model and model variables were determined at the 5% probability level ($\alpha = 0.05$). The software uses the quadratic model equation shown above to build response surfaces. Three-dimensional response surface plots and contour plots were generated by keeping one response variable at its optimal level and plotting that against two factors (independent variables). Response surface plots were determined for each response variable. The coded values of the experimental factors and factor levels used in the response surface analysis are shown in Table 1. The graphics and the RSM analysis were made by software Matlab version R2013b (The MathWorks Inc., Natick, MA, USA, 2013). All responses were determined in triplicate and are expressed as average ± standard deviation. The answers have a percentage deviation less than 10%.

2.7. Water-Oil Emulsions

Oil-in-water emulsions (20.2 g) were prepared by dissolving Tween-20 (1%) in acetate buffer (0.1 M, pH 5.4), either with or without protein, namely egg albumin (0.2% w/w) and avocado seeds extracts (0.45% w/w, 0.225% w/w, 0.1125% w/w). The emulsion was prepared by the dropwise addition of oil (sunflower oil) to the water phase, cooling it in an ice bath with continuous sonication with a Vibracell sonicator (Sonics and Materials, Newtown, CT, USA) for 4 min. All emulsions were stored in triplicate in 60 mL glass beakers in the dark (inside an oven) at 30 °C in an incubator. Two aliquots of each emulsion (0.005–0.1 g, depending on the extent of oxidation) were removed periodically for determination of peroxide value (PV).

2.8. Peroxide Value (PV)

PV was determined by the ferric thiocyanate method [24] (after calibrating the procedure with a series of oxidized oil samples analyzed using the AOCS Official Method Cd 8-53). Data from the PV measurements were plotted against time.

2.9. Meat Preparation

Fresh beef meat was purchased from a local processor 96 h postmortem. All subcutaneous and inter-muscular fat and visible connective tissue were removed from the fresh beef muscle. Lean meat was ground through Ø-4 mm plate using a meat grinder (PM-70, Mainca, Barcelona, Spain). The ground meat was divided into six portions for each experiment prior to the addition of the sodium chloride or different concentration of powder (freeze-dried extract of powder of avocado). The lyophilized avocado and the powder of direct avocado were mixed with the salt final concentration of 1.5% (w/w). Each portion of beef meat was mixed manually with each solid. Each mixed sample was divided into nine smaller portions (about 10 g each) and allocated onto trays. The meat was packed under MAP (20% CO_2 and 80% O_2) in polystyrene/EVOH/polyethylene trays, heat sealed with laminated barrier film and stored at 4 ± 1 °C for 8 days. Patties were evaluated for lipid oxidation.

2.10. Thiobarbituric Reactive Substances

Fat meat oxidation was determined by the concentration of thiobarbituric acid-reactive substances (TBARS) using the method described by Domenech Asensi (2013) [25] with some modifications. In the dark, 1 g of burger patty was dispensed in tubes and 1 mL of EDTA was added. The samples were homogenized for 5 min sin an Ultra-Turrax (Ika®-Werke, Staufen, Germany) with 5 mL of TBARS reactive (Trichloroacetic acid, 9.2%; Hydrochloric acid, 2%; Thiobarbituric acid, 0.22%, all w/w final). During homogenization, the tubes were placed in an ice bath to minimize the development of oxidative reactions. The sample tubes were heated at 90 °C in a boiling water bath for 20 min and then left to cool. Two milliliters of slurry was centrifuged (10,000 rpm for 10 min). The absorbance was measured at 531 nm in a Spectrophotometer Zuzi model 4201/20 (AUXILAB, SL, Navarra, Spain). The result is expressed in mg of MDA/kg sample.

3. Results and Discussion

3.1. Extraction Optimization

Experimental design was carried out to see the effects of temperature, solvent concentration (ethanol) and time in both TPC and radical scavenging (measured by ORAC). Several authors used ethanol/water as solvent to extract different raw material polyphenols, such as seeds, grape marc, fruits, among others [26–30]. Ethanol concentration with the highest polyphenols yield is in the range of 10%–60%. Ethanol, instead of methanol, is used when it is necessary to reduce the toxicity of extracts [18]. The time effect was measured between 5 and 45 min, because some research reported that it is enough to achieve the maximum amount of polyphenols [31,32]. Temperature bounds were taken between 40 and 80 °C, to achieve the maximum temperature that does not have a negative effect on the polyphenols stability [33]. All these parameters are collected in Table 1 which shows the experimental design for the variables temperature (T), ethanol concentration (% EtOH) and time (*t*), with responses of TPC and antiradical activity measured by ORAC.

Figure 1 shows the relationship between the variables *T*, % EtOH and t in polyphenol extraction. The process is favored by high temperatures and low concentrations of ethanol (in the studied range). This behavior can be attributed to the nature of the polyphenols present in the sample, mainly chlorogenic acid and protocatechuic acid [34] both highly soluble in water. The solvent plays an important role in mass transfer of the compounds; not all polyphenols show identical behavior in the extraction process, and the less polar polyphenols are favored by the highest concentration of ethanol [9].

The effect of temperature on the extraction is associated with the solubility of the components present in the avocado pit. This variable, *T*, has a marked influence on the diffusivity of the substances [30]. Solubility increases with temperature. Time has no influence in the extraction process. This means that from the beginning, the extraction is governed by the solubility and diffusion, and both are almost complete after 5 min.

Figure 2 shows the effect of the parameters on the antioxidant power measured by ORAC. The ORAC increases with temperature. In the investigated range, the ORAC is increased about

44% (Table 1). Furthermore, as stated above, it is in accordance with the higher polyphenols solubility at high temperature. This means that these kinds of polyphenols are thermo-resistant [20].

The effect observed for the percentage of ethanol is similar to that described in the TPC. An increase in the ethanol concentration causes a decrease in antioxidant activity. It is not a new fact, because similar results were described in other studies and were justified by the polarity of the compounds of the extract [18].

Figure 1. Response surface model plot: the variable is the total phenolic content (TPC) of the extract. % EtOH with temperature; temperature with time; % EtOH with time.

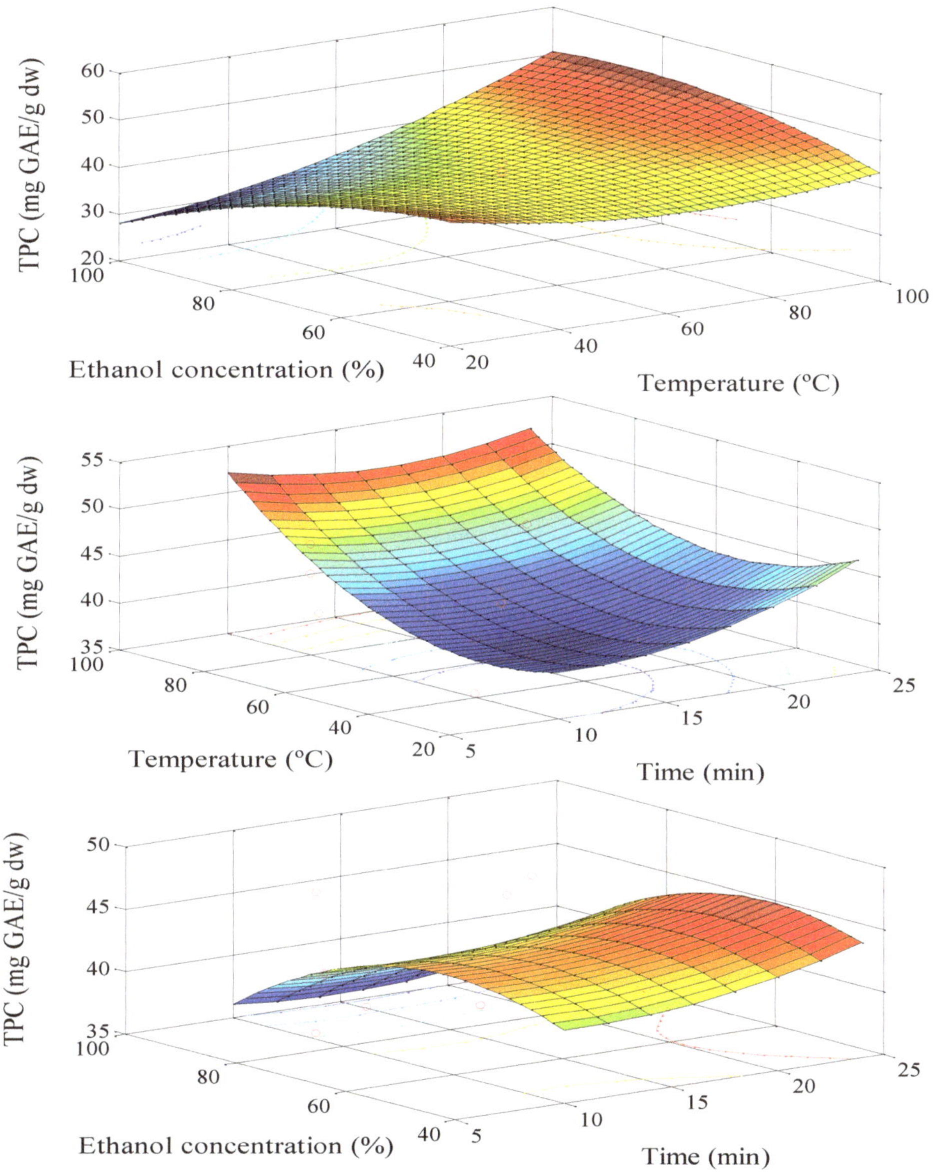

On TPC the variable t has no influence, while on ORAC small changes were observed, but all of them with similar final values. One possible explanation is that there are antioxidant compounds with slow solubilization and, therefore, the time promotes an increase in total extraction [35].

Table 2 shows the "*p* values" of the mathematical model for the coefficients, with the decoded variables. It starts with the complete model, taking the variables that have less influence, *i.e.*, with

$p > 0.05$. For TPC all those that are with % EtOH and t are involved. This means that the more important variable is T. However, on the ORAC, the variables that have more influence are % EtOH, *t*, and these quadratic terms. From the data, different iterations were made and less influential terms were eliminated; after which the values were recalculated. With these data the reduced model was obtained and provided a better fit. In ORAC the predicted R^2 becomes 77.88 which is within the range of a good set [36].

Figure 2. Response surface model plot: the variable is the Oxygen Radical Antioxidant Capacity (ORAC) of the extract. Temperature with time; % EtOH with temperature; % EtOH with time.

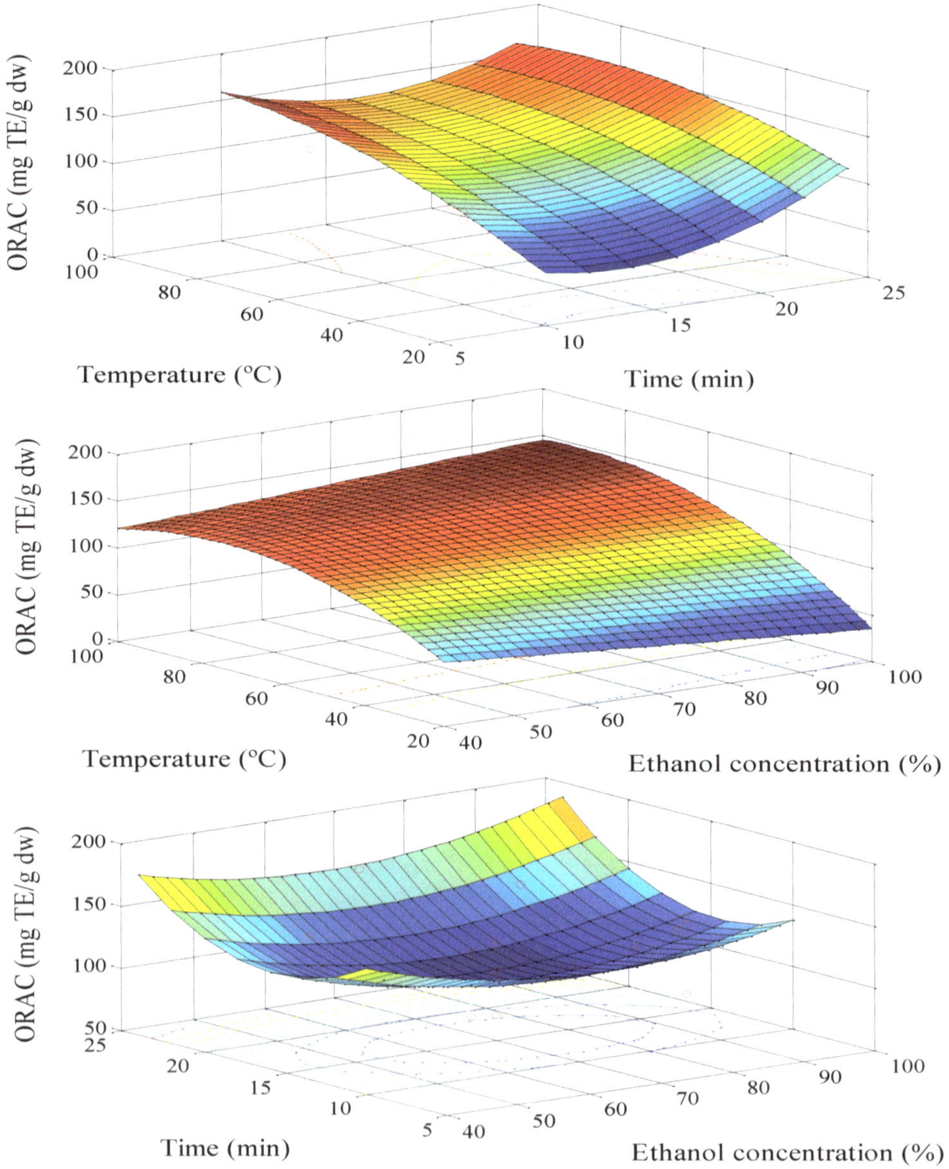

Therefore, with the exception of $T \times$ % EtOH for the TPC, all of the crossed terms disappear in the reduced model (which is used to adjust and to determine the optimal extraction conditions).

Additionally, the quadratic variables % EtOH × % EtOH and $t \times t$, as well as the linear variable t are eliminated for the TPC. The quadratic term $T \times T$ is eliminated from the model which determines the scavenging activity. This is summarized in Table 2.

Table 3 lists the completed model and the reduced model equations. The reduced model has a higher R^2 predicted which means that it is more reliable in estimating a response.

When the fitting was considered good enough, the experiment was performed in the laboratory to obtain the real value. Table 4 contains these values for the TPC and for the ORAC. The TPC is fitted with less than a 4% error (the predicted value is 43.6 mg GAE/g dw, compared to an experimental value of 45.01 mg GAE/g dw). This indicates that the initial hypothesis was correct, and demonstrates that T is the variable with the greatest influence on the maximum TPC extraction.

Table 2. *p*-Values for each of the constants in the equation of the mathematical model.

Term	*p*-Value	
	Response	
	TPC	ORAC
Complete Model		
Constant	0.001	0.006
Temperature (°C)	0.012	0.069
Ethanol (%)	0.291	0.022
Time (min)	0.804	0.001
Temperature (°C) × Temperature (°C)	0.014	0.135
Ethanol (%) × Ethanol (%)	0.622	0.046
Time (min) × Time (min)	0.068	0.000
Temperature (°C) × Ethanol (%)	0.003	0.186
Temperature (°C) × Time (min)	0.119	0.071
Ethanol (%) × Time (min)	0.610	0.435
Reduced Model		
Constant	0.000	0.000
Temperature (°C)	0.005	0.000
Ethanol (%)	0.001	0.031
Time (min)	-	0.000
Temperature (°C) × Temperature (°C)	0.029	-
Ethanol (%) × Ethanol (%)	-	0.033
Time (min) × Time (min)	-	0.000
Temperature (°C) × Ethanol (%)	0.004	-

TPC (mg GAE/g dw); ORAC (mg TE/g dw); GAE: Gallic Acid Equivalent; TE: trolox equivalent.

However, the values which maximize scavenging activity (ORAC) have a greater deviation. The value predicted by the reduced model was 200.66 mg TE/g dw, compared to an experimental value of 154.3 mg TE/g dw, which represents a deviation of 23.1%.

The best-fitting experimental conditions were then applied, *i.e.*, 23 min extraction with 56% EtOH and 63 °C. This extract was lyophilized and used in subsequent experiments.

Table 3. Mathematical equations from Response Surface Method (RSM) for each of the responses, with their respective value of R^2 and R^2-predicted.

Response	Equation	R^2 Value	
		R^2	R^2 Pred.
Complete Model			
TPC	$62.87 - 0.47\,T - 0.25\,[\%] - 0.14\,t + 0.003\,T^2 - 0.001\,[\%]^2 + 0.03\,t^2 + 0.006\,T \times [\%] - 0.007\,T \times t - 0.003\,[\%] \times t$	94.69	57.0
ORAC	$318.2 + 2.03\,T - 04.41\,[\%] - 019.5\,t - 00.009\,T^2 + 0.023\,[\%]^2 + 0.7\,t^2 + 0.012\,T \times [\%] - 00.053\,T \times t - 00.03\,[\%] \times t$	96.7	75.0
Reduced Model			
TPC	$69.7 - 00.53\,T - 00.39\,[\%] + 0.002\,T^2 - 00.006\,T \times [\%]$	85.7	66.76
ORAC	$345.7 + 1.01\,T - 03.92\,[\%] - 022.01\,t + 0.027\,[\%]^2 + 0.73\,t^2$	91.88	77.88

T: Temperature (°C); [%]: Ethanol concentration (%); t: Time (min); Pred.: response predicted by model. TPC in mg GAE/g dw and ORAC in mg TE/g dw.

Table 4. Optimal conditions for the extractions for TPC and ORAC, given by RSM.

Model	Conditions			Response		
	Temperature (°C)	Ethanol (%)	Time (min)	Predicted	Predicted RM	Experimental
TPC	63	56	23	51.75	43.6	45.01
ORAC	93.6	44.7	7	206.82	200.66	154.3

TPC in mg GAE/g dw; ORAC in mg TE/g dw.

3.2. Extract Optimized Effect in Oil-in-Water Emulsions (O/W)

Figure 3 shows the evolution of peroxide value over time. In this case, the possible synergy between the extract (with different concentrations) and egg albumin was determined. Firstly, it should be noted that both albumin and various concentrations of the extract of avocado produce significant protection against oxidation. For example, within the 400 h of the experiment the amount of hydroperoxides produced is 90% higher in the control than in any of the samples (20 mg hydroperoxides/kg of emulsion hydroperoxides *vs.* 38 mg/kg for the emulsion control). Notably, there were no significant differences ($p < 0.05$) for the three tested avocado concentrations (0.1125%, 0.225% and 0.45% w/w), as well as egg albumin (0.2% w/w). This fact could be explained by the solubility of the lyophilized extract in water and the ability to coat the oil drop generated in the emulsion and prevent oxidation thereof. The necessary concentration that allows this protection is already achieved with 0.1125% and the results do not improve if increased. Similar behavior has been published elsewhere [37–39].

In fact, putting together two different compounds (avocado pit extract and egg protein) allows greater protection against oxidation and further differentiates the two concentrations of the tested extract. For example, the time required to reach 15 mg hydroperoxides/kg emulsion goes from 180 h of the control group up to 480 h for the sample containing 0.45% extract + 0.2 egg protein. This is an increase of 260% superior durability. In the intermediate areas three avocado extract concentrations were tested, as well as the protein (an increase in the durability between 150% and 180%) and one that contains 0.225% of avocado and 0.2% protein, with an improvement of the durability of 220%. Almajano and Bonilo-Carbognin already published similar results of synergy [40,41]. As a summary, it can be said that increasing the concentration of the extract does not improve the durability. However, the incorporation of small amounts of protein allows significant differences to be found between the samples containing protein and those that do not contain it.

Figure 3. Peroxide values *vs.* time in the emulsions.

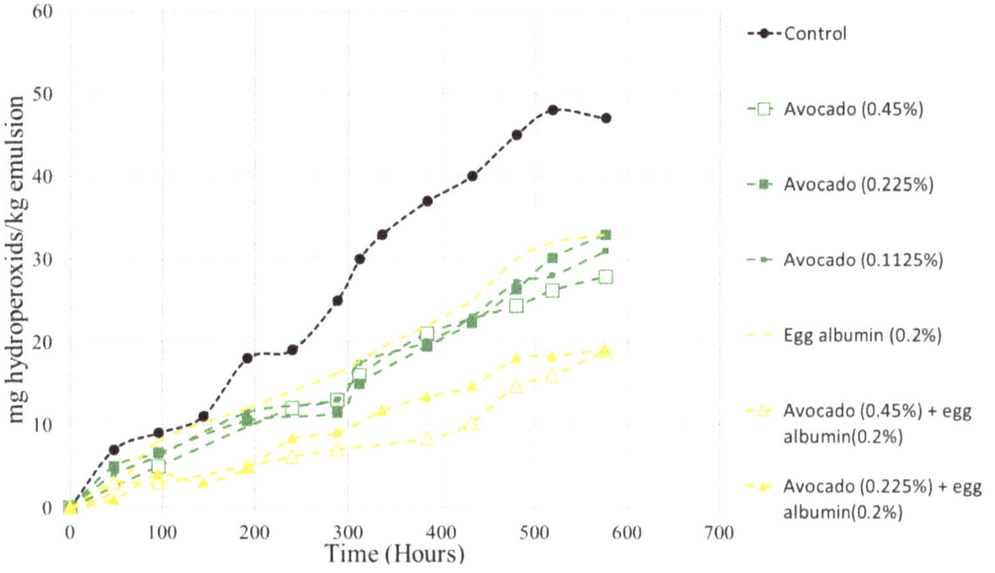

Avocado pits contain polyphenolic compounds (such as protocatechuic acid, chlorogenic acid, syringic acid and rutin), which are very strong antioxidants [34]. In 2010, Sasaki [42] studied the antioxidant power of chlorogenic acid in oil-water emulsions. The effects discovered are remarkable. The authors analyzed the presence of other compounds, which in that case were also polyphenolic compounds. Additionally, they demonstrated that the presence of several different compounds provided better results than the added individual effects.

As it was stated before, 1% of surfactant (Tween-20) was added to the emulsion prepared in the present work. This eases the dissolution of the polyphenolic compounds, thus increasing the antioxidant activity in the emulsion.

3.3. Effect of the Extract in Burger Meat

The TBARS method is widely used to determine the oxidation of fats and oils in foods [43–45]. In Figure 4, the evolution of TBARs *vs.* time for each of the studied beef burger meat patties is collected. Samples containing 0.1% lyophilized extract and 0.5% direct seed powder have no significant differences compared with the BHA (0.05%), but show a big difference compared with the control. The lower concentration (0.01% and 0.05% lyophilized extract powder direct seed) presented intermediate behavior, as expected. The duration of the experiment was 8 days and it was observed that the burger meat with 0.5% seed powder and 0.1% of lyophilized extract had no significant oxidation, or the protection is higher than 90%. These results are similar to those reported by Weiss *et al.* [46] for pork burgers. That study examined protecting fat oxidation also with excellent results [46]. Additional results along the same lines have avocado oil added directly to the pork burgers. This shows a positive effect on the conservation of the burger [47].

It is not the first time avocado pits have been used in meat products. Rodríguez-Carpena *et al.* (2011) [15], prepared pork meat pies and inserted the grinded avocado pits to protect the meat against lipid oxidation. The authors indicated that one of the factors might be the formation of chelates with the copper and iron cations. These cations, in their free ionic state, could cause the creation of free radicals.

Figure 4. The TBARS (Thiobarbituric acid reactive substances) values for the meat emulsions.

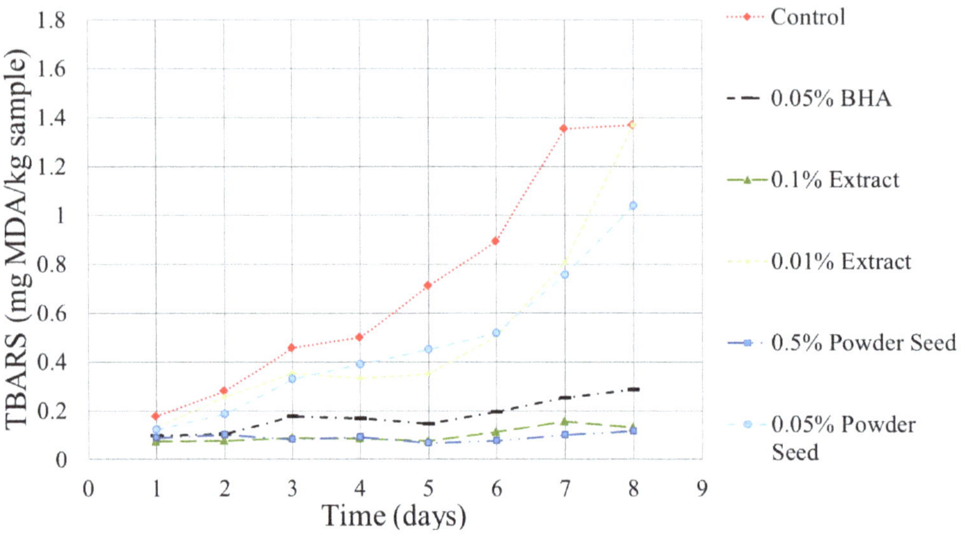

4. Conclusions

RSM was used to identify the best conditions for the extraction of compounds with an antioxidant activity from an organic residue: the avocado pit. The reduced model obtained provides parameters that fit with those of the TPC (with a 3.13% error when compared to the experimental value).

The lyophilized extract was used as protection from the oxidation of oils (oil-in-water emulsions) and fat (beef burgers) with excellent results, especially in meat, in which the durability of the burger meat is significantly increased relative to oxidation.

These studies should encourage further exploration in this area of study in order to obtain a byproduct of the natural antioxidants that currently as waste are worthless.

Acknowledgments

This research was conducted at the Department Chemical Engineering, University Polytechnic of Catalonia, Spain. The authors thank Universidad Experimental Politécnica “Antonio José de Sucre” and Carlos Barreiro, for their support in this research. The authors want to thank Mariona Vila for the support.

Author Contributions

Francisco Segovia and Sara Peiró have done the experimental section and the design, data acquisition, analysis and data interpretation; Maria Gabriela Gallego and Nurul Aini Mohd, participate in revising it critically for important intellectual content. Finally, María Pilar Almajano, as director, had the initial idea, the constant support, the English revision and the final approval of the version to be submitted and any revised version.

Conflicts of Interest

The authors declare no conflict of interest.

References

1. Martínez, R.; Torres, P.; Meneses, M.A.; Figueroa, J.G.; Pérez-Álvarez, J.A.; Viuda-Martos, M. Chemical, technological and *in vitro* antioxidant properties of mango, guava, pineapple and passion fruit dietary fibre concentrate. *Food Chem.* **2012**, *135*, 1520–1526.
2. Ignat, I.; Volf, I.; Popa, V.I. A critical review of methods for characterisation of polyphenolic compounds in fruits and vegetables. *Food Chem.* **2011**, *126*, 1821–1835.
3. Rubilar, M.; Pinelo, M.; Ihl, M.; Scheuermann, E.; Sineiro, J.; Nuñez, M.J. Murta leaves (*Ugni molinae* Turcz) as a source of antioxidant polyphenols. *J. Agric. Food Chem.* **2006**, *54*, 59–64.
4. Manach, C.; Scalbert, A.; Morand, C.; Rémésy, C.; Jime, L. Polyphenols: Food sources and bioavailability 1,2. *Am. J. Clin. Nutr.* **2004**, *79*, 727–747.
5. Perumalla, A.V.S.; Hettiarachchy, N.S. Green tea and grape seed extracts—Potential applications in food safety and quality. *Food Res. Int.* **2011**, *44*, 827–839.
6. Wettasinghe, M.; Shahidi, F. Phenolic acids in defatted seeds of borage (*Borago officinalis* L.). *Food Chem.* **2001**, *75*, 49–56.
7. Jordán, M.; Lax, V.; Rota, M.C.; Lora, S.; Sotomayor, J.A. Relevance of carnosic acid, carnosol, and rosmarinic acid concentrations in the *in vitro* antioxidant and antimicrobial activities of *Rosmarinus officinalis* (L.) methanolic extracts. *J. Agric. Food Chem.* **2012**, *60*, 9603–9608.
8. Ayala-Zavala, J.F.; Vega-Vega, V.; Rosas-Domínguez, C.; Palafox-Carlos, H.; Villa-Rodriguez, J.A.; Siddiqui, M.W.; Dávila-Aviña, J.E.; González-Aguilar, G.A. Agro-industrial potential of exotic fruit byproducts as a source of food additives. *Food Res. Int.* **2011**, *44*, 1866–1874.
9. Wijngaard, H.H.; Brunton, N. The optimisation of solid–liquid extraction of antioxidants from apple pomace by response surface methodology. *J. Food Eng.* **2010**, *96*, 134–140.
10. Lagha-Benamrouche, S.; Madani, K. Phenolic contents and antioxidant activity of orange varieties (*Citrus sinensis* L. and *Citrus aurantium* L.) cultivated in Algeria: Peels and leaves. *Ind. Crops Prod.* **2013**, *50*, 723–730.
11. Aguedo, M.; Kohnen, S.; Rabetafika, N.; Vanden Bossche, S.; Sterckx, J.; Blecker, C.; Beauve, C.; Paquot, M. Composition of by-products from cooked fruit processing and potential use in food products. *J. Food Compos. Anal.* **2012**, *27*, 61–69.

12. Wijngaard, H.; Hossain, M.B.; Rai, D.K.; Brunton, N. Techniques to extract bioactive compounds from food by-products of plant origin. *Food Res. Int.* **2012**, *46*, 505–513.
13. Balasundram, N.; Sundram, K.; Samman, S. Phenolic compounds in plants and agri-industrial by-products: Antioxidant activity, occurrence, and potential uses. *Food Chem.* **2006**, *99*, 191–203.
14. Guerrero, M.S.; Torres, J.S.; Nuñez, M.J. Extraction of polyphenols from white distilled grape pomace: Optimization and modelling. *Bioresour. Technol.* **2008**, *99*, 1311–1318.
15. Rodríguez-Carpena, J.G.; Morcuende, D.; Estévez, M. Avocado by-products as inhibitors of color deterioration and lipid and protein oxidation in raw porcine patties subjected to chilled storage. *Meat Sci.* **2011**, *89*, 166–173.
16. Soong, Y.-Y.; Barlow, P.J. Antioxidant activity and phenolic content of selected fruit seeds. *Food Chem.* **2004**, *88*, 411–417.
17. Moure, A.; Cruz, J.M.; Franco, D.; Domínguez, J.M.; Sineiro, J.; Domínguez, H.; Jose, M.; Parajo, J.C. Natural antioxidants from residual sources. *Food Chem.* **2001**, *72*, 145–171.
18. Saha, J.; Debnath, M.; Saha, A.; Ghosh, T.; Sarkar, P.K. Response surface optimisation of extraction of antioxidants from strawberry fruit, and lipid peroxidation inhibitory potential of the fruit extract in cooked chicken patties. *J. Sci. Food Agric.* **2011**, *91*, 1759–1765.
19. Díaz Reinoso, B.; Couto, D.; Moure, A.; Fernandes, E.; Domínguez, H.; Parajó, J.C. Optimization of antioxidants—Extraction from Castanea sativa leaves. *Chem. Eng. J.* **2012**, *203*, 101–109.
20. Pompeu, D.R.; Silva, E.M.; Rogez, H. Optimisation of the solvent extraction of phenolic antioxidants from fruits of *Euterpe oleracea* using Response Surface Methodology. *Bioresour. Technol.* **2009**, *100*, 6076–6082.
21. Wijngaard, H.; Brunton, N. The optimization of extraction of antioxidants from apple pomace by pressurized liquids. *J. Agric. Food Chem.* **2009**, *57*, 10625–10631.
22. Prior, R.L.; Wu, X.; Schaich, K. Standardized methods for the determination of antioxidant capacity and phenolics in foods and dietary supplements. *J. Agric. Food Chem.* **2005**, *53*, 4290–4302.
23. Casettari, L.; Gennari, L.; Angelino, D.; Ninfali, P.; Castagnino, E. ORAC of chitosan and its derivatives. *Food Hydrocoll.* **2012**, *28*, 243–247.
24. Singh, G.; Maurya, S.; DeLampasona, P.; Catalan, C. A comparison of chemical, antioxidant and antimicrobial studies of cinnamon leaf and bark volatile oils, oleoresins and their constituents. *Food Chem. Toxicol.* **2007**, *45*, 1650–1661.
25. Doménech-Asensi, G.; García-Alonso, F.J.; Martínez, E.; Santaella, M.; Martín-Pozuelo, G.; Bravo, S.; Periago, M.J. Effect of the addition of tomato paste on the nutritional and sensory properties of mortadella. *Meat Sci.* **2013**, *93*, 213–219.
26. Spigno, G.; de Faveri, D.M. Antioxidants from grape stalks and marc: Influence of extraction procedure on yield, purity and antioxidant power of the extracts. *J. Food Eng.* **2007**, *78*, 793–801.
27. Amendola, D.; de Faveri, D.M.; Spigno, G. Grape marc phenolics: Extraction kinetics, quality and stability of extracts. *J. Food Eng.* **2010**, *97*, 384–392.
28. Boussetta, N.; Vorobiev, E.; Le, L.H.; Cordin-falcimaigne, A.; Lanoisellé, J. Application of electrical treatments in alcoholic solvent for polyphenols extraction from grape seeds. *LWT-Food Sci. Technol.* **2012**, *46*, 127–134.

29. Rodríguez-Rojo, S.; Visentin, A.; Maestri, D.; Cocero, M.J. Assisted extraction of rosemary antioxidants with green solvents. *J. Food Eng.* **2012**, *109*, 98–103.
30. Cacace, J.E.; Mazza, G. Mass transfer process during extraction of phenolic compounds from milled berries. *J. Food Eng.* **2003**, *59*, 379–389.
31. Tubtimdee, C.; Shotipruk, A. Extraction of phenolics from *Terminalia chebula* Retz with water–ethanol and water–propylene glycol and sugaring-out concentration of extracts. *Sep. Purif. Technol.* **2011**, *77*, 339–346.
32. Qu, W.; Pan, Z.; Ma, H. Extraction modeling and activities of antioxidants from pomegranate marc. *J. Food Eng.* **2010**, *99*, 16–23.
33. Thoo, Y.Y.; Ho, S.K.; Liang, J.Y.; Ho, C.W.; Tan, C.P. Effects of binary solvent extraction system, extraction time and extraction temperature on phenolic antioxidants and antioxidant capacity from mengkudu (*Morinda citrifolia*). *Food Chem.* **2010**, *120*, 290–295.
34. Pahua-Ramos, M.E.; Ortiz-Moreno, A.; Chamorro-Cevallos, G.; Hernández-Navarro, M.D.; Garduño-Siciliano, L.; Necoechea-Mondragón, H.; Hernández-Ortega, M. Hypolipidemic effect of avocado (*Persea americana* Mill) seed in a hypercholesterolemic mouse model. *Plant Foods Hum. Nutr.* **2012**, *67*, 10–16.
35. Sun, Y.; Xu, W.; Zhang, W.; Hu, Q.; Zeng, X. Optimizing the extraction of phenolic antioxidants from kudingcha made frrom *Ilex kudingcha* C.J. Tseng by using response surface methodology. *Sep. Purif. Technol.* **2011**, *78*, 311–320.
36. Kahyaoglu, T. Optimization of the pistachio nut roasting process using response surface methodology and gene expression programming. *LWT-Food Sci. Technol.* **2008**, *41*, 26–33.
37. Kargar, M.; Spyropoulos, F.; Norton, I.T. The effect of interfacial microstructure on the lipid oxidation stability of oil-in-water emulsions. *J. Colloid Interface Sci.* **2011**, *357*, 527–533.
38. Poyato, C.; Navarro-blasco, I.; Isabel, M.; Yolanda, R.; Astiasarán, I.; Ansorena, D. Oxidative stability of O/W and W/O/W emulsions : Effect of lipid composition and antioxidant polarity. *Food Res. Int.* **2013**, *51*, 132–140.
39. Sun, C.; Gunasekaran, S. Effects of protein concentration and oil-phase volume fraction on the stability and rheology of menhaden oil-in-water emulsions stabilized by whey protein isolate with xanthan gum. *Food Hydrocoll.* **2009**, *23*, 165–174.
40. Almajano, M.P.; Delgado, M.E.; Gordon, M.H. Changes in the antioxidant properties of protein solutions in the presence of epigallocatechin gallate. *Food Chem.* **2007**, *101*, 126–130.
41. Bonoli-Carbognin, M.; Erretani, L.O.C.; Endini, A.L.B.; Goidanich, P.; Cesena, I. Bovine serum albumin produces a synergistic increase in the antioxidant activity of virgin olive oil phenolic compounds in oil-in-water emulsions. *J. Agric. Food Chem.* **2008**, *56*, 7076–7081.
42. Sasaki, K.; Alamed, J.; Weiss, J.; Villeneuve, P.; López, L.J.; Lecomte, J.; Figueroa-espinoza, M.; Decker, E.A. Relationship between the physical properties of chlorogenic acid esters and their ability to inhibit lipid oxidation in oil-in-water emulsions. *Food Chem.* **2010**, *118*, 830–835.
43. Mendes, R.; Cardoso, C.; Pestana, C. Measurement of malondialdehyde in fish: A comparison study between HPLC methods and the traditional spectrophotometric test. *Food Chem.* **2009**, *112*, 1038–1045.

44. Seljeskog, E.; Hervig, T.; Mansoor, M.A. A novel HPLC method for the measurement of thiobarbituric acid reactive substances (TBARS). A comparison with a commercially available kit. *Clin. Biochem.* **2006**, *39*, 947–954.
45. Wenjiao, F.; Yongkui, Z.; Yunchuan, C.; Junxiu, S.; Yuwen, Y. TBARS predictive models of pork sausages stored at different temperatures. *Meat Sci.* **2014**, *96*, 1–4.
46. Weiss, J.; Gibis, M.; Schuh, V.; Salminen, H. Advances in ingredient and processing systems for meat and meat products. *Meat Sci.* **2010**, *86*, 196–213.
47. Rodríguez-Carpena, J.G.; Morcuende, D.; Estévez, M. Avocado, sunflower and olive oils as replacers of pork back-fat in burger patties: Effect on lipid composition, oxidative stability and quality traits. *Meat Sci.* **2012**, *90*, 106–115.

Screening of Antioxidant Activity of *Gentian Lutea* Root and Its Application in Oil-in-Water Emulsions

Nurul Aini Mohd Azman [1,2], Francisco Segovia [1], Xavier Martínez-Farré [3], Emilio Gil [3] and María Pilar Almajano [1,*]

[1] Chemical Engineering Department, Technical University of Catalonia, Avda. Diagonal 647, 08028 Barcelona, Spain; E-Mails: ainiazman@gmail.com (N.A.M.A.); segoviafj@gmail.com (F.S.)

[2] Chemical and Natural Resources Engineering Faculty, University Malaysia Pahang, LebuhrayaTunRazak, Pahang 26300, Malaysia

[3] Agro-Food Engineering and Biotechnology Department, EsteveTerradas, 8, 08860 Castelldefels, Spain; E-Mails: Xavier.Martinez-Farre@upc.edu (X.M.-F.); Emilio.Gil@upc.edu (E.G.)

* Author to whom correspondence should be addressed; E-Mail: m.pilar.almajano@upc.edu

---

Abstract: *Gentiana Lutea* root (*G. Lutea*) is a medicinal herb, traditionally used as a bitter tonic in gastrointestinal ailments for improving the digestive system. The active principles of *G. Lutea* were found to be secoiridoid bitter compounds as well as many other active compounds causing the pharmacological effects. No study to date has yet determined the potential of *G. Lutea* antioxidant activity on lipid oxidation. Thus, the aim of this study was to evaluate the effects of an extract of *G. Lutea* on lipid oxidation during storage of an emulsion. *G. Lutea* extracts showed excellent antioxidant activity measured by DPPH scavenging assay and Trolox equivalent antioxidant capacity (TEAC) assays. An amount of 0.5% w/w *G. Lutea* lyophilise was able to inhibit lipid oxidation throughout storage ($p < 0.05$). A mixture of *G. Lutea* with 0.1% (w/w) BSA showed a good synergic effect and better antioxidant activity in the emulsion. Quantitative results of HPLC showed that *G. Lutea* contained secoiridoid-glycosides (gentiopicroside and sweroside) and post column analysis displayed radical scavenging activity of *G. Lutea* extract towards the ABTS radical. The results from this study highlight the potential of *G. Lutea* as a food ingredient in the design of healthier food commodities.

Keywords: *Gentiana Lutea*; lipid oxidation; antioxidant; HPLC

---

1. Introduction

Gentian Lutea root (*G. Lutea*), also known as Yellow Gentian, has over 300 species and is widely distributed in North America, Europe, Asia and some parts of South America [1]. The plant root was traditionally use as a medicinal plant to stimulate appetite and improve digestion [2]. Furthermore, the root is also well known for its bitter properties due to the existence of secoiridoid-glycosides (e.g., swertiamarin, gentiopicroside, amarogentin and sweroside) [3]. Many researchers have discovered the benefits of bitter tasting secoiridoid-glycoside through extensive pharmacological studies. These constituents are claimed to have many biological effects such as anti-apoptotic [4], anti-cancer [5], anti-fungi [6], anti-bacterial [7], anti-inflammatory [8], and hepatoprotective [9]. It has been reported recently by Nastasijevic *et al.* [10] that the *G. Lutea* extracts have potential to inhibit myeloperoxidase (MPO) activity which contributes to many disorders such as cardiovascular, inflammatory, neurodegenerative and immune-mediated diseases.

However, not only are the secoiridoids relevant for the plant pharmacological action, there are many active compounds in *G. Lutea* that also have relevant effects such as iridoidloganic acid, xanthone (e.g., gentisin and isogentisin) and xanthone glycosides (gentioside and its isomer) [3]. Iridoidloganic acid has shown potent activity as an anti-inflammatory and a number of studies demonstrated that xanthones and its derivatives have wide-ranging biological activities such as anti-inflammatory, anti-hepatotoxic, anti-tumor and anti-microbial [11]. Furthermore, xanthone, and its derivatives, are phenolic compounds with antioxidant properties which have attracted much attention recently [12]. There are few reports that have investigated the radical scavenging activities of secoiridoid-glycosides using DPPH assay [7,13], although some authors have measured total antioxidant capacity of *G. Lutea* but without sufficient assessment of individual antioxidants [10,14]. However, the observation of antioxidant activity of *G. Lutea* extract towards lipid oxidation has not been fully determined.

Lipid oxidation in high fat-containing food is a major cause of shelf life deterioration such as in meat products and emulsions. The oxidation process causes unsavoury alteration of flavor, texture, shelf life, appearance, and nutritional qualities [15].

Oil-in-water emulsion (o/w), is a food model which is highly susceptible to oxidation, besides it has also become one of the effective models to evaluate the antioxidant activity of natural plants towards lipids [16]. The first phase of lipid oxidation starts with the formation of unstable free radicals and hydroperoxides with further decomposition to secondary products like ketones, aldehydes, alcohols and acids [17]. The formation of peroxides in primary oxidation can be measured using peroxide value (PV) assay. Secondary oxidation can be measured by thiobarbituric acid reacting substances (TBARS), specifically aldehydes, and also leads to unpleasant taste, aroma and quality traits of the products.

Thus, our goal was to evaluate the potential antioxidant activity of *G. Lutea* by different methods (1) *in vitro* with such radicals as $ABTS^{\bullet+}$, DPPH and enzymatic activity and (2) in o/w emulsion.

2. Experimental Section

2.1. Plant Material

Commercially dried *G. Lutea* was kindly supplied by Manatial de la Salut (Barcelona, Spain), a registered herbal company. Reagents used were: thiobarbituric acid, 1,1-diphenyl-2-picrylhydrazyl (DPPH), Folin-Ciocalteu reagent, methanol, hydrogen chloride, aluminium oxide, ferrous chloride, anhydrous sodium carbonate, ethanol 96%, Phosphate Buffer Solution (PBS) and ammonium thiocyanate from Panreac (Barcelona, Spain). Gallic acid, 2,2′-azino-bis (3-ethylbenzothiazoline)-6-sulfonic acid diammonium salt (ABTS), (±)-6-hydroxy-2,5,7,8-tetramethylchromane-2–carboxylic acid (Trolox), NBT (nitrobluetetrazolium), Bovine Serum Albumin (BSA), Xanthine and Xanthine-oxidase from Sigma-Aldrich (Gillingham, UK).

2.2. Extraction of G. Lutea

Dried roots of *G. Lutea* were finely ground using a standard kitchen food processor. Ground *G. Lutea* (5 g) was extracted in two ways; (1) with 50:50 (v/v) methanol:water and (2) with water, always in the ratio 1:10 (w/v). The extraction was performed at 4 ± 1 °C for 24 h, in the dark with constant stirring. The extract solutions of *G. Lutea* were recovered by filtration using Whatman Filter paper, 0.45 μm. Part of the supernatant was taken for subsequent use to determine the antiradical capacity. The volume of the remaining supernatant was measured and the excess methanol was removed under vacuum using a rotary evaporator (BUCHI RE111, Postfachi, Switzerland) and kept frozen at −80 °C for 24 h. All extracts were dried in a freeze dryer (Unicryo MC2L −60 °C, Martinsried, Germany) under vacuum conditions at −60 °C for 3 days to remove moisture. Finally, *G. Lutea* lyophilize (freeze dried) were weighed to determine the concentration recovered (g/L) and the extraction yield (%) as Zhang *et al.* [18]. Samples were then weighed and kept protected from light in a desiccator until use.

2.3. Determination of the Total Phenolic Content (TPC)

The Folin-Ciocalteu method was used to determine the total phenolic content (TPC) as reported by Santas *et al.* [19]. The sample was diluted 1:25 (v/v) in order to be in the range of absorbance. The final concentration (v/v) for the mixture was; sample 7.7%, Folin reagent 4% and saturated sodium carbonate solution 30.8%. The mixture was finally diluted with Mili Q water, shaken and incubated in the dark for 1 h. Absorbance at 765 nm was measured using a microplate reader (Fluostar Omega, BMG Labtech, Ortenberg, Germany) against water as a blank. Gallic acid was used to prepare a standard calibration, and the results were expressed as mg of Gallic acid equivalents/g dry weight (mg GAE/g DW).

2.4. Determination of Free Radical Scavenging Activity Assays

2.4.1. TEAC Assay

The antioxidant capacities of *G. Lutea* were measured by using a modified TEAC assay, which was performed as described by Miller *et al.* [20]. The TEAC assay was based on the reduction of the $ABTS^{\bullet+}$ radical cation by the antioxidants present in the samples. $ABTS^{\bullet+}$ radical cation (7 mM, final concentration) was dissolved before adding potassium sulphate (2.45 mM, final concentration) and allowing the mixture to stand in the dark up to 16 h. Phosphate Buffer Solution (PBS, 10 mM) with the $ABTS^{\bullet+}$ radical cation was incubated at room temperature for 30 min before used. Then, the mixture of the $ABTS^{\bullet+}$ radical cation was adjusted to an absorbance of 0.73 ± 0.2 nm, using a microplate reader (Fluostar Omega, BMG Labtech, Ortenberg, Germany). The TEAC values for the different concentrations of each compound were interpolated from the trolox calibration curve and expressed as milligrams of trolox equivalent per gram of dry weight sample (mg TE/g DW sample).

2.4.2. DPPH Assay

The effect of the extracts on the scavenging of DPPH radical was determined according to the method adapted from Madhujith *et al.* [21] with slight modifications. The sample was diluted 1:20 (v/v) and DPPH radical in methanol (5.07 mM) was made for the study. Then, the sample (10% v/v) and DPPH solution (90% v/v) were added to the well of the microplate. The absorbance was measured at 517 nm over every 15 min for 75 min. The results were expressed as mg TE/g DW sample.

2.4.3. Superoxide Activity Xanthine/Xanthine Oxidase (X/XO)

The method was based on the developed method of Valentao *et al.* [22] and modified for application in microplates by Lopez *et al.* [23]. All test samples were dissolved in a 50 mM phosphate buffer to simulate the environment in which the reaction occurs in the body. The sample was mixed with 145 μM of a solution of xanthine, 50 μM of a solution of NBT and incubated in 37 °C. The sample extract was diluted from 1:10 to 1:100 (v/v) for the study. Finally, 0.29 U/mL of enzyme xanthine oxidase solution was added and the absorbance was recorded at 560 nm every 2 min. The value of IC_{50} was calculated to determine the inhibition rate of *G. Lutea* in the reaction.

2.5. Determination of Antioxidant Activity in o/w Emulsion

2.5.1. Removal of Tocopherols from Sunflower Oil

Alumina was placed in an oven at 200 °C for 24 h, and then removed and allowed to cool in a desiccator until it reached room temperature. Sunflower oil triacylglycerol was passed twice through the alumina in a column to remove the tocopherols as described by Yoshida *et al.* [24]. Finally, the filtered oil was stored at −80 °C until use.

2.5.2. Preparation of Emulsion

Oil in water emulsion was prepared by dissolving Tween-20 (1%, final concentration) in Milli Q water and adding oil (10%, final concentration). To form an emulsion, the oil was added drop wise to the solution of Tween-20 and water, which was kept cold, and sonication process was continued for 5 min. All samples were redissolved in ethanol-50% (v/v) to obtain the final concentration in the emulsion. The final samples were prepared either (i) control (no addition); (ii) 0.35% (w/w) Trolox (positive control); (iii) 0.1% (w/w) BSA; (iv) 0.5% (w/w) lyophilise *G. Lutea*; (v) 0.5% (w/w) lyophilise *G. Lutea* mixed with 0.1% (w/w) BSA; (vi) 0.2% (w/w) lyophilise *G. Lutea* and (vii) 0.2% (w/w) lyophilise *G. Lutea* mixed with 0.1% (w/w) BSA. The emulsion for each sample was prepared in quadruplicate, obtaining a total of 28 samples and stored in the dark and allowed to oxidize at 37 °C. The pH of the samples was measured four times for each sample (pH meter GLP21, Crison Instruments, Barcelona, Spain) as a parameter to investigate its correlation with PV.

2.5.3. Determination of Peroxide Value (PV)

The primary oxidation products were measured using peroxide value (PV) according to the thiocyanate method of the Association of Official Analytical Chemists (AOAC) 8195 [25]. Ferrous chloride solution was prepared in hydrochloric acid (1 M) with the addition of iron chloride (II) (2 mM, final concentration). Ammonium thiocyanate solution was prepared in water (2 mM, final concentration). The assay was performed with a drop of emulsion in the range from 0.007 to 0.01 g, diluted with ethanol. From this solution the required amount of sample, varying according to the degree of oxidation, was taken in a cuvette and ethanol (96%) was added. Ferrous chloride and ammonium thiocyanate solutions were added, each in a proportion of 1.875% (v/v), final concentration. The absorbance was measured spectrophotometrically at $\lambda = 500$ nm. The results are expressed as meq hydroperoxides/kg of emulsion.

2.5.4. Determination of Secondary Oxidation by Thiobarbituric Acid Reactive Substances (TBARS)

The TBARS method was adapted from Gallego *et al.* [26]. The TBARS reagent was prepared (15% w/v trichloroacetic acid, 0.375% w/v thiobarbituric acid and hydrochloric acid 2.1% v/v). One mL of each emulsion was taken and the TBARS reagent was added in the ratio 1:5 (v/v). Immediately the samples were added to an ultrasonic bath (5 min) and after immersing in a water bath preheated to 95 °C (20 min) the samples were centrifuged and the absorbance of the supernatant was measured at $\lambda = 531$ nm. The results are expressed as mg malondialdehyde (MDA)/kg of emulsion.

2.6. Statistical Analysis

Statistical analysis was performed using a one-way analysis of variance ANOVA using Minitab 16 software program (Minitab, Inc., Paris, France). When a statistically significant difference was found, Tukey's tests were performed and the statistical significance was set at $p < 0.05$. The results were presented as mean values ($n \geq 3$).

2.7. HPLC and Post-Column HPLC-ABTS$^{\bullet+}$ Radical Scavenging Method

The method for identification of peaks with antioxidant activity was that used by Koleva *et al.* [27] with some modifications. The instrument was a Waters 2695 separations module (Meadows Instrumentation Inc., Bristol, USA) system with a photodiode array detector Waters 996 (Meadows Instrumentation Inc., Bristol, USA). The column used was a Kinetex C18 100A, (100 × 4.6 mm, Phenomenex, Torrence, CA, USA). Solvents used for separation were 0.1% acetic acid in water (v/v) (eluent A) and 0.1% acetic acid in methanol (v/v) (eluent B). The gradient used was isocratic, 75% A. The flow rate was 0.6 mL/min. Detection wavelength was 230 nm (to see the peaks) and 734 nm (to see the ABTS radical). The sample injection volume was 10 µL. The chromatographic peaks of gentiopicroside and sweroside were confirmed by comparing their retention times and diode array spectra with that of their reference standards. The pump for ABTS post-column injection was a Merk-Hitachi HPLC gradient pump (Model L-6200, Hitachi High Technologies America Inc., Schaumburg, Illinois, IL, USA) with a 0.2 mL/min flow; ABTS concentration was of 0.03% (w/v).

3. Results and Discussion

3.1. Analysis of Total Polyphenols and Free Radical Activity Assays

On average, from 5 g of dried *G. Lutea* extracted with aqueous methanol 50:50 (v/v) and water alone, it was possible to recover 1.5 ± 0.05 g and 1.0 ± 0.04 g of lyophilised, respectively. The concentration recovered was proportional to the extraction yield shown in Table 1. Previous studies reported that gentiopicroside compound, an active compound that signifies the main bitter principle in *G. Lutea* was still preserved at almost 83.5% after drying [28].

Table 1. Extraction yield, total phenolic content (TPC), 1,1-diphenyl-2-picrylhydrazyl (DPPH), Trolox equivalent capacity assay (TEAC) and enzymatic activity of *G. Lutea*.

Activity *G. Lutea*	**Extraction Solvent**	
	H_2O	**50:50 MeOH:H_2O**
Extraction yield (%)	20.00 ± 0.9	29.10 ± 0.3
Total phenolic content (g GAE/g DW)	3.79 ± 1.7	12.03 ± 1.8
DPPH (µmol of TE/g DW)	12.34 ± 1.5	15.89 ± 0.5
TEAC (µmol of TE/g DW)	33.28 ± 1.5	48.90 ± 1.8
Superoxide activity (mg/mL)	30.00 ± 2.8	23.21 ± 2.8

Mean value $n = 3$ and the standard deviation for each assay is less than 5%. Gallic Acid Equivalent (GAE), Trolox Equivalent (TE), Dry Weight (DW).

The concentration of total polyphenols and the value of antioxidant activity assays were determined and the results are shown in Table 1. The extract of *G. Lutea* in methanol-50% showed higher phenolic content and antioxidant activity than the water extract. The total phenolic content of *G. Lutea* extracts allowed the estimation of all phenolic acids, flavonoids, anthocyanins, nonflavonoids and many classes of polyphenol compounds present in the samples. On the other hand, Nastasijevic *et al.* determined the total polyphenol content of *G. Lutea*in in water extract as being slightly higher compared to different concentrations of aqueous ethanol and methanol extracts [10].

Water extracts of *G. Lutea* showed the lowest activity in free $DPPH^{\bullet}$ scavenging activity compared to methanol-50% extract, similar to previous research from Kintzios *et al.* [14]. It is not the first time that the antioxidant activity of *G. Lutea* by DPPH method has been carried out. However, variation in results may be due to the plant age, solvent, method and system used throughout the experiment [10,14,29].

For the TEAC assay, the finding was consistent with the DPPH method where the aqueous methanol extract showed higher activity than the water extract. TEAC assay indicated the extract potency used as a source of antioxidants based on the ability of the antioxidant compound to scavenge the long-life radical cation $ABTS^{\bullet+}$. In the results shown in Table 1 it can be appreciated that the methanol-50% extract has a higher capacity to scavenge $ABTS^{\bullet+}$ radicals and consequently shows a higher antioxidant activity than DPPH assay. To the best of our knowledge, this is the first report of the antioxidant activity of extracts from *G. Lutea* roots assessed using the TEAC methods.

Some of the previous reports showed that the antioxidant activity of plant extracts correlates with the phenolic content [30] and the yield of phenolic components from herbs is higher with methanol-50% rather than with water as extract. The mixtures of alcohol and water have been more efficient in extracting compounds and give a better yield than the corresponding mono-component solvent system. Xanthones such as isogentisin and gentisin and its derivatives are one of main sources of phenolic compounds in *G. Lutea* and are expected to be more soluble in aqueous alcohol. A study showed good correlation between phenolic content and antioxidant activity [31] whereas another found no correlation [32].

In the present work, an effective antioxidant activity in *G. Lutea* was found. Methanol-50% extract exhibited $O_2^{\bullet-}$ scavenging activity, measured using the X/XO system (Table 1), with an IC_{50} at 23.21 ± 2.8 mg/mL. Water extract of *G. Lutea* showed lower scavenging activity than methanol aqueous extract, with IC_{50} = 30.00 ± 2.8 mg/mL. These results are consistent with Kusar *et al.* [29], who demonstrated the effect of superoxide activity of *G. Lutea* leaf and root in methanol extracts, with IC_{50} inhibition value of 11.1 mg/mL and 8.2 mg/mL, respectively. Kusar and co-workers' findings were accomplished by X/XO reaction mixture with DEPMO-OOH scavenger that transformed the reaction to a stable radical measured by electro spin resonance (ESR). Valentao *et al.* [22] observed the phenolic acids (*p*-coumaric acid, ferulic acid, sinapic acid and kaempferol) exhibited superoxide scavenger activity and an inhibitory effect on XO. Considering the results obtained from TPC assay, it may be anticipated that *G. Lutea* extract has antioxidant activity achieved by the scavenging of superoxide radical and XO inhibition.

3.2. Antioxidant Effect in Stored o/w Emulsion

Many strategies on laboratory scale have been developed to improve the stability of shelf life in food models including adding a minimum amount of natural plants to delay the oxidation rate. The effect of *G. Lutea* on inhibiting lipid oxidation in oil-in-water (o/w) emulsion as a food model has not been described. In this study it also has been carried out to determine the synergic effect of *G. Lutea* with BSA in o/w emulsions. The oxidation in o/w emulsions was measured in two stages of oxidation; primary oxidation product (Peroxide Value) and secondary oxidation products (TBARS). In addition the change in pH was monitored, since pH tends to fall during oxidation.

3.2.1. Evolution of Peroxide Value (PV)

Figure 1 shows the evolution of PV *vs.* time. The control (without extract added) showed the highest oxidation throughout the storage time followed by the emulsion with only BSA (0.1%). The sample containing Trolox (0.35%, positive control) and the samples containing extracts, were not oxidized during the first 10 days. They show significant difference from the control ($p < 0.05$). The time required for the emulsions to reach a peroxide value of 10 meq hydroperoxides/kg of emulsion was determined as a standard to measure the stability of emulsion. The limits of fat product (animal, plant and anhydrous) margarine and fat preparation were set <10 meq hydroperoxides/kg as a guarantee of the product quality [33]. When the peroxide value of the sample is measured as greater than 15 meq hydroperoxide/kg, the sample is considered rancid, which may alter the color, taste and nutritional quality due to the deterioration of the lipid. The control was the first sample to reach 10 meq hydroperoxides/kg of emulsion which occurred rapidly in two days. The emulsion with BSA exhibited a similar deterioration rate to the control, revealing that BSA, in this concentration of 0.1%, does not provide any antioxidant effect in the emulsions. Positive control samples (Trolox) showed good antioxidant effect over 11 days and begin to oxidize rapidly after 15 days, reaching 88 meq hydroperoxides/kg emulsion on the final days of the experiment.

Figure 1. Change of peroxide value over time stored at 37 °C (each value is expressed as mean ($n = 3$)).

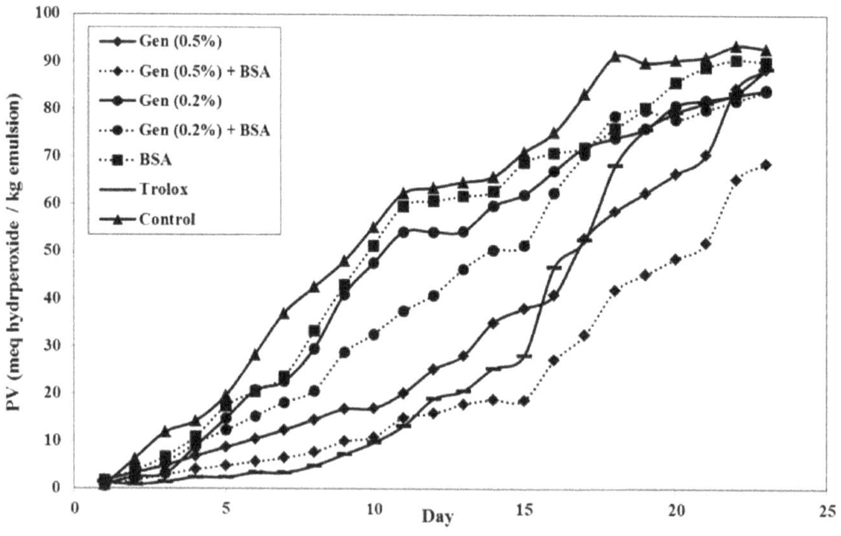

Adding 0.2% *G. Lutea* to the emulsion, with or without adding BSA, did not result in any relevant effect towards oxidation in the first stage (PV <10 meq hydroperoxides/kgin time <3 days). There is a significant difference between 0.2% antioxidant sample with BSA and in its absence ($p < 0.05$). The sample with 0.5% *G. Lutea* showed antioxidant activity towards lipid degradation first at 15 days and gradually oxidized after 15 days ($p < 0.05$). Finally, the sample with *G. Lutea* 0.5% and BSA 0.1%

displayed the lowest PV, with significant differences with the other samples throughout storage time ($p < 0.05$); it took almost 10 days to reach above 10 meq hydroperoxides/kg of emulsion.

Almajano *et al.* [34] reported that some antioxidant compounds such as EGCG and caffeic acid mixed with BSA cause a marked increase of the antioxidant activity in an emulsion. Since BSA is known to be surface active [35], the increase of antioxidant activity in emulsions containing a mixture of antioxidant and BSA could be due to BSA binding with the antioxidant and transporting it to the oil water interface, where it is highly effective in reducing the rate of oxidation. The authors also stated that the antioxidant molecule had bound to the BSA protein, proved by TEAC assay, and showed a progressive increase in the radical scavenging $ABTS^{\bullet+}$ with the storage time over several days [34]. The mixture of 0.1% BSA and 0.5% *G. Lutea* in emulsion exhibit the lowest oxidation rate compared to all samples shown in this experiment. These results showed for the first time the important effect of *G. Lutea* extract on lipid oxidation with synergic effect to BSA tested in an o/w emulsion. This concentration of 0.5%, demonstrated the best antioxidant effect throughout the storage period.

3.2.2. Evolution of pH over Time

Since decomposition of hydroperoxide measured in PV assay is acidic, the pH change in the sample is considered inversely proportional to the PV. Thus, the pH measured is a parameter of which its correlation with PV can be investigated. Antioxidant activity in food models is less effective under low pH conditions. However, some antioxidant compounds such as carnosic acid and carnosol (found in rosemary) have been reported to have high antioxidant activity at lower pH, which is at pH 4–5 [36]. Figure 2 shows the changes of pH value on emulsions over 23 days storage. Overall, the decrease in pH value throughout storage was of a similar order with increased primary oxidation measured in the PV assay. These results agreed with the studies done by Frankel *et al.* [16]. They found that the lipid oxidation in emulsion is slower at higher pH, and decreased when oxidation is accelerated. All samples started nearly at neutral pH and 0.5% *G. Lutea* with 0.1% BSA showed the highest value throughout storage. Similar to PV, the pH of the sample with 0.2% *G. Lutea* with or without BSA showed higher pH than control but the value was not significant throughout storage ($p > 0.05$). The pH of 0.5% *G. Lutea* with BSA and 0.5% *G. Lutea* alone, showed significant differences compared to all samples during storage period ($p < 0.05$). The behavior of pH of the sample with 0.5% *G. Lutea* with BSA was stable until 19 days before it started to decrease.

Skowyra *et al.* [37] demonstrated that the pH and PV have the best correlation with $R^2 = 0.9648$. Our results are in agreement with them and it can be described that the antioxidant activity in o/w emulsion which is stable at pH 6 showed an inverse relationship at a lower PV value. Some authors also reported a similar agreement of pH change which was inversely proportional to the lipid oxidation [26,38,39]. Meanwhile, Mancuso *et al.* [40,41] suggested the initial oxidation of emulsion depended on pH, by varying the effect of emulsifier. The authors observed a higher oxidation rate occurring at pH 7 rather than pH 3 o/w emulsion. Results may be due to the iron solubility increasing at low pH and allowing iron to be partitioned into the continuous phase, whereas insoluble iron at high pH may precipitate onto the emulsion droplet surface resulting in an increase in the lipid oxidation.

Figure 2. Change of pH over time, stored at 37 °C (each value is expressed as mean ($n = 3$)).

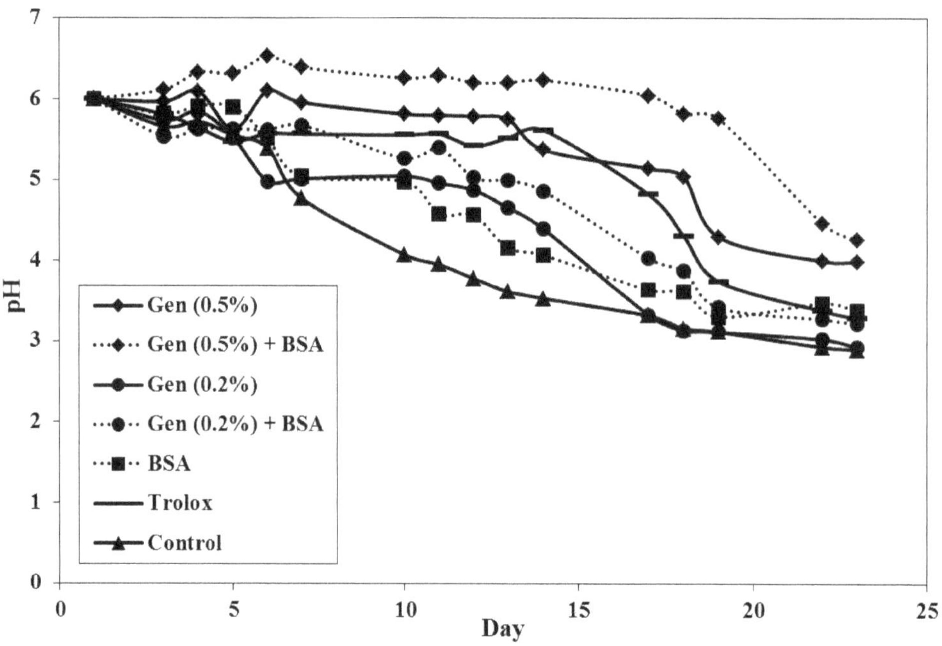

3.2.3. Evolution of Thiobarbituric Acid Reactive Substances (TBARS)

One of the compounds produced from secondary oxidation in lipids is MDA (malondialdehyde) which can be measured by the TBARS method. The secondary lipid oxidation is responsible for the alteration of flavor, rancid odor and the undesirable taste in foods [42]. Secondary oxidation products were monitored by TBARS assay and are shown in Figure 3. Similar to PV, the control had the most rapid increase in TBARS followed by the BSA sample. TBARS values for samples treated with 0.5% gentian powder, with and without BSA, experienced below 1.2 mg MDA/kg sample over the first 21 days and showed prominently lower than positive control up to 4 weeks ($p < 0.05$). The sample with 0.2% of *G. Lutea* alone does not display significant delay in lipid oxidation ($p > 0.05$), meanwhile 0.2% *G. Lutea* with BSA and positive control showed significant different during 20 days ($p \leq 0.05$). From the TBARS results exhibited, the synergic effect between both the concentration of *G. Lutea* and BSA in the emulsion during the storage time is demonstrated and the samples with both (gentian and BSA) show the lowest oxidation rate compared to all samples.

After 23 days, all samples are oxidized with above 1.2 mg malondialdehyde/kg sample even though *G. Lutea* with the BSA mixture showed minimum rise and had the best antioxidant effect in the emulsion. This behavior is not new. It has been previously reported that artificial antioxidants such as Trolox, epigallocatechingallate (EGCG), caffeic acid are more stable in emulsions during storage in the presence of BSA than in its absence [34].

G. Lutea has compounds which of family of iridoids, scoiridoids and xanthones [3]. It could be developed as antioxidant agent and radical scavengers and may contribute to the decrease of lipid oxidation [43]. The water soluble antioxidant molecular structures differ in the number of phenolic hydroxyl groups, their location and the carboxylic acid group [34]. Thus, the range of structure presented may allow any possible interaction with the BSA to be detected.

Figure 3. Change of Thiobarbituric Acid Reactive Substances (TBARS) over time, stored at 37 °C (each value is expressed as mean ($n = 3$)).

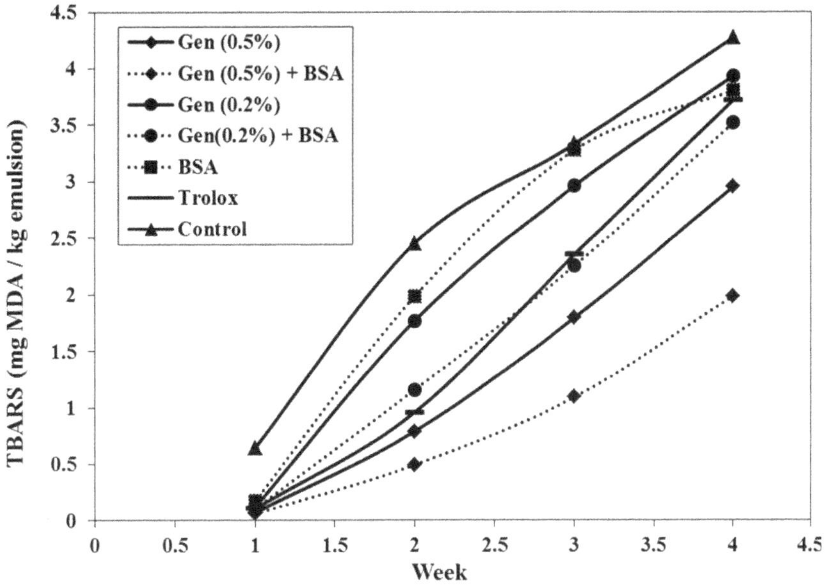

Many findings determined such amount of natural plants work as antioxidant efficiently; depended on variation of system tested and plant preparation. However, the more important is to identify the minimum concentration required to reduce lipid oxidation significantly throughout storage time. Adding minimum amount of natural plant not only can delay lipid oxidation, but also to avoid some changes on sensorial quality and flavoured. Analysis on our data showed for *G. Lutea* required as minimum 0.5% (w/w) would be beneficial for reducing the velocity of lipid oxidation significantly in both primary and secondary oxidation in emulsion system ($p < 0.05$). Adding also 0.1% (w/w) of BSA gave better effect of the antioxidant activity towards emulsion, compare to sample with *G. Lutea* alone. The study confirmed the potential of edible *G. Lutea* to prevent oxidation of emulsion.

3.3. HPLC Analysis of G. Lutea and the Total Antioxidant Activity Based on Post-Column On-Line Coupling ABTS$^{\bullet+}$

The HPLC analysis is targeted to identifysecoiridoid-glycoside, the bitter constituent occurred in the extract shown in (a) and (b). Results in Table 2 presentsecoiridoid glycoside; gentiopicroside, swerosideand amarogentin were the important compounds in the *G. Lutea*. The highest content of secoiridoidin *G. Lutea* extract in methanol-50% weregentiopicroside (1805 ± 62 mg/L extract) and the amount of sweroside found was 72 ± 4 mg/L extract. It was not possible to identifyamarogentin in the extract, meanwhile only low traces of amarogentinwere found in various commercial *G. Lutea* (less than 0.09%) [3]. Carnat *et al.* [28] also reported that amarogentin can only be found in some fresh root of *G. Lutea*. There are comprehensive studies measuring the active compound in *G. Lutea* using bioassay-guided fractionation such as HPLC [3], CapilaryElectrophoris [44] and Thin Layer Chromatography [45]. From the literature study, there is constituent of secoiridoidthat has not been identified in this study (swertiamarin)are believed to be existed in the extract [2]. This is most likely due to the objective of the method was not optimize the quantification of the traces but to measure the antioxidant activity of each compound in the *G. Lutea* extract.

Table 2. Amount of secoiridoid-glycoside quantified by HPLC.

Sample	Concentration (mg/L)
Gentiopocroside	1805 ± 62
Sweroside	72 ± 4
Amarogentin	n.d

n.d = not detected.

Aberham *et al.* [3] analyzed the active compound of 12 commercial samples of *G. Lutea* root. They found that swertiamarin was shown to have consistent occurrence between 0.21% and 0.45% and gentiopicroside was the most dominant compound in the sample up to 9.53%. Meanwhile, Ando *et al.* [46] reported that gentiopicroside was not detected from the fresh roots of 3-year-old *G. Lutea*. In contrast, Hayashi *et al.* [47] described that one year root contains high amounts of gentiopicroside and amarogentin and decreases over 5 years.

The Gentianaceae family is well known for its intensive bitter root used as a tonic for the digestive system with many pharmacological benefits. There are other plants such as *Swertiachirayita* [48] and *Lonicera japonica* [49] that also possess similar compounds of secoiridoid-glycosides (swertiamarin and sweroside).

Investigation of the main individual compounds in *G. Lutea* root was previously developed and optimized by Aberham *et al.* [2,3]. Furthermore, a substantial number of studies have demonstrated the effect of the secoiridoid group on the scavenging function to generate free radicals [7,13,48]. Wei *et al.* [13] reported that five secoiridoids, including gentiopicroside, sweroside, swertiamarin and sweroside, did not show any scavenging ability towards free radicals by *in vitro* DPPH assay. However, taking into account their report, it was desirable to explore more individual extracted compounds by isolating the compounds followed by a biochemical assay, such as ABTS radical, to measure their activity. Online post-column methods are very dependable because they combine systems for investigating different features of the sample simultaneously. Our initial observation of using *in vitro* ABTS assay showed that gentiopicroside and sweroside displayed no scavenging activity towards ABTS radicals (data not shown) while the activity of these compounds is similar towards DPPH radicals. However, our finding showed an activity of amarogentin analyzed by ABTS *in-vitro* assay (644.5 ± 17.5 mg eq TE/L sample). However, the amarogentin was not identified in the extract, as discussed above. This is in contrast with Phoboo *et al.* [48] who observed no scavenging activity of amarogentin towards DPPH radical.

The HPLC separated analytes reacted with ABTS radical post column (see Figure 4) and the reduction was detected as a negative peak at 734 nm. In Figure 5, the chromatographic analysis showed gentiopicroside (a) and sweroside (b) detected in *G. Lutea* extract and unknown compounds, (c) and (d), detected as negative peaks using the ABTS radical assay, which indicated that these components had free radical scavenging activity. The result of antioxidant response peaks (negative peak) of the *G. Lutea* compounds, expressed as mg galic acid equivalent (GAE)/L extract, indicates their relative contribution to the antioxidant activity of the extract with concentration taken into account. Analysis of *G. Lutea* extract of total antioxidant activity had been reported several times mainly measured by DPPH *in vitro* assay [10,14]. Even though we were unable to identify the compound relevant to the scavenging activity in the post-column ABTS assay, the results showed that the antiradical capacity of *G. Lutea* is not related to gentiopicroside and sweroside. There are many other compounds that maybe related to the scavenging activity in the extract such as xanthone-glycosides.

Figure 4. Scheme of HPLC-ABTS for screening of antioxidant compounds in *G. Lutea* root extract.

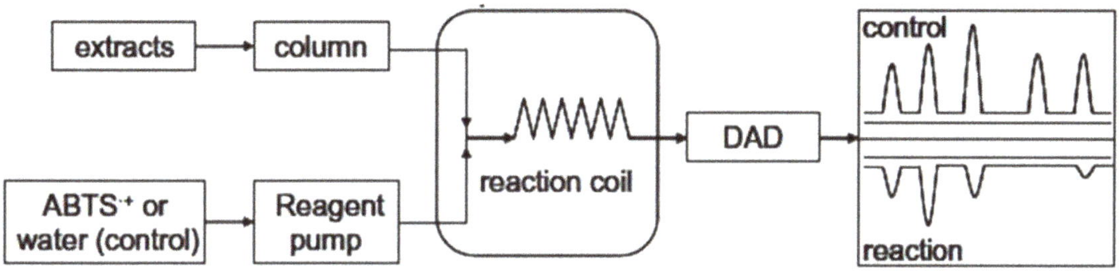

Figure 5. Chromatogram of *G. Lutea* root extract obtained direct and from the post-column HPLC-ABTS$^{\bullet+}$ radical scavenging method. (**a**) Gentiopicroside; (**b**) Sweroside; (**c**) antiradical activity by unknown compound; 31.33 ± 1.16 mg GAE/L and (**d**) antiradical activity by unknown compound; 8.30 ± 0.12 mg GAE/L.

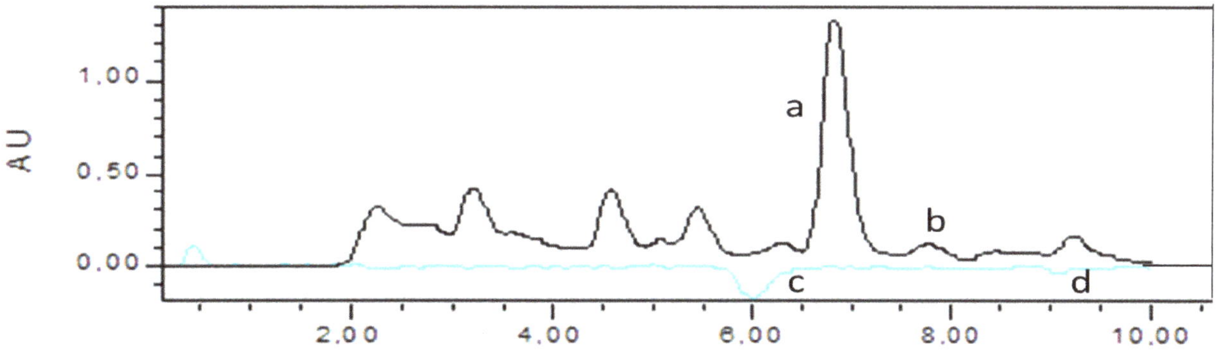

Nevertheless, our finding showed that the 0.5% *G. Lutea* extracts is able to delay the process of oxidation and give better storage stability as emulsion. However, this activity is not due to the bitter compounds (gentiopicroside amarogentin and sweroside) presented in our extract. Studies (from the revised literature) proved that xanthone compounds such as gentioside, gentisin and isogentisin found in *G. Lutea* have antiradical activity even though their constituents possess remarkable activity in pharmacological study

4. Conclusions

G. Lutea has valuable pharmacological properties due to the bitter properties of its secoiridoid-glycosides; their extraction using methanol-water mixture was better than water alone. *G. Lutea* extract showed excellent antioxidant activity in aqueous methanol measured by DPPH scavenging activity assay and Trolox equivalent capacity assay (TEAC) methods (15.89 and 48.90 µmol of TE/g DW, respectively). *G. Lutea* lyophilize can be applied as antioxidants in oil-in-water emulsions. 0.2% (w/w) of lyophilize. *G. Lutea* does not inhibit lipid oxidation significantly.

An amount of 0.5% (w/w) *G. Lutea* lyophilise exhibited antioxidant activity towards primary and secondary oxidation in an o/w emulsion. Adding 0.1% (w/w) BSA with *G. Lutea* in an emulsion showed a synergic effect and better activity in delaying lipid oxidation.

Gentiopicroside and sweroside found in HPLC analysis do not show any antiradical capacity in *G. Lutea* aqueous methanol extract. However, total antiradical capacities shown in post-column measurements presented activities of 31.33 ± 1.16 mg GAE/L and 8.30 ± 0.12 mg GAE/L towards the ABTS free radical. This study confirmed that *G. Lutea* roots as a source of edible natural antioxidants have potential to be used by the food industry.

Acknowledgments

The authors acknowledge the special fund from The Technical University of Catalonia, University Malaysia Pahang and Malaysia Government for the financial support of this research.

Author Contributions

Nurul Aini Mohd Azman and Francisco Segovia did the experimental section and the first writing; Xavier Martínez-Farré and Emilio Gil helped in the writing and correction of English; María Pilar Almajano has been the supervisor and manager of all the process.

Conflicts of Interest

The authors declare no conflict of interest.

References

1. Balijagić, J.; Janković, T.; Zdunić, G.; Bosković, J.; Savikin, K.; Godevac, D.; Stanojković, T.; Jovancević, M.; Menković, N. Chemical profile, radical scavenging and cytotoxic activity of yellow gentian leaves (*Genitaneae luteaefolium*) grown in northern regions of Montenegro. *Nat. Prod. Commun.* **2012**, *7*, 1487–1490.
2. Aberham, A.; Pieri, V.; Croom, E.M.; Ellmerer, E.; Stuppner, H. Analysis of iridoids, secoiridoids and xanthones in *Centaurium erythraea, Frasera caroliniensis* and *Gentiana lutea* using LC-MS and RP-HPLC. *J. Pharm. Biomed. Anal.* **2011**, *54*, 517–525.
3. Aberham, A.; Schwaiger, S.; Stuppner, H.; Ganzera, M. Quantitative analysis of iridoids, secoiridoids, xanthones and xanthone glycosides in *Gentiana lutea* L. roots by RP-HPLC and LC-MS. *J. Pharm. Biomed. Anal.* **2007**, *45*, 437–442.
4. Lian, L.H.; Wu, Y.L.; Wan, Y.; Li, X.; Xie, W.X.; Nan, J.X. Anti-apoptotic activity of gentiopicroside in D-galactosamine/Lipopolysaccharide-induced murine fulminant hepatic failure. *Chem. Biol. Interact.* **2010**, *188*, 127–133.
5. Pal, D.; Sur, S.; Mandal, S.; Das, A.; Roy, A.; Das, S.; Panda, C.K. Prevention of liver carcinogenesis by amarogentin through modulation of G 1/S cell cycle check point and induction of apoptosis. *Carcinogenesis* **2012**, *33*, 2424–2431.
6. Tan, R.X.; Kong, L.D.; Wei, H.X. Secoiridoid glycosides and an antifungal derivative from gentian 4 tibetica. *Phytochemistry* **1998**, *47*, 1223–1226.

7. Kumarasamy, Y.; Nahar, L.; Sarker, S. Bioactivity of gentiopicroside from the aerial parts of Centaurium erythraea. *Fitoterapia* **2003**, *74*, 151–154.
8. Jia, N.; Li, Y.; Wu, Y.; Xi, M.; Hur, G.; Zhang, X.; Cui, J.; Sun, W.; Wen, A. Comparison of the anti-inflammatory and analgesic effects of *Gentiana macrophylla* Pall. and *Gentiana straminea* Maxim., and identification of their active constituents. *J. Ethnopharmacol.* **2012**, *144*, 638–645.
9. Mihailović, V.; Mihailović, M.; Uskoković, A.; Arambašić, J.; Mišić, D.; Stanković, V.; Katanić, J.; Mladenović, M.; Solujić, S.; Matić, S. Hepatoprotective effects of *Gentiana asclepiadea* L. extracts against carbon tetrachloride induced liver injury in rats. *Food Chem. Toxicol.* **2013**, *52*, 83–90.
10. Nastasijević, B.; Lazarević-Pašti, T.; Dimitrijević-Branković, S.; Pašti, I.; Vujačić, A.; Joksić, G.; Vasić, V. Inhibition of myeloperoxidase and antioxidative activity of *Gentiana lutea* extracts. *J. Pharm. Biomed. Anal.* **2012**, *66*, 191–196.
11. Jiang, D.J.; Dai, Z.; Li, Y.J. Pharmacological effects of xanthones as cardiovascular protective agents. *Cardiovasc. Drug Rev.* **2004**, *22*, 91–102.
12. Rana, V.S.; Rawat, M.S.M. A new xanthone glycoside and antioxidant constituents from the rhizomes of *Swertia speciosa*. *Chem. Biodivers.* **2005**, *2*, 1310–1315.
13. Wei, S.; Chen, G.; He, W.; Chi, H.; Abe, H.; Yamashita, K.; Yokoyama, M.; Kodama, H. Inhibitory effects of secoiridoids from the roots of *Gentiana straminea* on stimulus-induced superoxide generation, phosphorylation and translocation of cytosolic compounds to plasma membrane in human neutrophils. *Phytother. Res.* **2012**, *26*, 168–173.
14. Kintzios, S.; Papageorgiou, K.; Yiakoumettis, I.; Baricevic, D.; Kusar, A. Evaluation of the antioxidants activities of four Slovene medicinal plant species by traditional and novel biosensory assays. *J. Pharm. Biomed. Anal.* **2010**, *53*, 773–776.
15. Estévez, M.; Ventanas, S.; Cava, R. Food chemistry and toxicology protein oxidation in frankfurters with increasing levels of added rosemary essential oil: Effect on color and texture deterioration. *J. Food. Sci.* **2005**, *70*, 427–432.
16. Frankel, E.N.; Huang, S.; Aeschbach, R. Antioxidant activity of green teas in different lipid systems. *J. Am. Oil Chem. Soc.* **1997**, *74*, 1309–1315.
17. Kiokias, S.; Dimakou, C.; Oreopoulou, V. Activity of natural carotenoid preparations against the autoxidative deterioration of sunflower oil-in-water emulsions. *Food Chem.* **2009**, *114*, 1278–1284.
18. Zhang, S.; Bi, H.; Liu, C. Extraction of bio-active components from *Rhodiola sachalinensis* under ultrahigh hydrostatic pressure. *Sep. Purif. Technol.* **2007**, *57*, 277–282.
19. Santas, J.; Carbo, R.; Gordon, M.; Almajano, M. Comparison of the antioxidant activity of two Spanish onion varieties. *Food Chem.* **2008**, *107*, 1210–1216.
20. Miller, N.J.; Sampson, J.; Candeias, L.P.; Bramley, P.M.; Rice-Evans, C.A. Antioxidant activities of carotenes and xanthophylls. *FEBS Lett.* **1996**, *384*, 240–242.
21. Madhujith, T.; Shahidi, F. Optimization of the extraction of antioxidative constituents of six barley cultivars and their antioxidant properties. *J. Agric. Food Chem.* **2006**, *54*, 8048–8057.
22. Valentão, P.; Fernandes, E.; Carvalho, F.; Andrade, P.B.; Seabra, R.M.; Bastos, M.L. Antioxidant activity of *Centaurium erythraea* infusion evidenced by its superoxide radical scavenging and xanthine oxidase inhibitory activity. *J. Agric. Food Chem.* **2001**, *49*, 3476–3479.

23. López-Cruz, R.I.; Zenteno-Savín, T.; Galván-Magaña, F. Superoxide production, oxidative damage and enzymatic antioxidant defenses in shark skeletal muscle. *Comp. Biochem. Physiol.* **2010**, *156*, 50–56.
24. Yoshida, H.; Kajimoto, G.; Emura, S. Antioxidant effects of D-tocopherols at different concentrations in oils during microwave heating. *J. Am. Oil Chem. Soc.* **1993**, *70*, 989–995.
25. American Oil Chemists' Society. *AOCS Official Method Cd 8-53*; Firestone, D., Ed.; Official Methods and Recommended Practices of the American Oil Chemists' Society: Champaign, IL, USA, 1997.
26. Gallego, M.G.; Gordon, M.H.; Segovia, F.J.; Skowyra, M.; Almajano, M.P. Antioxidant properties of three aromatic herbs (rosemary, thyme and lavender) in oil-in-water emulsions. *J. Am. Oil Chem. Soc.* **2013**, *90*, 1559–1568.
27. Koleva, I.I.; Niederländer, H.A.; van Beek, T. Application of ABTS radical cation for selective on-line detection of radical scavengers in HPLC eluates. *Anal. Chem.* **2001**, *73*, 3373–3381.
28. Carnat, A.; Fraisse, D.; Carnat, A.-P.; Felgines, C.; Chaud, D.; Lamaison, J.-L. Influence of drying mode on iridoid bitter constituent levels in gentian root. *J. Sci. Food Agric.* **2005**, *85*, 598–602.
29. Kusar, A.; Zupancic, A.; Sentjurc, M.; Baricevic, D. Free radical scavenging activities of yellow gentian (*Gentiana lutea* L.) measured by electron spin resonance. *Hum. Exp. Toxicol.* **2006**, *25*, 599–604.
30. Almajano, M.P.; Carbó, R.; Jiménez, J.A.L.; Gordon, M.H. Antioxidant and antimicrobial activities of tea infusions. *Food Chem.* **2008**, *108*, 55–63.
31. Deighton, N.; Brennan, R.; Finn, C.; Davies, H.V. Antioxidant properties of domesticated and wild Rubus species. *J. Sci. Food Agric.* **2000**, *80*, 1307–1313.
32. Gazzani, G.; Papetti, A.; Massolini, G.; Daglia, M. Anti- and prooxidant activity of water soluble components of some common diet vegetables and the effect of thermal treatment. *J. Agric. Food Chem.* **1998**, *46*, 4118–4122.
33. Nollet, L.M.L.; Toldra, F. *Handbook of Analysis of Edible Animal By-Products*; CRC Press: Gent, Belgium, 2011; p. 471.
34. Almajano, M.P.; Gordon, M.H. Synergistic effect of BSA on antioxidant activities in model food emulsions. *Am. Oil Chem. Soc.* **2004**, *81*, 275–280.
35. Rampon, V.; Lethuaut, L.; Mouhous-Riou, N.; Genot, C. Interface characterization and aging of bovine serum albumin stabilized oil-in-water emulsions as revealed by front-surface fluorescence. *J. Agric. Food Chem.* **2001**, *49*, 4046–4051.
36. Frankel, E.N.; Huang, S.-W.; Aeschbach, R.; Prior, E. Antioxidant activity of a rosemary extract and its constituents, carnosic acid, carnosol, and rosmarinic acid, in bulk oil and oil-in-water emulsion. *J. Agric. Food Chem.* **1996**, *44*, 131–135.
37. Skowyra, M.; Falguera, V.; Gallego, G.; Peiró, S.; Almajano, M.P. Antioxidant properties of aqueous and ethanolic extracts of tara (*Caesalpinia spinosa*) pods *in vitro* and in model food emulsions. *J. Sci. Food Agric.* **2013**, *94*, 911–918.
38. Sørensen, A.-D.M.; Haahr, A.-M.; Becker, E.M.; Skibsted, L.H.; Bergenståhl, B.; Nilsson, L.; Jacobsen, C. Interactions between iron, phenolic compounds, emulsifiers, and pH in omega-3-enriched oil-in-water emulsions. *J. Agric. Food Chem.* **2008**, *56*, 1740–1750.

39. Donnelly, J.L.; Decker, E.A.; McClements, D.J. Iron-catalyzed oxidation of menhaden oil as affected by emulsifiers. *J. Food Sci.* **1998**, *63*, 997–1000.
40. Mancuso, J.R.; McClements, D.J.; Decker, E.A. Ability of iron to promote surfactant peroxide decomposition and oxidize alpha-tocopherol. *J. Agric. Food Chem.* **1999**, *47*, 4146–4149.
41. Mancuso, J.R.; McClements, D.J.; Decker, E.A. The effects of surfactant type, pH, and chelators on the oxidation of salmon oil-in-water emulsions. *J. Agric. Food Chem.* **1999**, *47*, 4112–4116.
42. Pangloli, P.; Melton, S.L.; Collins, J.L.; Penfield, M.P.; Saxton, A.M. Flavor and storage stability of potato chips fried in cottonseed and sunflower oils and palm olein/sunflower oil blends. *J. Food Sci.* **2002**, *67*, 97–103.
43. Wu, Q.-X.; Li, Y.; Shi, Y.-P. Antioxidant phenolic glucosides from Gentiana piasezkii. *J. Asian Nat. Prod. Res.* **2006**, *8*, 391–396.
44. Citová, I.; Ganzera, M.; Stuppner, H.; Solich, P. Determination of gentisin, isogentisin, and amarogentin in *Gentiana lutea* L. by capillary electrophoresis. *J. Sep. Sci.* **2008**, *31*, 195–200.
45. Skrzypczak, L.; Wesolowska, M.; Bajaj, Y.P.S. *Medicinal and Aromatic Plants IV*; Springer Verlag: Berlin, Germany, 1993; Volume 21, pp. 172–186.
46. Ando, H.; Hirai, Y.; Fujii, M.; Hori, Y.; Fukumura, M.; Niiho, Y.; Nakajima, Y.; Shibata, T.; Toriizuka, K.; Ida, Y. The chemical constituents of fresh *Gentian* root. *J. Nat. Med.* **2007**, *61*, 269–279.
47. Hayashi, T.; Minamiyama, Y.; Miura, T.; Yamagishi, T.; Kaneshina, H. Cultivation of *Gentiana lutea* and chemical evaluation evaluation of gentiana radix. *Hokkaidoritsu eisei Kenkyushoho* **1990**, *40*, 103–106.
48. Phoboo, S.; Pinto, M.D.S.; Barbosa, A.C.L.; Sarkar, D.; Bhowmik, P.C.; Jha, P.K.; Shetty, K. Phenolic-linked biochemical rationale for the anti-diabetic properties of *Swertia chirayita* (Roxb. ex Flem.) Karst. *Phytother. Res.* **2013**, *27*, 227–235.
49. Oku, H.; Ogawa, Y.; Iwaoka, E.; Ishiguro, K. Allergy-preventive effects of chlorogenic acid and iridoid derivatives from flower buds of *Lonicera japonica*. *Biol. Pharm. Bull.* **2011**, *34*, 1330–1333.

Changes of Major Antioxidant Compounds and Radical Scavenging Activity of Palm Oil and Rice Bran Oil during Deep-Frying

Azizah Abdul Hamid [1,*], Mohd Sabri Pak Dek [1], Chin Ping Tan [2], Mohd Asraf Mohd Zainudin [1] and Evelyn Koh Wee Fang [1]

[1] Department of Food Science, Faculty of Food Science and Technology, Universiti Putra Malaysia, UPM Serdang, Selangor 43400, Malaysia; E-Mails: mpakdek@uoguelph.ca (M.S.P.D.); asraf_zainudin@yahoo.com (M.A.M.Z); evelyn_kwf@hotmail.com (E.K.W.F.)

[2] Department of Food Technology, Faculty of Food Science and Technology, Universiti Putra Malaysia, UPM Serdang, Selangor 43400, Malaysia; E-Mail: tancp@upm.edu.my

* Author to whom correspondence should be addressed; E-Mail: azizahah@upm.edu.my

---

Abstract: Changes in antioxidant properties and degradation of bioactives in palm oil (PO) and rice bran oil (RBO) during deep-frying were investigated. The alpha (α)-tocopherol, gamma (γ)-tocotrienol and γ-oryzanol contents of the deep-fried oils were monitored using high performance liquid chromatography, and antioxidant activity was determined using 2-diphenyl-1-picryl hydrazyl (DPPH) radical scavenging activity. Results revealed that the antioxidant activity of PO decreased significantly ($p < 0.05$), while that of RBO was preserved after deep-frying of fries. As expected, the concentration of α-tocopherol in PO and γ-tocotrienol in both PO and RBO decreased significantly ($p < 0.05$) with increased frying. Results also showed that γ-tocotrienol was found to be more susceptible to degradation compared to that of α-tocopherol in both PO and RBO. Interestingly, no significant degradation of α-tocopherol was observed in RBO. It is suggested that the presence of γ-oryzanol and γ-tocotrienol in RBO may have a protective effect on α-tocopherol during deep-frying.

Keywords: vegetable oils; deep-frying; kinetic degradation; free radical scavenging; oil antioxidants

1. Introduction

Vegetable oils are generally known to be healthy, due to being high in unsaturated fatty acid and other various phytochemical compounds. Palm oil (PO) is produced from the fruit bunches of the tree oil palm (*Elaeis guineensis*) and has been used for food preparations, as well as other commercial applications in many parts of the world. Yellow palm olein is well known for its high concentration of lipophilic antioxidants, in particular, tocopherols and tocotrienols, as compared to that of other vegetable oils, but lacking in carotenoids, which are removed during the refining process [1]. In addition, PO contains an equal ratio of unsaturated to saturated fatty acids, making it a versatile oil for commercial usage without involving major modification processes [2].

Rice bran oil (RBO) has been used as cooking oil in some parts of the world, particularly in Japan and India [3]. Rice bran oil has been reported to be an excellent source of antioxidants, such as tocopherols, tocotrienols and oryzanols [4]. Generally, RBO consists of up to 1.5% of oryzanol depending on the variety of the rice and is well known for its stability as a frying oil [5,6]. The role played by both tocols and oryzanol in improving lipid stability and human health is fascinating, although the use of rice bran oil is not as popular as that of other vegetables oils [7,8]. Studies have shown that tocopherols, tocotrienol, carotenoids and oryzanols exhibited potent antioxidant properties in both *in vivo* and *in vitro* systems [9,10].

Deep-frying is a convenient and rapid process to produce fried foods with intense flavor [11]. In this process, food is immersed in edible oil at a maintained temperature (150 to 200 °C) for a specific period of time [12]. However, in certain conditions, instead of the desirable intensive flavor, undesirable off-odor and toxic compounds from lipid peroxidation can also be produced during deep-frying, which is normally associated with the type of oil used. The alteration of flavor and oil quality during deep-frying can occur via the hydrolysis, oxidation and polymerization of the oil used as a result of the long exposure to high temperature [13]. Several *in vivo* studies revealed the existence of the relationship between deep-frying oil quality intake with oxidative stress level. The intake of such an altered oil quality could affect both the plasma and mitochondrial membrane [14,15].

Although oils, like PO and RBO, contain bioactive compounds that can protect lipids from oxidation, the reaction can still occur during deep-frying, which results from the degradation of the bioactive compounds, as affected by exposure to high temperature [3,13]. Therefore, it is important to evaluate the stability and degradation pattern of bioactive compounds in the oils during food preparations, in an effort to get a better picture on effective oil usage before complete depletion of bioactive compounds occur. The consumer can then gain the beneficial effects from such oils, including the lowering of serum cholesterol and blood pressure, reducing the risk of diabetic necropathy and cardio-reperfusion injury [8,16,17]. Chiou *et al.* [18] reported that during frying, foods absorb oil, the composition of which is similar to that remaining in the frying pan. Therefore, it is vital to investigate the quality of the oils during and after the frying process.

A number of studies have provided information on the quality deterioration of frying oils, but only a few studies focused on the stability of bioactive compounds during deep-frying [19–21]. The information regarding the antioxidant degradation kinetic and the changes of these bioactive compounds during deep-frying is relatively scarce. Hence, this study was conducted to evaluate the radical scavenging capacity and prominent bioactive compounds (α-tocopherol, γ-tocotrienol and γ-oryzanol) in PO and RBO, as well as to examine their degradation kinetic during the deep-frying of French fries.

2. Experimental Section

2.1. Materials

Yellow palm olein oil (Buruh Brand, Shah Alam, Malaysia), rice bran oil (Green Love Brand, Amorchai Co., Ltd., Bangkok, Thailand) and French fries (KG Brand, Shah Alam, Malaysia) were purchased from a local supermarket in Serdang, Selangor, Malaysia. Alpha-tocopherol, γ-tocotrienol and 2,2-diphenyl-1-picrylhydrazyl (DPPH) were purchased from Sigma-Aldrich, Steinheim, Germany. The gamma-oryzanol standard was purchased from Wako Pure Chemical Industries Ltd., Beijing, China. Hexane, isopropyl alcohol, acetonitrile, butanol, acetic acid and isooctane were purchased from Fisher Scientific, U.K. All other reagents and chemicals used were of analytical or HPLC grade.

2.2. Deep-Frying Model

A deep-frying model as described by Schroeder *et al.* [22] was adopted with slight modification. Frying oil with an initial volume of 3 L was heated to 180 ± 5 °C in a kitchen fryer (Philux, Model DF30AIT, Libertronic Sdn Bhd, Seri Kembangan Selangor). French fries (100 g) were then fried in the oil for 2 min, denoted as first batch deep-frying. The fried French fries were then removed, and the deep-frying was repeated with a new batch of French fries without any time lag. This procedure was repeated until the 60th frying cycle. An aliquot of the oil (30 mL) was sampled after each tenth batch of French fries deep-frying to be analyzed and compared with that of fresh oil.

2.3. Determination of Free Radical Scavenging Capacity

The antioxidant capacity of fresh PO and RBO and after deep-frying (20th, 40th and 60th frying cycles) were determined using the 2,2-diphenyl-1-picryl hydrazyl (DPPH) radical scavenging assay as described by Lee *et al.* [23]. Briefly, 0.7 grams of oil samples were dissolved in 10 mL isooctane as a stock solution. Dissolved oil (0.25 mL) was then added to 1.75 mL DPPH solution (25 mg·L^{-1} prepared in isooctane). The mixture was then vortexed and incubated at ambient temperature for 30 min. The absorbance was measured at 515 nm using a UV-Vis spectrophotometer (EL800, Biotek, Winooski, VT, USA). The radical scavenging activity was calculated using the following formula:

$$Radical\ scavenging\ activity\ (\%) = \frac{(Abs\ Control - Abs\ Sample)}{Abs\ Control} \times 100\%$$

where Abs control is the absorbance of the control reaction (containing all reagents except the test compound) and Abs sample is the absorbance of the test sample. Synthetic antioxidant butylated hydroxyanisole (BHA) and α-tocopherol were used as positive controls.

2.4. Determination of α-Tocopherol and γ-Tocotrienol

Alpha-tocopherol and γ-tocotrienol were determined according to the American Oil Chemists' Society (AOCS) Ce 8-89 [24] method with slight modifications. Oil was dissolved with 10 mL of hexane:isopropyl alcohol (ratio 98:2) and filtered through a 0.45 μm (Whatman) nylon membrane filter prior to injecting into a HPLC. An aliquot of sample (20 μL) was analyzed using a normal phase HPLC system equipped with chromatography software (Millennium LC-6A, Shimadzu, Kyoto, Japan), a UV detector (SPD-6A, Shidmazu, Kyoto, Japan) and pumps. The separation was performed with an ACE 5 Silica column (250 × 4.6 mm, 5 μm) operated at ambient temperature. The mobile phase consisted of hexane:isopropyl alcohol at a ratio of 98:2 (v/v) with a flow rate of 1 $mL \cdot min^{-1}$, and the detector was set at 295 nm. The identification of unknown tocopherol and tocotrienol isomers was based upon matching their retention time with those of pure standards, while the concentrations calculated using the peak area based on standard calibration curves.

2.5. Determination of γ-Oryzanol in Rice Bran Oil

The content of γ-oryzanol was determined according to the method of Xu and Godber [25]. An aliquot (0.1 g) of oil was dissolved in 10 mL hexane:isopropyl alcohol at a ratio of 98:2 (v/v). The dissolved oil sample was then filtered through a 0.45 μM (Whatman) nylon membrane filter prior to injecting into a HPLC. The filtered sample (20 μL) was analyzed using an HPLC system (Millennium LC-6A, Shimadzu, Tokyo, Japan) chromatography manager, a UV detector (SPD-6A, Shidmazu, Tokyo, Japan) and pumps. The separation was carried using a Waters reverse-phase (RP) μBondapak C18 column (250 × 4.6 mm, 5 μM; Waters Corp., Milford, MA, USA) at ambient temperature. The mobile phase consisted of acetonitrile/butanol/acetic acid/water at a ratio of 94:3:2:1 (v/v/v/v). The flow rate was kept at 1 $mL \cdot min^{-1}$ with the isocratic mode, and the detector was set at 330 nm. Identification of the γ-oryzanol component in the RBO was done by comparing its retention with that of pure standard, and quantification was done based on the standard calibration curve.

2.6. Determination of Degradation Kinetics Using the Order of Reaction Equation

According to Taoukis *et al.* [26], the degradation kinetics of many compounds in foods at constant temperature follows the first-order kinetics model, which can be expressed as follows:

$$-dC/dt = kC \tag{1}$$

Integration of Equation 1 gives:

$$\ln C = \ln Co - kt \tag{2}$$

$$\ln C/Co = -kt \tag{3}$$

where C is the concentration of the compound, Co is the initial concentration of the compound, t is time and k is the reaction rate constant. By plotting the logarithm of the concentration of α-tocopherol and γ-tocotrienol, respectively, over the initial concentration against a number of frying cycles during the degradation process, the degradation rate constant for each compound was calculated from the slope of the simple linear regression line.

2.7. Statistical Analysis

All frying experiments and analyses were conducted in triplicate. Data was analyzed statistically through one-way ANOVA by Duncan's multiple range tests using the commercially available software, the SPSS 16 software program (SPSS Inc., Chicago, IL, USA). A *p*-value of less than 0.05 ($p < 0.05$) was considered to denote statistical significance.

3. Results and Discussion

3.1. Antioxidant Capacity of the Oils

In this study, free radical scavenging activity utilizing the DPPH radical was done in determining the antioxidant capacity of the oils, due to its simplicity and wide use in antioxidant research. The free radical scavenging capacity of PO and RBO after the 20th, 40th and 60th frying cycle is shown in Figure 1. Results showed that there is no significant difference in the antioxidant capacity between the fresh and the 20th frying cycle PO. However, a significant ($p < 0.05$) decrease in the activity was observed for the oil from the 40th and 60th frying cycle. On the other hand, there was no significant difference in the antioxidant activity exhibited by all the batches of RBO. This indicates that RBO is more stable compared to that of PO upon deep-frying. It was also noted that RBO derived from the 40th and 60th frying cycle exhibited significantly ($p < 0.05$) higher radical scavenging activity than that of PO.

The radical scavenging assay is one of the common methods that can be used to determine the antioxidant capacity of vegetable oils [27]. The results of the study showed that both PO and RBO exhibited good antioxidant capacity. However, the antioxidant capacity of PO decreased with deep-frying. This is probably attributed to the degradation of bioactive compounds as a result of the exposure to high temperature used during deep-frying. Results presented here are in line with that of a previous study conducted by Gomez-Alonso *et al.* [28], where the reduction of antioxidant activity of olive oil correlated well with the number of deep-frying cycles. However, similar the destruction of some of these bioactive compounds may not be occurring in RBO, as the antioxidant activity in the oil was preserved throughout the frying process.

3.2. Degradation of α-Tocopherol and γ-Tocotrienol during Deep-Frying

The degradation of α-tocopherol in PO and RBO during deep-frying is shown in Figure 2. As expected, results showed that the concentration of α-tocopherol in PO decreased significantly with deep-frying. Similar to that of its antioxidant activity, α-tocopherol content decreased from 1.29 $mg{\cdot}g^{-1}$ to 0.59 $mg{\cdot}g^{-1}$ after the 60th frying cycle, depicting a loss of 54%. Surprisingly, α-tocopherol in RBO was found to be more stable and remained almost unaffected upon deep-frying up to the 60th frying cycle. The contents of α-tocopherol in fresh RBO and the 60th frying cycle were 0.75 mg·g^{-1} and 0.73 mg·g^{-1}, respectively, a loss of only 2%. This is reflected in the antioxidant activity exhibited by the oil. The degradation of α-tocopherol in PO is in agreement with that of previous literature [14,15]. Battino *et al.* [14] found that α-tocopherol in olive oil was degraded up to 28% from the initial value after frying for 60 min.

Figure 1. The percentage of radical-scavenging activity in palm oil and rice bran oil over the 60th batch of deep-frying at a concentration of 70 mg/mL. The values given are means ± standard deviation of a triplicate analysis. The values marked with the same letters are not significantly different at $p < 0.05$ analyzed using the Duncan multiple range test. PO, palm oil; RBO, rice bran oil.

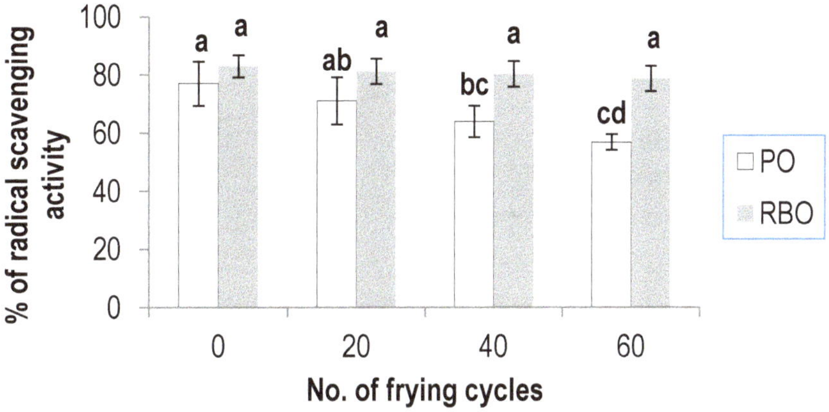

Figure 2. The degradation of α-tocopherol in palm oil and rice bran oil during deep-frying. The values given are means ± standard deviation of triplicate analysis. The values marked with the same letters are not significantly different at $p < 0.05$ analyzed using the Duncan multiple range test. A,B, the values with different capital letters indicate a significant difference between the type of oil at $p < 0.05$. a,b, the values with different small letters indicate a significant difference between frying cycle numbers at $p < 0.05$.

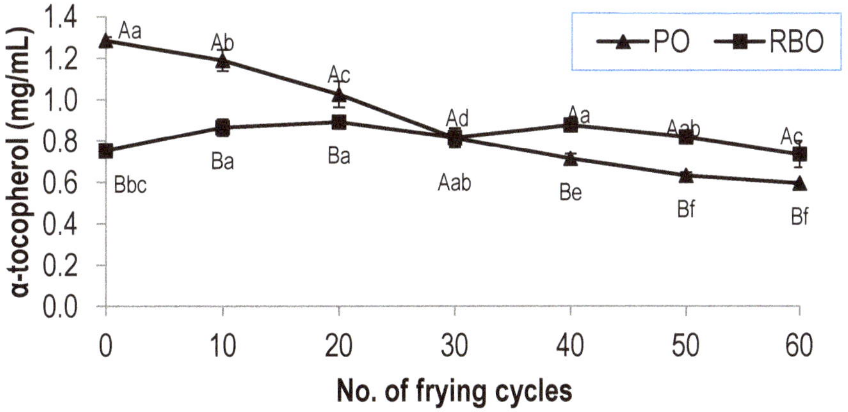

Figure 3 showed the degradation of γ-tocotrienol in deep-fried PO and RBO. There was a similar trend on the loss of γ-tocotrienol in both oils during the deep-frying of French fries, where it decreased significantly with frying. However, interestingly, γ-tocotrienol's content in RBO was always higher than that of PO in all batches of oils. The fresh PO and the 60th frying cycle PO consisted of 0.79 $mg \cdot g^{-1}$ and 0.28 $mg \cdot g^{-1}$ tocotrienol, respectively depicting a loss of 65%. Whereas, the content of γ-tocotrienol in fresh RBO and the 60th frying cycles of RBO were 1.34 $mg \cdot g^{-1}$ and 0.77 $mg \cdot g^{-1}$, respectively, showing a lower loss of 42%.

Figure 3. The degradation of γ-tocotrienol in palm oil and rice bran oil during deep-frying. The values given are means ± standard deviation of a triplicate analysis. The values marked with the same letters are not significantly different at $p < 0.05$ analyzed using the Duncan multiple range test. A,B, the values with different capital letters indicate a significant difference between the type of oil at $p < 0.05$. a,b, the values with different small letters indicate a significant difference between frying cycle numbers at $p < 0.05$.

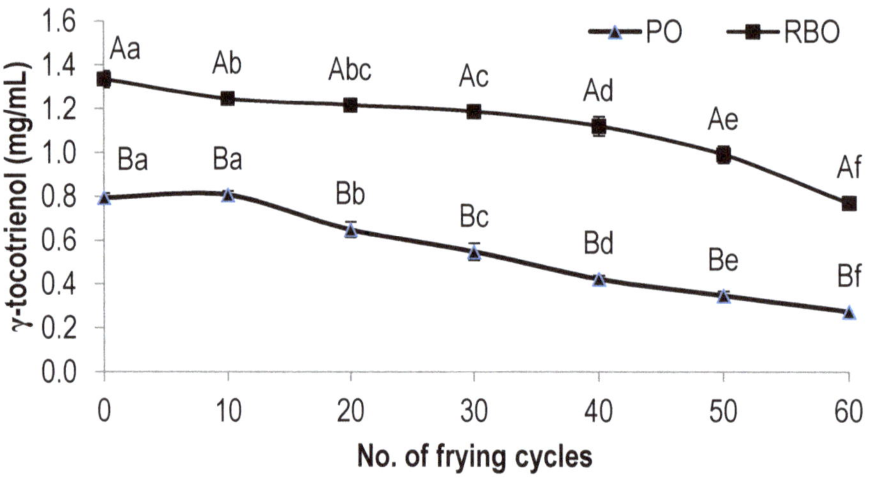

During the deep-frying of fries, the decrease in the concentration of tocopherols and tocotrienols in PO might have occurred due to their protective role against oxidation and degradation upon exposure to high temperature. A similar observation has been reported by Hoffmann [29], where at a high temperature, tocopherols and tocotrienols tend to protect the oil by oxidizing themselves to quinones and dimers. The present result also indicated that α-tocopherol and γ-tocotrienol in PO degraded at a faster rate than that of RBO. Generally, PO consisted of an equal amount of saturated fatty acid to unsaturated fatty acid, whereas RBO contained a higher amount of unsaturated fatty acid, usually up to 74%. A higher proportion of unsaturated fatty acids in RBO, relative to PO, could be correlated to the greater number of double bonds in the former oil [21,22]. Therefore, it can be suggested that during the deep-frying of RBO, these unsaturation might have competed with tocopherols and tocotrienols as substrates for oxidation, resulting in a slower degradation of these antioxidants. In addition, a higher degree of unsaturation, as a basis of more competitiveness of the fatty acids towards oxidation, may explain the high degradation rate of α-tocopherol and γ-tocotrienol in PO as compared with that of RBO [30]. This is in agreement with that reported by Hoffmann [29], who showed that during the propagation phase of oxidation, the fatty acid peroxy free radicals may react preferentially with the phenolic hydrogen molecule of tocopherol. Seppanen *et al.* [31] reported that the antioxidant activities of tocopherols and tocotrienols in oils are mainly due to their abilities to donate phenolic hydrogen to reactive lipid free radicals, thus retarding the normal steps of autocatalytic lipid peroxidation.

Tocopherols can effectively scavenge peroxy radicals and yield relatively stable products, which interrupt the propagation stage of the oxidative chain reaction, thereby preventing the destruction of lipid molecules [32,33].

It is interesting to note that the concentration of α-tocopherol in RBO was maintained during the deep-frying of the fries exhibiting only a low percent loss (Figure 2). In contrast, the content of γ-tocotrienol in RBO decreased considerably with increased frying. In RBO, γ-tocotrienol is expected to possess higher antioxidant activity than that of α-tocopherol. This was also applied to PO, which exerts the same trend [22]. Therefore, it can be suggested that γ-tocotrienol was firstly oxidized in order to protect other weak antioxidants. This is in agreement with that reported by Rossi *et al.* [30], where among vitamin E homologs, γ-tocotrienol was found to be the least stable and easily oxidized during deep-frying, which might be attributed mainly to its protective role and self-degradation in preserving other homologues, such as tocopherols. Apart from that, the initial content of γ-tocotrienol in fresh RBO was significantly higher than that in PO, although the concentration of both decreased with an increase in the number of frying cycles. Therefore, with a higher concentration, the protective effect of γ-tocotrienol in preserving α-tocopherol was evident.

3.3. Degradation of γ-Oryzanol in Rice Bran Oil during Deep-Frying

Figure 4 showed the degradation of γ-oryzanol in RBO over the 60th frying cycles. The concentration of γ-oryzanol in RBO was seen to decrease significantly ($p < 0.05$) in accordance to the number of frying batches. Fresh oil, containing 3.0 $mg{\cdot}g^{-1}$ γ-oryzanol, was reduced to 2.8 $mg{\cdot}g^{-1}$ at the 20th cycles, 2.6 $mg{\cdot}g^{-1}$ after 40 cycles and 2.4 $mg{\cdot}g^{-1}$ after 60 batches. The remaining γ-oryzanol content in RBO after the 60th frying cycles was estimated to be 80% of the original amount in the fresh oil.

RBO naturally contains α-tocopherol, γ-tocotrienol and γ-oryzanol, making it a suitable model to study the synergistic interaction between these bioactive compounds in protecting lipids from oxidation. The ability of a compound to inhibit the oxidation process is via several mechanisms. Gamma-oryzanol is one of the most potent antioxidant compounds reported. It is interesting to note that besides acting as an antioxidant to prevent oil from oxidation, γ-oryzanol is also associated with decreasing cholesterol in plasma, platelet aggregation and cholesterol absorption [34–36]. In addition, γ-oryzanol has also been used to treat nerve imbalance and disorders of menopause in women [37].

The results in this study indicated that γ-oryzanol in RBO was degraded with the number of frying cycles, and its degradation is parallel to that of γ-tocotrienol. However, the α-tocopherol content in RBO did not decrease significantly ($p < 0.05$) until the 60th frying cycles (Figure 2). Thus, it can be assumed that, in RBO, γ-oryzanol and γ-tocotrienol synergistically protect α-tocopherol from degradation during deep-frying. The presence of γ-oryzanol in RBO may confer a protective effect for α-tocopherol, thus slowing down its degradation compared to that in PO. This observation was in agreement with that of Kochhar [38], who reported that γ-oryzanol has substantial synergistic effects with tocopherols. Similarly, the higher amount of γ-oryzanol, due to the addition of RBO to soybean oil, was related to the synergistic antioxidant effects in preserving tocopherol degradation during the frying of dough [39].

Figure 4. The degradation of γ-oryzanol in RBO during deep-frying. The values given are means ± standard deviation of a triplicate analysis. The values marked with the same letters are not significantly different at $p < 0.05$ analyzed using the Duncan multiple range test.

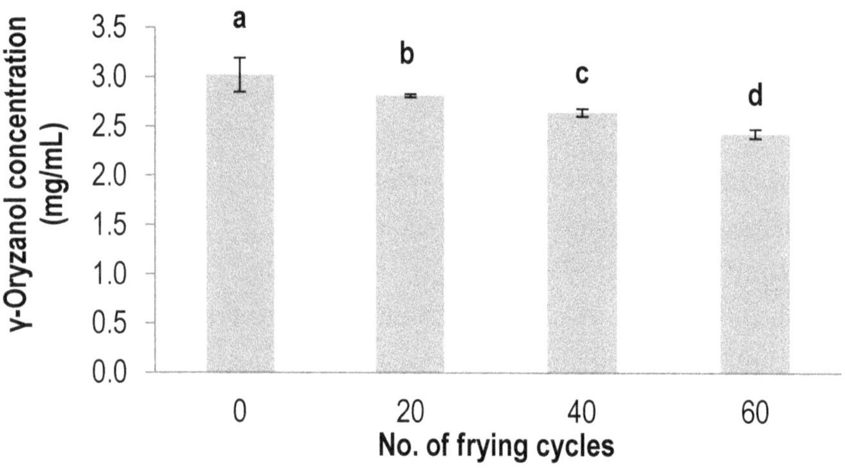

RBO usually contains oryzanol up to 1.5% and is well known for its good stability as a frying oil [6]. Even though RBO contained a higher percent of γ-oryzanol in comparison to that of α-tocopherol and γ-tocotrienol, the main reaction that caused the degradation of γ-oryzanol during deep-frying is due to the oxidation process [40]. In the present study, the antioxidant capacity of PO as measured by DPPH decreased significantly ($p < 0.05$) after deep-frying, similar to its α-tocopherol and γ-tocotrienol degradation. In contrast, the antioxidant capacity of RBO did not decrease significantly over 60 frying cycles. The slow degradation of γ-tocotrienol and the retention of α-tocopherol in RBO was probably attributed to the preservation of its antioxidant activity. This is in agreement with the findings of Nanua *et al.* [41], who reported that γ-oryzanol can improve the stability of RBO and acts as a natural antioxidant to improve the stability of several fried foods. Hence, it is suggested that blending PO with RBO might be helpful to preserve the tocol content during deep-frying.

3.4. Degradation Kinetic of α-Tocopherol and γ-Tocotrienol during Deep-Frying

The first order kinetic plots of the degradation of α-tocopherol and γ-tocotrienol in PO and RBO during deep-frying of French fries are presented in Figures 5 and 6, respectively. The kinetic constant of degradation (k) of α-tocopherol and γ-tocotrienol and the determination of the coefficient (R^2) for PO and RBO during deep-frying is shown in Table 1.

The results showed that the first-order kinetics model can be applied to approximately describe the degradation reaction of α-tocopherol in PO, but not in RBO. It is reported that the degradation kinetics of many compounds in foods at constant temperature will follow the first-order kinetics model [26]. This is in good agreement with the findings of Sabliov *et al.* [42], where heat during frying caused the degradation of free α-tocopherol followed first order kinetics and holding oils at 180 ± 5 °C showed the greatest degradation rate.

Figure 5. First order kinetic plots of the degradation of α-tocopherol in palm oil and rice bran oil during deep-frying.

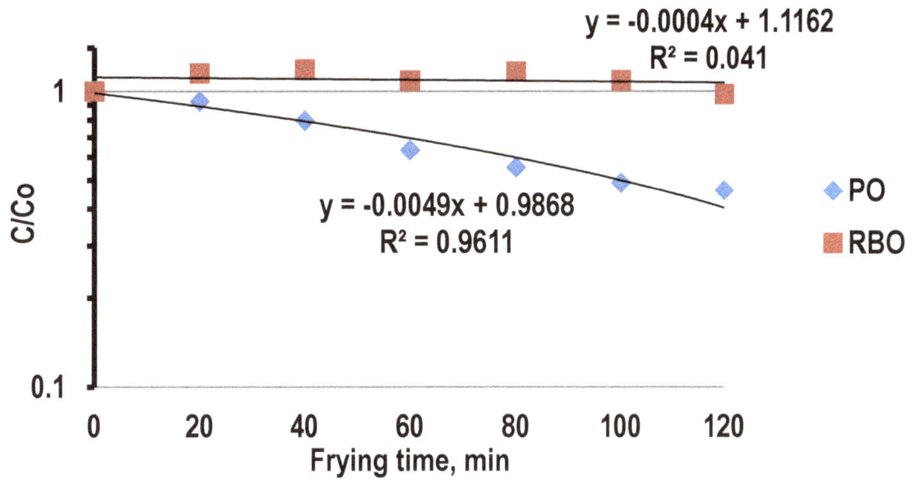

Figure 6. First order kinetic plots of the degradation of γ-tocotrienol in palm oil and rice bran oil during deep-frying.

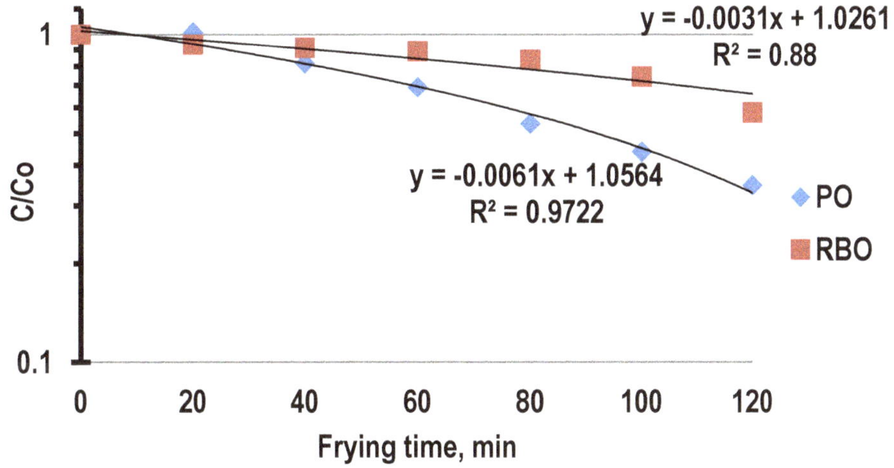

Table 1. Kinetic constant (k) of degradation of α-tocopherol and γ-tocotrienol and the coefficient of determination (r^2) of palm oil and rice bran oil during deep-frying.

Variable	k (min^{-1})	Coefficient of Determination, (r^2)
α-tocopherol in palm oil	0.004	0.961
α-tocopherol in rice bran oil	0.000	0.041
γ-tocotrienol in palm oil	0.006	0.972
γ-tocotrienol in rice bran oil	0.003	0.880

In the present study, the reaction rate constant for the degradation of both α-tocopherol and γ-tocotrienol upon deep-frying can be obtained using the order of reaction equation. The reaction rate constants of the degradation of α-tocopherol in PO and RBO were found to be 0.009 min^{-1} and 0 min^{-1}, respectively and their correlation coefficients (r^2) were found to be 0.961 and 0.041, respectively.

Similarly, γ-tocotrienol in PO and RBO obeyed the first-order kinetics with the reaction rate constants of degradation being 0.0061 min^{-1} and 0.0031 min^{-1}, respectively, and their correlation coefficients were found to be 0.972 and 0.88, respectively. The results are in line with that of previous studies that showed that tocopherols and tocotrienols decrease as a function of time and temperature, following first-order kinetics [43]. Results from the present study showed that γ-tocotrienol in PO (k = 0.0061 min^{-1}) degraded faster than that of α-tocopherol in PO (k = 0.0049 min^{-1}). In addition, it was reported that the antioxidant property of γ-tocotrienol is better than that of α-tocopherol [22]. Similarly, the antioxidant capacity of γ-tocotrienol under heating conditions in stripped oils is better than that of α-tocopherol [44]. Therefore, it could be suggested that the faster degradation of γ-tocotrienol could be due to its greater antioxidant activity.

4. Conclusions

Based on the present study, it is concluded that PO and RBO are good sources of natural antioxidants that include tocols and oryzanols. These compounds exhibited degradation at a faster rate in PO compared to that of RBO, with the degradation kinetics obeying the first-order kinetics. It was also noted that α-tocopherol does not degrade to a significant extent in RBO by virtue of the presence of γ-oryzanol, indicating the synergy between the bioactive compounds in maintaining the nutritional value of the oil.

Acknowledgments

The authors would like to acknowledge The Ministry of Science, Technology and Innovation (MOSTI) of Malaysia for financing the project and the Faculty of Food Science and Technology, Universiti Putra Malaysia, for providing the laboratory facilities.

Author Contributions

This project was conducted by Evelyn Koh Wee Fang with the guidance from Mohd Sabri Mohd Pak Dek who also technically assisted in statistical analysis. Meanwhile, the kinetic study was assisted by Mohd Asraf Mohd Zainudin. Azizah Abd Hamid, senior lecturer in Functional Food and Food Chemistry subjects was the main supervisor for this project and co-supervised by Chin Ping Tan who is expert in Fats and Oil field. All authors contributed to and have approved the final manuscript.

Conflicts of Interest

The authors declare no conflict of interest.

References

1. Szydłowska-Czerniak, A.; Trokowski, K.; Karlovits, G.; Szłyk, E. Effect of refining processes on antioxidant capacity, total contents of phenolics and carotenoids in palm oils. *Food Chem.* **2011**, *129*, 1187–1192.

2. Lam, M.K.; Tan; K.T.; Lee, K.T.; Mohamed, A.R. Malaysian palm oil: Surviving the food *versus* fuel dispute for a sustainable future. *Renew. Sustain. Energy Rev.* **2009**, *13*, 1456–1464.
3. Sugano, M.; Tsuji, E. Rice bran oil and cholesterol metabolism. *J. Nutr.* **1997**, *127*, S521–S524.
4. Xu, Z.; Godber, J.S. Purification and identification of components of gamma-oryzanol in rice bran oil. *J. Agric. Food Chem.* **1999**, *47*, 2724–2728.
5. Iqbal, S.; Bhanger, M.I.; Anwar, F. Antioxidant properties and components of some commercially available varieties of rice bran in Pakistan. *Food Chem.* **2005**, *93*, 265–272.
6. Norton, R.A. Quantitation of steryl ferulate and *p*-coumarate esters from corn and rice. *Lipids* **1995**, *30*, 269–275.
7. Lerma-García, M.J.; Herrero-Martínez, J.M.; Simó-Alfonso, E.F.; Mendonça, C.R.B.; Ramis-Ramos, G. Composition, industrial processing and applications of rice bran γ-oryzanol. *Food Chem.* **2009**, *115*, 389–404.
8. Siddiqui, S.; Rashid Khan, M.; Siddiqui, W.A. Comparative hypoglycemic and nephroprotective effects of tocotrienol rich fraction (TRF) from palm oil and rice bran oil against hyperglycemia induced nephropathy in type 1 diabetic rats. *Chem. Biol. Interact.* **2010**, *188*, 651–658.
9. Müller, L.; Theile, K.; Böhm, V. *In vitro* antioxidant activity of tocopherols and tocotrienols and comparison of vitamin E concentration and lipophilic antioxidant capacity in human plasma. *Mol. Nutr. Food Res.* **2010**, *54*, 731–742.
10. Chelchowska, M.; Ambroszkiewicz, J.; Gajewska, J.; Laskowska-Klita, T.; Leibschang, J. The effect of tobacco smoking during pregnancy on plasma oxidant and antioxidant status in mother and newborn. *Eur. J. Obstet. Gynecol. Reprod. Biol.* **2011**, *155*, 132–136.
11. Sakurai, H.; Yoshihashi, T.; Nguyen, H.T.T.; Pokorn, J. A new generation of frying oils. *Czech. J. Food Sci.* **2003**, *21*, 145–151.
12. Yamsaengsung, R.; Moreira, R.G. Modeling the transport phenomena and structural changes during deep fat frying. Part II: Model solution & validation. *J. Food Eng.* **2002**, *53*, 11–25.
13. Choe, E.; Min, D.B. Chemistry of deep-fat frying oils. *J. Food Sci.* **2007**, *72*, R77–R86.
14. Battino, M.; Quiles, J.L.; Huertas, J.R.; Ramirez-Tortosa, M.C.; Cassinello, M.; Mañas, M.; Lopez-Frias, M.; Mataix, J. Feeding fried oil changes antioxidant and fatty acid pattern of rat and affects rat liver mitochondrial respiratory chain components. *J. Bioenerg. Biomembr.* **2002**, *34*, 127–134.
15. Quiles, J.L.; Huertas, J.R.; Battino, M.; Ramìrez-Tortosa, M.C.; Cassinello, M.; Mataix, J.; Lopez-Frias, M.; Mañas, M. The intake of fried virgin olive or sunflower oils differentially induces oxidative stress in rat liver microsomes. *Br. J. Nutr.* **2002**, *88*, 57–65.
16. Raederstorff, D.E.; Elste, V.; Aebischer, C.; Weber, P. Effect of either γ-tocotrienol or a tocotrienol mixture on plasma lipid profile in hamster. *Ann. Nutr. Metab.* **2002**, *46*, 17–23.
17. Esterhuyse, J.S.; van Rooyen, J.; Strijdom, H.; Bester, D.; du Toit, E.F. Proposed mechanisms for red palm oil induced cardioprotection in a model of hyperlipidaemia in the rat. *Prostaglandins Leukot. Essent. Fatty Acids* **2006**, *75*, 375–384.
18. Chiou, A.; Kalogeropoulos, N.; Salta, F.N.; Efstathiou, P.; Andrikopoulos, N.K. Pan-frying of French fries in three different edible oils enriched with olive leaf extract: Oxidative stability and fate of microconstituents. *Food Sci. Technol.* **2009**, *42* 1090–1097.

19. Juárez, M.D.; Osawa, C.C.; Acuña, M.E.; Sammán, N.; Gonçalves, L.A.G. Degradation in soybean oil, sunflower oil and partially hydrogenated fats after food frying, monitored by conventional and unconventional methods. *Food Control* **2011**, *22*, 1920–1927.
20. Bou, R.; Navas, J.A.; Tres, A.; Codony, R.; Guardiola, F. Quality assessment of frying fats and fried snacks during continuous deep-fat frying at different large-scale producers. *Food Control* **2012**, *27*, 254–267.
21. Debnath, S.; Rastogi, N.K.; Gopala Krishna, A.G.; Lokesh, B.R. Effect of frying cycles on physical, chemical and heat transfer quality of rice bran oil during deep-fat frying of poori: An Indian traditional fried food. *Food Bioprod. Process.* **2012**, *90*, 249–256.
22. Schroeder, M.T.; Becker, E.M.; Skibsted, L.H. Molecular mechanism of antioxidant synergism of tocotrienols and carotenoids in palm oil. *J. Agric. Food Chem.* **2006**, *54*, 3445–3453.
23. Lee, Y.L.; Huang, G.W.; Liang, Z.C.; Mau, J.L. Antioxidant properties of three extracts from *Pleurotus citrinopileatus*. *LWT-Food Sci. Technol.* **2007**, *40*, 823–833.
24. AOCS Ce 8-89. Determination of tocopherols and tocotrienols in vegetable oils and fats by HPLC. In *Official Methods and Recommended Practices of the American Oil Chemists' Society*, 5th ed.; American Oil Chemists' Society: Champaign, IL, USA, 1998.
25. Xu, Z.; Godber, J.S. Antioxidant activities of major components of γ-oryzanol from rice bran using a linoleic acid model. *J. Am. Oil Chem. Soc.* **2001**, *78*, 645–649.
26. Taoukis, P.S.; Labuza, T.P.; Saguy, I.S. Kinetics of food deterioration and shelf-life prediction. In *Handbook of Food Engineering Practice*; Valentas, K.J., Rotstein, E., Singh, R.P., Eds.; CRC Press: Boca Raton, FL, USA, 1997; pp. 261–402.
27. Ramadan, M.F.; Amer, M.M.A.; Sulieman, A.E.M. Correlation between physicochemical analysis and radical scavenging activity of vegetable oil blends as affected by frying of French fries. *Eur. J. Lipid Sci. Technol.* **2006**, *108*, 670–678.
28. Gomez-Alonso, S.; Fregapane, G.; Salvador, M. Changes in phenolic composition and antioxidant activity of virgin olive oil during frying. *J. Agric. Food Chem.* **2003**, *51*, 667–672.
29. Hoffmann, G. The chemistry of edible fats. In *The Chemistry and Technology of Edible Oils and Fats and Their High Fat Products*; Hoffmann, G., Ed.; Academic Press: London, UK, 1998; pp. 1–28.
30. Rossi, M.; Alamprese, C.; Ratti, S. Tocopherols and tocotrienols as free radical-scavengers in refined vegetable oils and their stability during deep-fat frying. *Food Chem.* **2007**, *102*, 812–817.
31. Seppanen C.M.; Song Q.H.; Csallany A.S. The antioxidant functions of tocopherol and tocotrienol homologues in oils, fats, and food systems. *J. Am. Oil Chem. Soc.* **2010**, *87*, 469–481.
32. Fang, X.; Wada, S. Enhancing the antioxidant effect of α-tocopherol with rosemary in inhibiting catalyzed oxidation caused by Fe^{2+} and hemoprotein. *Food Res. Int.* **1993**, *26*, 405–411.
33. Michael, H.; Kourimska, G.; Kourimska, L. Effect of antioxidants on losses of tocopherols during deep-fat frying. *Food Chem.* **1995**, *52*, 175–177.
34. Lichenstein, A.H.; Ausman, L.M.; Carrasco, W.; Gualtieri, L.J.; Jenner, J.L.; Ordovas, J.M.; Nicolosi, R.J.; Goldin, B.R.; Schaefer, E.J. Rice bran oil consumption and plasma lipid levels in moderately hypercholesterolemic humans. *Arterioscler. Thromb.* **1994**, *14*, 549–556.
35. Seetharamaiah, G.S.; Krishnakantha, T.P.; Chandrasekhara, N. Influence of oryzanol on platelet aggregation in rats. *J. Nutr. Sci. Vitaminol.* **1990**, *36*, 291–297.

36. Rong, N.; Ausman, L.M.; Nicolosi, R.J. Oryzanol decreases cholesterol absorption and aortic fatty streaks in hamsters. *Lipids* **1997**, *32*, 303–309.
37. Nakayama, S.; Manabe, A.; Suzuki, J.; Sakamoto, K.; Inagake, T. Comparative effects of two forms of γ-oryzanol in different sterol compositions on hyperlipidemia induced by cholesterol. *Jpn. J. Pharmacol.* **1987**, *44*, 135–143.
38. Kochhar, S.P. Stabilisation of frying oils with natural antioxidative components. *Eur. J. Lipid Sci. Technol.* **2000**, *102*, 552–559.
39. Chotimarkorn, C.; Silalai, N. Addition of RBO to soybean oil during frying increases the oxidative stability of the fried dough from rice flour during storage. *Food Res. Int.* **2008**, *41*, 308–317.
40. Khuwijitjaru, P.; Taengtieng, N.; Changprasit, S. Degradation of gamma-oryzanol in rice bran oil during heating: An analysis using derivative UV-spectrophotometry. *Silpakorn Univ. Int. J.* **2004**, *4*, 154–165.
41. Nanua, J.N.; McGregor, J.U.; Godber, J.S. Influence of high-oryzanol RBO on the oxidative stability of whole milk powder. *J. Dairy Sci.* **2000**, *83*, 2426–2431.
42. Sabliov, C.M.; Fronczek, C.; Astete, C.E.; Khachaturyan, M.; Khachatryan, L.; Leonardi, C.; Brazzoli, I. Effects of temperature and *uv* light on degradation of a-tocopherol in free and dissolved form. *J. Am. Oil Chem. Soc.* **2009**, *86*, 895–902.
43. Alyssa, H.; Andrea, B.; Carlo, P. Kinetics of tocols degradation during the storage of einkorn (*Triticum monococcum* L. ssp. *Monococcum*) and breadwheat (*Triticum aestivum* L. ssp. *aestivum*) flours. *Food Chem.* **2009**, *116*, 821–827.
44. Feng, H.P. Preparative Techniques for Isolation of Vitamin E Homologs and Evaluation of Their Antioxidant Activities. Ph.D. Thesis, Atherns, University of Georgia, Athens, GA, USA, 1996.

Endogenous Phenolics in Hulls and Cotyledons of Mustard and Canola: A Comparative Study on Its Sinapates and Antioxidant Capacity

Shyamchand Mayengbam [1], Ayyappan Aachary [1,2] and Usha Thiyam-Holländer [1,2,]*

[1] Department of Human Nutritional Sciences, University of Manitoba, Winnipeg, MB R3T 2N2, Canada; E-Mails: Ummayens@cc.umanitoba.ca (S.M.); Ayyappan.Appukuttan@ad.umanitoba.ca (A.A.)

[2] Richardson Centre for Functional Foods and Nutraceuticals & Department of Human Nutritional Sciences, University of Manitoba, Winnipeg, MB R3T 2N2, Canada

* Author to whom correspondence should be addressed; E-Mail: Usha.Thiyam@umanitoba.ca

---

Abstract: Endogenous sinapic acid (SA), sinapine (SP), sinapoyl glucose (SG) and canolol (CAN) of canola and mustard seeds are the potent antioxidants in various lipid-containing systems. The study investigated these phenolic antioxidants using different fractions of canola and mustard seeds. Phenolic compounds were extracted from whole seeds and their fractions: hulls and cotyledons, using 70% methanol by the ultrasonication method and quantified using HPLC-DAD. The major phenolics from both hulls and cotyledons extracts were SP, with small amounts of SG, and SA with a significant difference of phenolic contents between the two seed fractions. Cotyledons showed relatively high content of SP, SA, SG and total phenolics in comparison to hulls ($p < 0.001$). The concentration of SP in different fractions ranged from 1.15 ± 0.07 to 12.20 ± 1.16 mg/g and followed a decreasing trend- canola cotyledons > mustard cotyledons > mustard seeds > canola seeds > mustard hulls > canola hulls. UPLC-tandem Mass Spectrometry confirmed the presence of sinapates and its fragmentation in these extracts. Further, a high degree of correlation ($r = 0.93$) was noted between DPPH scavenging activity and total phenolic content.

Keywords: sinapic acid derivatives; sinapine; sinapoyl glucose; mustard; canola; hulls; cotyledons; antioxidant activity

---

1. Introduction

Over the last two decades, the positive effects of endogenous bioactive phenolic compounds from plants and oilseeds have received considerable attention due to the role of phenolic antioxidants in human nutrition and health [1,2]. Rapeseed (*Brassica napus* L. spp. *oleifera*), an excellent source of phenolic antioxidants, is one among the 100 species in the *Brassica genus*. Additionally, canola that is another variety of *B. napus* contains various minor constituents such as tocopherols, carotenoids, phytic acid, sinapic acid (SA), and its derivatives (SADs) namely sinapine (SP), the choline ester of SA and sinapoyl glucose (SG), the glucose ester of SA and low levels of erucic acid (~2%) and glucosinolates (<30 μg/g). The relatively high concentration of both free and esterified SA was reported in press cakes and proteins of rapeseed [3]. The SA and the decarboxylation product canolol (CAN) are considered to be potent antioxidants as demonstrated recently by various *in vitro* assays in various food products. The endogenous bioactive principles of canola and rapeseeds have great potential as therapeutic agents to maintain and improve human health and well-being, and can be incorporated into many food and non-food products. There are, however, several questions that still need to be answered with respect to SADs distribution in various fractions of canola and mustard seeds to be able to understand their antioxidant efficacy in bulk oil and emulsion systems.

Oilseeds are potential sources for various bioactive molecules such as phenolics and glucosinolates and most of them are retained in oilseed processing by-products (meal, press cakes, and hulls) in significant amounts. The isolation of these bioactive molecules is justified in the value addition perspective of these by-products. With respect to various factors, especially the genetics of the rapeseed, and the processes of oil extraction, the contents of SADs in rapeseed meal vary significantly (6–18 mg/g) [4,5]. Among various SADs of rapeseed meal, the glucose ester of SA (SG) is highly potent in terms of its antioxidant efficacy [6]. SA is a potential peroxyl radical scavenger [7], and it had been demonstrated to retard oxidation process in many emulsion systems including bulk methyl linoleate (MeLo), emulsified MeLo, sunflower oil methyl esters and low-density lipoprotein [7–10]. Wanasundara *et al.* proved that, for the oxidation of liposomes and low-density lipid particles, SP is the principal contributor towards the antioxidative potential of phenolic extracts from rapeseed [11]. Various phenolic bioactive constituents of rapeseed meal and crude oil were also shown to have antioxidative properties [4,6,12,13]. A significant reduction (>90%) in the oxidation of LDL particles by rapeseed phenolics was reported [14]. The antioxidant effectiveness of canola hulls extracts in methanol and acetone was comparable to butylated hydroxyanisole in model systems of β-carotene-linoleate [15]. The rapeseed phenolics were better antioxidants towards liposomes membrane oxidation and the radical scavenging activities of such phenolics from rapeseed oil were significantly high [16]. The authors also suggested the possible use of rapeseed phenolics in functional food product development, owing to its abundance and potent bioactive attributes.

Wakamatsu *et al.* isolated 4-Vinylsyringol or CAN from crude canola oil [17] and it was reported that the decarboxylation of SA via roasting treatments could increase CAN content of rapeseed [18]. CAN is a highly effective 1,1-diphenyl-2-picrylhydrazyl (DPPH) radical scavenger [13,14,19] and it inhibits oxidative degradation of lipids and proteins [16]. Kuwahara *et al.* studied scavenging capacity of CAN with respect to peroxynitrite, which is an endogenous mutagen and reported that CAN effectively suppressed peroxynitrite-induced bactericidal action [20].

Recently Bala and Singh developed a Near Infra-Red spectroscopy method allowing rapid and non-destructive detection of total phenolics in mustard using whole seed [21]. However, high-performance liquid chromatography-diode array detector (HPLC-DAD) is proven to be the most convenient, efficient, and reliable technique to quantify phenolics [22–26]. The quantitative profile of SADs in commercial canola and mustard products and their extracts need to be established. There is a lack of information on the concentration of phenolic antioxidants of commercial oil seeds. Recently, Yang *et al.* estimated the concentration of some of the minor components of rapeseed oil produced via cold-pressing of a few varieties cultivated in Yangtze River Valley, China [27]. However, the study was not extensively carried out on the phenolic constituents but on the total phenolic concentration which was reported to be 36 mg/100 g of sample.

The phenolic compounds non-uniformly distributed in different fractions of oilseed, where in certain fractions have more phenolics than others. However, there are no data on the *in-situ* distribution of phenolics in different sections of an oilseed, particularly canola or mustard. This information will further strengthen the rationale of value addition of oilseed by-products such as hulls, meals, and press cakes for recovering phenolics. In this scenario, the purposes of our study were to identify, quantify, and characterize the antioxidant phenolics of hulls and cotyledons of canola and mustard seeds. The study encompassed the identification and quantification of SADs by reversed-phase HPLC–DAD at 330 and 275 nm to understand the distribution of these phenolics between hulls and cotyledons. The study also used the quantitative comparison of total phenolic contents following two methods: HPLC and Folin-Ciocalteu's method to ascertain the applicability of these methods in canola/rapeseed phenolics. Additionally, the major phenolics were characterized using ultra-high performance liquid chromatography-mass spectrometry (UPLC-MS), and the potential DPPH scavenging of these various residues of canola and mustard seeds was also investigated. This knowledge is essential to contribute to the optimized extraction of canolol from sinapic acid and other precursors.

2. Experimental Section

2.1. Chemicals and Materials

Analytical grade chemicals were used. Standards of sinapic acid and sinapine were procured from Sigma-Aldrich (St. Louis, MO, USA) and EPL Bioanalytical Services (Niantic, IL, USA) respectively. Dr. A. Baumert kindly donated standard of sinapoyl glucose. Sinapinaldehyde was a product of ChromaDex Inc. (Irvine, CA, USA). An authentic standard of 4-Vinylsyringol (canolol) was kindly donated by Amy Logan of CSIRO Animal Food and Health Sciences, Werribee, Australia. Mustard seeds were procured from G.S. Dunn Limited, Ontario, Canada and a local store. Dow AgroSciences (Calgary, AB, Canada) supplied Nexera, a variety of canola.

2.2. Phenolic Extraction from Hulls and Cotyledons of Canola and Mustard

Hulls and cotyledon of canola and mustard were manually separated. Whole seeds, hulls and cotyledons were defatted with *n*-hexane using Soxtec 2050 and extracted as per Thiyam *et al.* 2004 [28]. Briefly, 1 g of defatted canola or mustard fractions were extracted thrice in aqueous methanol (70%) assisted by ultra-sonication (60 s) followed by refrigerated centrifugation at 5000 rpm for 10 min. The filtrates from the three extractions, obtained through the filtration of methanolic layers using Whatman No. 1 filter paper from Sigma-Aldrich (St. Louis, MO, USA), were combined and made up to a known volume (25 mL). All the extractions were conducted in triplicates.

2.3. Total Phenolic Content of Canola and Mustard Seed Fractions

Phenolic contents of different fractions of canola and mustard were characterized by reversed phase HPLC-DAD [29]. Folin-Ciocalteu's reagent based assay was used to estimate the total phenolics [30] with slight modifications. Briefly, the extracts, were appropriately diluted (2.5 fold) with distilled water, and 500 μL of this was thoroughly mixed with Folin-Ciocalteau's phenol reagent (1:1 ratio). After a specified reaction period (3 min), 1 mL of 19% Na_2CO_3 was added, followed by monitoring of absorption at 750 nm after 60 min in a DU 800 UV/Visible Spectrophotometer (Beckman Coulter Inc., Mississauga, ON, Canada). The analysis was carried out in duplicates and compared with a calibration graph of SA and the results were expressed as SA equivalents (SAE).

2.4. DPPH Scavenging Activity

Different fractions of canola and mustard were assessed for their DPPH radical scavenging activities following Schwarz *et al.* method with slight modifications [31]. In covered test tubes (three for each sample), 100 μL of the phenolic extracts were combined with methanolic DPPH (2.9 mL, 0.1 mM). The tubes were vortexed thoroughly and placed in a dark cabinet for exactly 10 min before measuring the absorption values at 516 nm. The absorbance of control (A_c) and absorbance of the sample (A_s) were used to calculate scavenging effect (%), which is the percentage change in absorbance (A_c–A_s) with respect to A_c. To calculate the EC_{50} concentration, different concentrations of 100 μL sample were used (20, 40, 60, 80, 100 μL of the sample and all of them were made to 100 μL using 80, 60, 40, 20 & 0 μL of methanol).

2.5. HPLC-DAD Analysis of Phenolic Compounds in Canola and Mustard Extracts

The phenolic profile of canola and mustard extracts was established following a reversed-phase HPLC-DAD (Ultimate 3000; Dionex, Sunnyvale, CA, USA) analysis [29]. Solvent A, 90% methanol (aqueous) acidified with *o*-phosphoric acid (1.2%) and solvent B, 100% methanol acidified with o-phosphoric acid (0.1%) were used as mobile phases in a gradient elution, where in the concentration of mobile phase B (%, indicated in brackets) changed in the following sequences at specified time periods (min) 0 (10), 7 (20), 20 (45), 25 (70), 28 (100), 31 (100) and 40 (10). Synergi 4 μ Fusion-RP 80 Å; 150 × 4.0 mm- 4 micron (Phenomenex, Torrance, CA, USA) column was used for SADs separation. Both the mobile phases and canola and mustard extracts were passed through syringe filters (0.45 μm). The following conditions of analysis were maintained: flow rate (1 mL/min), column compartment temperature (25 °C) and wavelengths of analysis (275 nm and 330 nm). Version 6.8 of Chromeleon software (Dionex Corporation, Sunnyvale, CA, USA) was used to acquire the HPLC data. Standards of SP, SG, SA, and CAN were also analyzed for comparison purpose based on retention time. Triplicate samples were analyzed for statistical validation of results.

2.6. UPLC-MS Analysis of Phenolics from Hulls and Cotyledons

A Waters Acquity UPLC system coupled to a Quattro micro API tandem mass spectrometer (Milford, MA, US) was used for the confirmation of SA and SADs. A Phenomenex Synergi 4 μ Fusion-RP column (150 × 4 mm, Torrance, CA, USA) was employed for the separation of SADs. 5 mM of ammonium acetate (pH 3.2 with acetic acid) and 100% methanol were the two mobile phases, A and B respectively. A constant flow rate (0.5 mL/min) was maintained throughout the gradient elution, which consisted the following sequence of solvent mixing: initially, phase B was set at 25% (1 min), then the concentration was changed to 95% over 10 min in a linear manner and maintained at this condition for 2 min. Column was re-equilibrated for 3 min after each injection. The column temperature was kept constant at 35 °C. Samples were stored at 4 °C throughout the analysis, and 10 μL of the sample was injected. The PDA detector was set at a range between 210 nm and 400 nm with 2-channel monitoring at 275 nm and 330 nm.

The tandem mass spectrometer consisted of an atmospheric pressure ionization (API) probe and for analysis of SP and CAN, positive ion mode (ES+) and for SA and SG, negative ion mode (ES−) were selected, with the condition tuned based on each authentic standard for identification purpose. The general MS/MS parameters were as follows: cone gas (N_2) flow, 50 L/h; source temperature, 100 °C; desolvation temperature, 400 °C; capillary voltage, 3.00 kV; desolvation gas (N_2) flow, 400 L/h; cone voltage, 25 V except for SP (22 V) and collision energy, 15 eV. The precursor to product ion transition was monitored using Multiple Reactions Monitoring (MRM) mode: SP, *m/z* 310 > 251; CL, *m/z* 181 > 121; SA, *m/z* 223 > 208 and SG, *m/z* 385 > 205. Daughter ion mode was used to obtain MS/MS spectrum of their precursor ions (also molecular ions except *m/z* 310 for SP). Mass resolution was set at maximum.

2.7. Data Expression and Analysis

Means and standard deviations were based on triplicate values. Data on phenolic content of mustard and canola fractions and their antioxidant activity were statistically interpreted using one factor ANOVA. For multiple comparisons, Tukey mean separation was followed using the Statistical Analysis System Program (SAS Institute, Carey, NC, USA), where in $p \leq 0.05$ was fixed as level of significance.

3. Results and Discussion

The information on extraction and analysis of phenolics from canola and mustard was focused on either whole seed or extracted-oil or the meal. There is limited data on the comparative profile of phenolics of hulls and cotyledons of canola and mustard. Phenolics are known for its heterogeneous distribution in various fractions of oil seeds. There are very few studies available on the *in-situ* distribution of phenolics in different sections of an oilseed. Krygier *et al.* examined the distribution of phenolics in rapeseed hulls and de-hulled flour [32]. However, the results might not be very suitable for comparisons as the authors used an alkaline treatment method which might have affected the structural attribute of the original phenolics and thereby, its quantification. Similarly, Liu *et al.* investigated the distribution of soluble and insoluble phenolics in certain varieties of rapeseeds [33].

In the present study, we avoided destructive and harsh methods of phenolic extraction but rather followed solvent extractions at optimum conditions for better stability of phenolics. The various by-products of canola and mustard have been previously suggested as a potential substrate for phenolics and the results of our study further strengthen the understanding of phenolic distribution prior to the objective of value addition.

A large number of hydroxycinnamic acid derivatives, with varied structural attributes such as sinapoyl, caffeoyl, coumaroyl, hydroxyferulolyl and ferulolyl esters, are found in mustard greens [34]. The phenolic profile of mustard seeds is less complex, and most of the hydroxycinnamic acids except SADs were not reported. In the extracts of crude mustard seeds, SP was the principal phenolic, while the free SA was detected only in trace amount. Previously, SP, SG and free SA were reported in mustard meal [25]. Data's on the phenolic profile and antioxidant activity of hulls and cotyledons of mustard and canola are scarce. Thus, the current study was conducted to investigate the antioxidant properties of SA and its derivatives present in different fractions of canola and mustard. Methanolic (70%) extracts of defatted fractions were analyzed for the total phenolic contents following two methods—the Folin-Ciocalteau assay, and HPLC profiling based on diode array detection.

3.1. Phenolic Profile of Extracts from Canola and Mustard Seed Fractions

Hydroxycinnamate conjugates are characteristic of brassicaceous plants [34–36]. The shikimate/phenylpropanoid pathway is responsible for the production of SA and its conversion to *O*-ester conjugates via a system of multiple enzymes [37]. From the taxonomic point of view, the seed-specific SP can be used as a biomarker to group the members of family Brassicaceae [35]. For the biosynthesis of phosphatidylcholine, SP (a choline ester) might work as a storage vehicle, whereas sinapoyl esters might include UV protection of plants [37]. In the present study, the distribution of these phenolics in various fractions of canola and mustard seeds was investigated. Previously, the efficiency of different solvents (70% v/v) such as methanol, iso-propanol and ethanol for canola phenolic extraction was evaluated following HPLC-DAD and found that the methanol was more efficient to obtain SADs [29]. The methanolic extracts from canola seed, in comparison to ethanol or iso-propanol extracts, had a higher total phenolic content, which was mainly contributed by its phenolics (SP, SG and SA).

The total phenolic content of canola samples were assessed using HPLC and Folin-Ciocalteu method (Table 1). The total phenolic contents (mg/g) were 10.60 (canola seeds), 4.50 (canola hulls), 16.89 (canola cotyledons), 10.31 (mustard seeds), 6.24 (mustard hulls), and 10.60 (mustard cotyledons) when analyzed by Folin-Ciocalteu method. However, the HPLC analysis showed relatively higher total phenolics values than Folin-Ciocalteu method, except in canola hulls and mustard hulls (3.57 and 5.67 mg/g respectively). The total phenolic contents with regards to the HPLC method (mg/g) were 14.06 (canola seeds), 20.20 (canola cotyledons), 11.12 (mustard seeds), and 11.45 (mustard cotyledons). Phenolic content estimated by both methods indicated that cotyledons are a richer source of phenolics than the hulls. The results are comparable with those of other researchers who reported a total phenolic content ranges from 10 to 18 mg/g [28,29,32,38]. Khattab *et al.* observed a total phenolic content (mg/g) of 17.71 (defatted canola seeds), 15.83 (canola meals) and 18.48 (canola seed press cakes) [29]. Kozlowska *et al.* emphasized that both the variety and processing methods affect the total phenolic content of canola meal (6.4–18.4 mg/g) [39]. Moreover, Cai and Arntfield indicated an insignificant difference between the total phenolics in the methanolic extracts of canola flour based on two estimation methods in which 22.90 and 22.58 mg/g were reported for Folin-Ciocalteu and HPLC methods respectively [23]. The varietal genetics, environment and the extent of maturation of seeds will determine the profile of phenolic constituents and thereby the content of total phenolics.

Table 1. Profile of sinapic acid and its derivatives in different fractions of mustard and canola with their EC_{50} values.

Samples	Sinapoyl Glucose	Sinapine	Sinapic Acid	Total Phenolics (HPLC) *	Total Phenolics Folin-Ciocalteu	EC_{50}
Canola Cotyledon	8.71 ± 0.76^{a}	12.20 ± 1.16^{a}	0.22 ± 0.02^{b}	20.20 ± 1.85^{a}	16.89 ± 0.69^{a}	1.78
Canola Seeds	5.45 ± 0.35^{b}	8.35 ± 0.44^{c}	0.15 ± 0.01^{d}	14.06 ± 0.71^{b}	10.60 ± 0.81^{b}	2.31
Canola Hulls	1.34 ± 0.07^{c}	1.15 ± 0.07^{e}	0.04 ± 0.00^{e}	3.57 ± 0.20^{d}	4.50 ± 0.16^{d}	5.82
Mustard Cotyledon	0.67 ± 0.01^{c}	10.62 ± 0.08^{b}	0.18 ± 0.00^{c}	11.45 ± 0.05^{c}	10.60 ± 0.24^{b}	2.36
Mustard Seeds	0.66 ± 0.01^{c}	10.17 ± 0.27^{b}	0.19 ± 0.02^{c}	11.12 ± 0.39^{c}	10.31 ± 0.32^{b}	2.54
Mustard Hulls	0.41 ± 0.02^{c}	4.74 ± 0.28^{d}	0.90 ± 0.01^{a}	5.67 ± 0.32^{d}	6.24 ± 0.40^{c}	4.40

All the values (mg/g) except for EC_{50} are average and SD, while EC_{50} values are expressed in mg/mL, $n = 3$, values with different superscripts were significantly different at $p \leq 0.05$. * Expressed as sinapic acid equivalents (SAE).

The HPLC-DAD was carried out at 330 nm (SADs) and 270 nm (CAN). The concentration of major SADs in canola and mustard extracts are shown in Table 1. Previously, Khattab *et al.* reported that SP solely represented about 69%–87% of the total phenolics of various canola fractions (seeds, meal and press cakes) [29]. A similar pattern was observed in the present study with respect to the canola variety analyzed (Nexera). The canola hulls had the lowest SP (1.15 mg/g) in comparison with canola seeds (8.35 mg/g), and canola cotyledon (12.20 mg/g). Interestingly, the concentration of SA was significantly less (0.04–0.22 mg/g) and this corresponds to 0.8%–1.09% of the total phenolics. Figure 1 represents a typical HPLC-DAD chromatogram (330 nm) of phenolic extracts from canola cotyledons showing the peaks of SG, SP and SA. It is well known that the rapeseeds have the highest amount of phenolics among various oilseeds of commercial origin, and such phenolics are leached into the oil during pressing the oil from rapeseed. Depending on the processing parameters, the amount of such phenolics in rapeseed oil will vary, for example, cold-pressed rapeseed oil and refined rapeseed oils contain 3–4 mg/kg and 16 mg/kg (caffeic acid equivalents) phenolics respectively [13,40]. The analysis of phenolics from Nexera canola variety indicated that the major phenolic was SP, with trace amounts of SA. Interestingly, the seeds and cotyledons of this canola variety had a significantly higher concentration of SG. Moreover, it was also found that the total phenolic concentration of cotyledon was relatively higher than other fractions of canola or mustard.

Figure 1. A representative high-performance liquid chromatography-diode array detector (HPLC-DAD) chromatogram (330 nm) of phenolic extracts from canola cotyledons showing the principal component sinapine (RT 11.3 min).

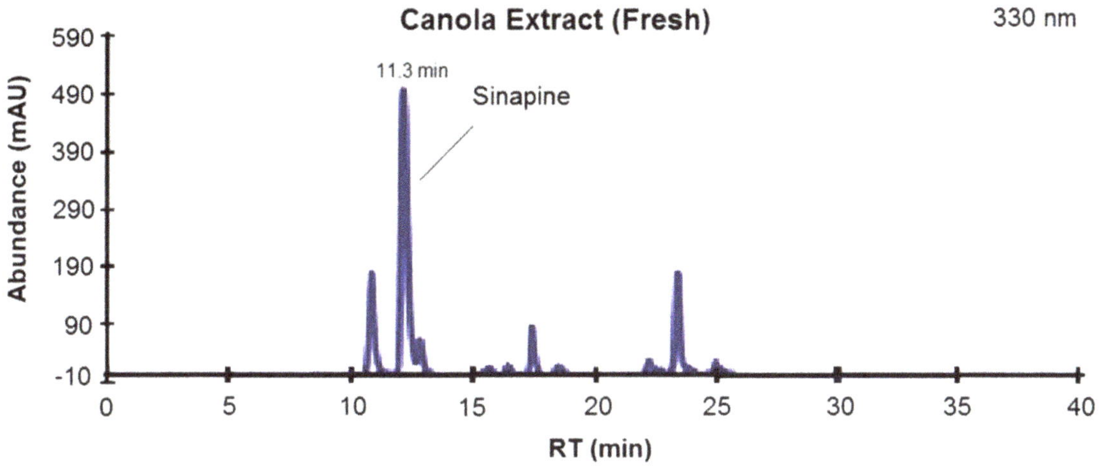

Recently, Siger *et al.* [41] identified and quantified SADs in the crude extracts of *Brassica napus* L. seeds as well as extracts after acidic and alkaline hydrolysis and indicated a high content of total phenolics (1577–1705 mg/100 g SA equivalent) in the crude extracts. 1-*O*-β-D-glucopyranosyl sinapate was the major SADs with the highest antioxidant capacity [41]. However, the concentrations of SG (4% of the total phenolics) as well as the total phenolics were much lower in *Brassica juncea* seeds in comparison to *Brassica napus*. Interestingly, SP and SA contents were insignificantly different. Both the mustard and canola cotyledons showed high content of oil than seeds or hulls (Table 2).

Table 2. Oil and moisture contents of seeds, hulls and cotyledons of mustard and canola.

Samples	Oil content % (Dry wt)	Moiture %
Canola Cotyledon	53.82 ± 0.66^{a}	3.17 ± 0.00^{f}
Canola Seeds	41.70 ± 0.62^{b}	6.45 ± 0.09^{c}
Mustard Seeds	38.35 ± 1.08^{c}	5.10 ± 0.11^{e}
Mustard Hulls	27.69 ± 0.07^{d}	5.53 ± 0.13^{d}
Canola Hulls	18.00 ± 0.09^{e}	6.71 ± 0.04^{b}
Mustard Cotyledon	42.18 ± 0.00^{b}	7.06 ± 0.05^{a}

All the values are average and SD, $n = 3$, values with different superscripts were significantly different at $p \leq 0.05$.

In the case of mustard also, the major phenolic was SP, with the content of SP been significantly higher in both seeds and cotyledon in comparison to hulls (Table 1). In all of these fractions, both SG and SA were present only in minute quantities. Like canola hulls, the mustard hulls also showed the lowest SP (4.74 mg/g) in comparison with mustard seeds (10.17 mg/g) and mustard cotyledon (10.62 mg/g). There was no significant difference between the SP content of seeds and cotyledons. Interestingly, the concentration of SA was significantly higher (0.9 mg/g) in mustard hulls than mustard seed or cotyledon. The SG content of mustard seeds and cotyledons were almost same (0.66 and 0.67 mg/g respectively) and is higher than mustard hulls (0.41 mg/g). Between canola and mustard, the SG content was significantly higher in canola with a maximum of 8.71 mg/g in cotyledon followed by 5.45 and 1.34 mg/g in seeds and hulls respectively.

In our study, CAN was not detected in any of the samples, probably because CAN would be produced only through decarboxylation of SA during roasting of mustard seed and canola [13,17,18,42]. Since the CAN synthesis is majorly dependent on the partial hydrolysis of other esterified SADs while roasting, the relatively low content of free SA in the unroasted oilseed is not sufficient enough to produce it [42,43].

3.2. UPLC-MS Analysis of Phenolics from Mustard and Canola Cotyledons and Hulls

For UPLC-MS analysis, methanolic extracts of canola and mustard phenolics were used and compared with the fragmentation pattern of standard SP, SA and SG. Since the type of molecules in the phenolic extracts of hulls and cotyledons of canola and mustard and their basic fragmentation pattern are similar, only a typical UPLC-MS of phenolic extracts obtained from mustard cotyledons is presented (Figure 2). SP was the predominant substance in the polyphenolic fractions of seed, cotyledons, and hulls of canola as well as mustard as indicated by HPLC-DAD. Based on UPLC-MS data, ions at *m/z* 254, and its breakdown product at *m/z* 119 tentatively identified SP as the choline ester of SA (Figure 2). While SP is generally regarded as the major phenolic compound in most of the brassicaceous species, data on the fragmentation pattern using mass spectrometry is scarce. The characteristic major (*m/z* 251) and minor (*m/z* 207, 175, 147, 119 and 91) fragments of standard SP were observed (spectrum not shown). The fragmentation pattern noticed in the present investigation was comparable with earlier observations for phenolics obtained from extracts of canola seed [44,45]. Even though, the signal at *m/z* 309, which corresponded to the $[M - H]^-$ ion was observed, a signal at *m/z* 311 ($[M + H]^+$) ion was absent. With respect to the ionization of phenolic compounds from mustard and canola fractions, our study indicated that a positive mode is best suited for detecting SP (*m/z* 310) in its molecular ion form.

Figure 2. A typical ultra-high performance liquid chromatography-mass spectrometry (UPLC)-Mass spectrum of phenolics in the extracts of mustard cotyledons showing major fragments of sinapates. (SG: sinapoyl glucose, SP: sinapine, SA: sinapic acid, FSG: fragments of SG, FSP: fragments of SP, FSA: fragments of SA).

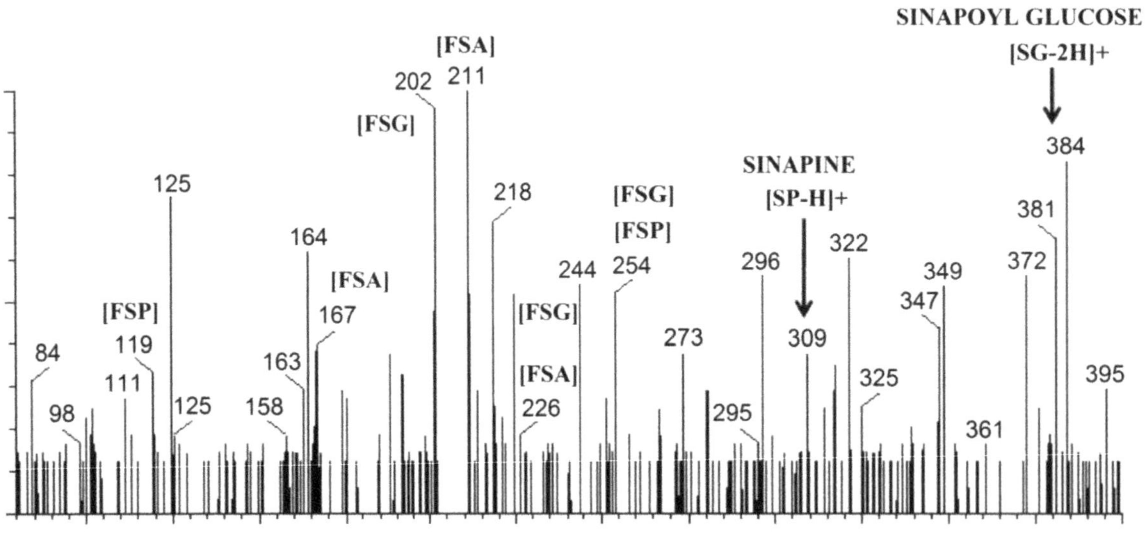

SA in the canola and mustard fractions was readily identified based on MS spectra, UV data, and comparison with respective reference standard. Hydroxycinnamic acids produce two fragments (*m/z* 208 and 193) by losing two methyl groups and are characteristics of hydroxycinnamic acid fragmentation. The standard SA exhibited its characteristic fragmentations at *m/z* 223 and 164 (spectrum not shown). The major fragments of SA were found at *m/z* 223, 208, 179, 164 and 149 while the minor fragments were observed at *m/z* 193 and 147. In the present study, when the canola and mustard phenolics were analyzed, some of these fragments were detected at *m/z* 226, 211 and 167, corresponding to $[M + 3H]^+$ ions of fragments of SA standard. Sinapate esters, which showed the similar fragmentation pattern, were also characterized. The MS analysis of the current study matched with the results by Engels *et al.* [45].

Presence of $[M - 2H]^+$ ion at m /z 384 confirmed the presence of sinapoyl hexose, possibly SG (Figure 2). Typical fragmentation pattern of SG consists of fragments at *m/z* 251, 223, 205, 190, 179 and 164. When mustard and canola phenolics were analyzed, two of these fragments (*m/z* 254 and 226) corresponding to $[M + 3H]^+$ ions of fragments and two signals (*m/z* 384 for the parent molecule and 202) corresponding to $[M - 3H]^+$ ions of fragments of standard SG were observed.

3.3. DPPH Scavenging Effects of Various Phenolic Extracts

Among various methods used to evaluate the antioxidant activities of plant phenolics, the DPPH radical assay is very common and reliable [46–48]. The ratio of reduction in absorbance (517 nm) of DPPH solution in the presence and absence of phenolics is widely used as an estimate for radical-scavenging activity of an antioxidant [47]. This procedure was later modified to consider various kinetic properties of antioxidants [46]: however, the modification was not apt to assess the antioxidant potential of phenolic extracts (crude), because the idea of structural characteristics of molecules is imperative. In the present study, to overcome this limitation, EC_{50} was suggested as an appropriate expression of radical-scavenging potential. EC_{50} was defined as the quantity of phenolic extract (crude) needed for a half decrease in the concentration of DPPH radicals during assay [31]. A low EC_{50} value corresponds to a strong radical-scavenging activity.

Previously, Khattab *et al.* investigated the antioxidative efficacy of minor components (phytic acid, chlorophyll, and condensed tannin) of a few varieties of canola (*Brassica napus* L.) seeds, meals and cakes in comparison to one Indian mustard (*Brassica juncea* L.) [49]. In this study, the authors did not focus on the scavenging effects of SADs. In our study, we observed that phenolic extracts of canola cotyledon is significantly more potent than other extracts in terms of its effectiveness to scavenge DPPH (Figure 3). Among the various phenolic extracts from mustard, the cotyledon extract showed more scavenging activity than extracts of whole seed or hulls while the scavenging efficacy of mustard hull phenolics is comparatively lower than others. The result also indicated that the EC_{50} concentration (mg/mL) of these phenolics followed a decreasing order of scavenging activity: canola cotyledon > canola seed > mustard cotyledon > mustard seed > mustard hulls > canola hulls (Table 1). A high degree of correlation was established between DPPH scavenging activity (%) and total phenolic content ($r = 0.93$). Utilization of canola hulls as a potential substrate for anti-oxidants extraction was discussed previously [50]; however, there were no comparative data for hulls and cotyledons in comparisons with canola and mustard. The present study is relevant in this perspective.

Figure 3. 1,1-diphenyl-2-picrylhydrazyl (DPPH) scavenging activity (%) of phenolic extracts from seeds, cotyledons and hulls of Nexera canola and mustard. (Values with different letters were significantly different at $p < 0.001$).

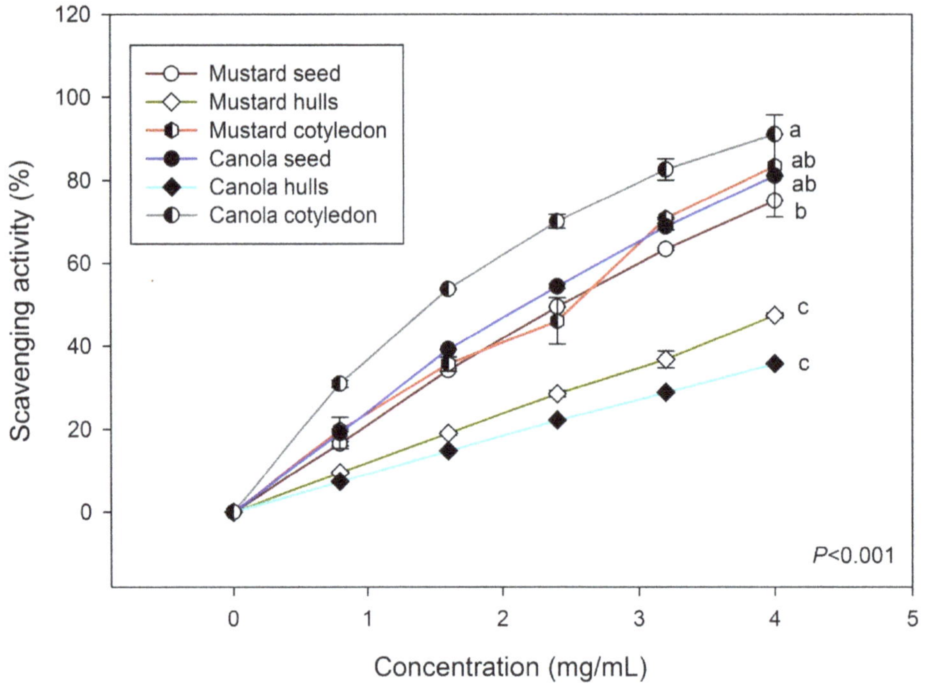

4. Conclusions

Aqueous methanol (70%) was the best extraction solvent to recover phenolics, especially SADs from canola and mustard seeds and their fractions (hulls and cotyledon). The major phenolic compound in both hull and cotyledon extracts was SP with a relatively smaller amount of SA and SG, with significant variation between the two seed fractions. Canolol was not found in the crude extracts investigated in this study. Further, the DPPH scavenging activities of the methanolic extracts indicated a co-relation with their total phenolic content. Based on the quantitative profile of SADs, the major contributor towards the antioxidant potential is SP. Even though UPLC-MS confirmed the presence of SG, SP and SA in the extracts, detailed studies on their fragmentation pattern are needed to establish the structural changes and formation of novel antioxidants like canolol from SADs. In order to translate the results of this study to food development and applications, for example, the effects of various phenolic fractions of canola and mustard seeds in bulk oil and emulsion systems, more research is warranted. The study also implies that most of the phenolics are accumulated in cotyledons of canola and mustard and only a fraction of it is found in hulls. This information on the quantitative distribution of SADs is certainly a catalyst to add value to by-products for recovery of endogenous phenolics either to add back to oils or to fortify other lipid-systems.

Acknowledgments

The National Sciences and Engineering Research Council of Canada (NSERC), the Canola Council of Canada, and Syngenta Inc., Guelph, ON, Canada, funded this work through a Collaborative Research and Development grant. Amy Logan of CSIRO Animal Food and Health Sciences, Werribee, Australia is kindly acknowledged for the authentic canolol standard synthesized in her laboratory. The authors are thankful to Hai Feng, Department of Human Nutritional Sciences, the University of Manitoba for technical assistance with UPLC-MS.

Author Contributions

Usha Thiyam-Hollander is the principal investigator of this project and supervised the trainees. Shyamchand Mayengbam conducted the research, analyzed data and wrote manuscript. Ayyappan Aachary wrote manuscript. Usha Thiyam-Hollander reviewed and revised manuscripts for the final content contributed to this manuscript.

Conflicts of Interest

The authors declare no conflict of interest.

References

1. Stevenson, D.E.; Hurst, R.D. Polyphenolic phytochemicals—Just antioxidants or much more? *Cell Mol. Life Sci.* **2007**, *64*, 2900–2916.
2. Aachary, A.A.; Thiyam-Hollander, U. An update on characterization and bioactivities of sinapic acid derivatives. In *Canola and Rapeseed: Production, Processing, Food Quality, and Nutrition*; Thiyam-Hollander, U., Eskin, N.A.M., Matthaus, B., Eds.; CRC Press: Boca Raton, FL, USA, 2012; pp. 21–38.
3. Thiyam, U.; Claudia, P.; Jan, U.; Alfred, B. De-oiled rapeseed and a protein isolate: Characterization of sinapic acid derivatives by HPLC-DAD and LC-MS. *Eur. Food Res. Tech.* **2009**, *229*, 825–831.
4. Matthäus, B. Antioxidant activity of extracts obtained from residues of different oilseeds. *J. Agric. Food Chem.* **2002**, *50*, 3444–3452.
5. Baumert, A.; Milkowski, C.; Schmidt, J.; Nimtz, M.; Wray, V.; Strack, D. Formation of a complex pattern of sinapate esters in *Brassica napus* seeds, catalyzed by enzymes of a serine carboxypeptidase-like acyltransferase family? *Phytochemistry* **2005**, *66*, 1334–1345.
6. Wanasundara, U.N.; Shahidi, F. Canola extract as an alternative natural antioxidant for canola oil. *J. Am. Oil Chem. Soc.* **1994**, *71*, 817–822.
7. Natella, F.; Nardini, M.; di Felice, M.; Scaccini, C. Benzoic and cinnamic acid derivatives as antioxidants: Structure-activity relation. *J. Agric. Food Chem.* **1999**, *47*, 1453–1459.
8. Cuvelier, M.E.; Richard, H.; Berset, C. Comparison of the antioxidative activity of some acid-phenols: Structure-activity relationship. *Biosci. Biotechnol. Biochem.* **1992**, *56*, 324–325.
9. Marinova, E.M.; Yanishlieva, N.V. Effect of lipid unsaturation on the oxidative activity of some phenolic acids. *J. Am. Oil Chem. Soc.* **1994**, *71*, 427–434.

10. Pekkarinen, S.S.; Stöckmann, H.; Schwarz, K.; Heinonen, I.M.; Hopia, A.I. Antioxidant activity and partitioning of phenolic acids in bulk and emulsified methyl linoleate. *J. Agric. Food Chem.* **1999**, *47*, 3036–3043.
11. Wanasundara, U.N.; Amarowicz, R.; Shahidi, F. Partial characterization of natural antioxidants in canola meal. *Food Res. Int.* **1995**, *28*, 525–530.
12. Amarowicz, R.; Raab, B.; Shahidi, F. Antioxidant activity of phenolic fractions of rapeseed. *J. Food Lipids*. **2003**, *10*, 51–62.
13. Koski, A.; Pekkarinen, S.; Hopia, A.; Wähälä, K.; Heinonen, M. Processing of rapeseed oil: Effects on sinapic acid derivative content and oxidative stability. *Eur. Food Res. Technol.* **2003**, *217*, 110–114.
14. Vuorela, S.; Meyer, A.S.; Heinonen, M. Impact of isolation method on the antioxidant activity of rapeseed meal phenolics. *J. Agric. Food Chem.* **2004**, *52*, 8202–8207.
15. Naczk, M.; Amarowicz, R.; Zadernowski, R.; Shahidi, F. Antioxidant capacity of phenolics from canola hulls as affected by different solvents. In *Phenolic Compounds in Foods and Natural Health Products*; Shahidi, F.C.-H., Ed.; ACS Symposium Series: Washington, DC, USA, 2005; Volume 909, pp. 57–66.
16. Vuorela, S.; Kreander, K.; Karonen, M.; Nieminen, R.; Hämäläinen, M.; Galkin, A.; Laitinen, L.; Salminen, J.; Moilanen, E.; Pihlaja, K.; *et al.* Preclinical evaluation of rapeseed, raspberry, and pine bark phenolics for health related effects. *J. Agric. Food Chem.* **2005**, *53*, 5922–5931.
17. Wakamatsu, D.; Morimura, S.; Sawa, T.; Kida, K.; Nakai, C.; Maeda, H. Isolation, identification, and structure of a potent alkyl-peroxyl radical scavenger in crude canola oil, canolol. *Biosci. Biotechnol. Biochem.* **2005**, *69*, 1568–1574.
18. Spielmeyer, A.; Wagner, A.; Jahreis, G. Influence of thermal treatment of rapeseed on the canolol content. *Food Chem.* **2009**, *112*, 944–948.
19. Galano, A.; Francisco-Márquez, M.; Alvarez-Idaboy, J.R. Canolol: A promising chemical agent against oxidative stress. *J. Phys. Chem. B* **2011**, *115*, 8590–8596.
20. Kuwahara, H.; Kanazawa, A.; Wakamatu, D.; Morimura, S.; Kida, K.; Akaike, T.; Maeda, H. Antioxidative and antimutagenic activities of 4-vinyl-2,6-dimethoxyphenol (canolol) isolated from canola oil. *J. Agric. Food Chem.* **2004**, *52*, 4380–4387.
21. Bala, M.; Singh, M. Non destructive estimation of total phenol and crude fiber content in intact seeds of rapeseed-mustard using FTNIR. *Ind. Crop. Prod.* **2013**, *42*, 357–362.
22. Escarpa, A.; González, M.C. Evaluation of high-performance liquid chromatography for determination of phenolic compounds in pear horticultural cultivars. *Chromatographia* **2000**, *51*, 37–43.
23. Cai, R.; Arntfield, S.D. A rapid high-performance liquid chromatographic method for the determination of sinapine and sinapic acid in canola seed and meal. *J. Am. Oil Chem. Soc.* **2001**, *78*, 903–910.
24. Mattila, P.; Kumpulainen, J. Determination of free and total phenolic acids in plant-derived foods by HPLC with diode-array detection. *J. Agric. Food Chem.* **2002**, *50*, 3660–3667.
25. Thiyam, U.; Stöckmann, H.; Zum Felde, T.; Schwarz, K. Antioxidative effect of the main sinapic acid derivatives from rapeseed and mustard oil by-products. *Eur. J. Lipid Sci. Technol.* **2006**, *108*, 239–248.

26. Wen, D.; Li, C.; Di, H.; Liao, Y.; Liu, H. A universal HPLC method for the determination of phenolic acids in compound herbal medicines. *J. Agric. Food Chem.* **2005**, *53*, 6624–6629.
27. Yang, M.; Zheng, C.; Zhou, Q.; Huang, F.; Liu, C.; Wang, H. Minor components and oxidative stability of cold-pressed oil from rapeseed cultivars in China. *J. Food Com. Anal.* **2013**, *29*, 1–9.
28. Thiyam, U.; Kuhlmann, A.; Stöckmann, H.; Schwarz, K. Prospects of rapeseed oil by-products with respect to antioxidative potential. *Comptes Rendus Chim.* **2004**, *7*, 611–616.
29. Khattab, R.; Eskin, M.; Aliani, M.; Thiyam, U. Determination of sinapic acid derivatives in canola extracts using high-performance liquid chromatography. *J. Am. Oil Chem. Soc.* **2010**, *87*, 147–155.
30. Schanderl, S.H. Tannins and related phenolics. In *Food Science. A Series of Monograps*; Stewart, G.F., Chichester, C.O., Galliver, G.B., Morgan, A.I., Eds.; Academic Press: New York, NY, USA, 1970; pp. 701–725.
31. Schwarz, K.; Bertelsen, G.; Nissen, L.R.; Gardner, P.T.; Heinonen, M.I.; Hopia, A.; Huynh-Ba, T.; Lambelet, P.; McPhail, D.; Skibsted, L.H.; *et al.* Investigation of plant extracts for the protection of processed foods against lipid oxidation. Comparison of antioxidant assays based on radical scavenging, lipid oxidation and analysis of the principal antioxidant compounds. *Eur. Food Res. Technol.* **2001**, *212*, 319–328.
32. Krygier, K.; Sosulski, F.; Hogge, L. Free, esterified, and insoluble-bound phenolic acids. 2. Composition of phenolic acids in rapeseed flour and hulls. *J. Agric. Food Chem.* **1982**, *30*, 334–336.
33. Liu, Q.; Wu, L.; Pu, H.; Li, C.; Hu, Q. Profile and distribution of soluble and insoluble phenolics in Chinese rapeseed (*Brassica napus*). *Food Chem.* **2012**, *135*, 616–622.
34. Lin, L.Z.; Sun, J.; Chen, P.; Harnly, J. UHPLC-PDA-ESI/ HRMS/MS^{n} analysis of anthocyanins, flavonol glycosides, and hydroxycinnamic acid derivatives in red mustard greens (*Brassica juncea* coss variety). *J. Agric. Food Chem.* **2011**, *59,* 12059–12072.
35. Bouchereau, A.; Hamelin, J.; Lamour, I.; Renard, M.; Larher, F. Distribution of sinapine and related compounds in seeds of *Brassica* and allied genera. *Phytochemistry* **1991**, *30*, 1873–1881.
36. Cartea, M.E.; Francisco, M.; Soengas, P.; Velasco, P. Phenolic compounds in *Brassica* vegetables. *Molecules* **2011**, *16*, 251–280.
37. Milkowski, C.; Strack, D. Sinapate esters in brassicaceous plants: Biochemistry, molecular biology, evolution and metabolic engineering. *Planta* **2010**, *232*, 19–35.
38. Shahidi, F.; Naczk, M. An overview of the phenolics of canola and rapeseed: Chemical, sensory and nutritional significance. *J. Am. Oil Chem. Soc.* **1992**, *69*, 917–924.
39. Kozlowska, H.; Naczk, M.; Shahidi, F.; Zadernowski, R. Phenolic acids and tannins in rapeseed and canola. In *Canola and Rapeseed; Production, Chemistry, Nutrition and Processing Technology*; Shahidi, F., Ed.; Van Nostrand Reinhold: New York, NY, USA, 1990; pp. 193–210.
40. Koski, A.; Psomiadou, E.; Tsimidou, M.; Hopia, A.; Kefalas, P.; Wähälä, K.; Heinonen, M. Oxidative stability and minor constituents of virgin olive oil and cold-pressed rapeseed oil. *Eur. Food Res. Technol.* **2002**, *214*, 294–298.
41. Siger, A.; Czubinski, J.; Dwiecki, K.; Kachlicki, P.; Nogala-Kalucka, M. Identification and antioxidant activity of sinapic acid derivatives in *Brassica napus* L. seed meal extracts. *Eur. J. Lipid Sci. Technol.* **2013**, *115*, 1130–1138.

42. Shrestha, K.; Stevens, C.V.; de Meulenaer. B. Isolation and identification of a potent radical scavenger (canolol) from roasted high erucic mustard seed oil from nepal and its formation during roasting. *J. Agric. Food Chem.* **2012**, *60*, 7506–7512.
43. Shrestha, K.; Gemechu, F.G.; de Meulenaer, B. A novel insight on the high oxidative stability of roastedmustard seed oil in relation to phospholipid, Maillard type reaction products, tocopherol and canolol contents. *Food Res. Int.* **2013**, *54*, 587–594.
44. Marles, M.A.S.; Gruber, M.Y.; Scoles, G.J.; Muir, A.D. Pigmentation in the developing seed coat and seedling leaves of *Brassica carinata* is controlled at the dihydroflavonol reductase level. *Phytochemistry* **2003**, *62*, 663–672.
45. Engels, C.; Schieber, A.; Gänzle, M.G. Sinapic acid derivatives in defatted Oriental mustard (*Brassica juncea* L.) seed meal extracts using UHPLC-DAD-ESI-MSn and identification of compounds with antibacterial activity. *Eur. Food Res. Technol.* **2012**, *234*, 535–542.
46. Brand-Williams, W.; Cuvelier, M.E.; Berset, C. Use of a free radical method to evaluate antioxidant activity. *LWT-Food Sci. Technol.* **1995**, *28*, 25–30.
47. Chen, C.W.; Ho, C.T. Antioxidant properties of polyphenols extracted from green and black teas. *J. Food Lipids*. **1995**, *2*, 35–46.
48. Amarowicz, R.; Naczk, M.; Shahidi, F. Antioxidant activity of crude tannins of canola and rapeseed hulls. *J. Am. Oil Chem. Soc.* **2000**, *77*, 957–961.
49. Khattab, R.; Goldberg, E.; Lin, L.; Thiyam, U. Quantitative analysis and free-radical-scavenging activity of chlorophyll, phytic acid, and condensed tannins in canola. *Food Chem.* **2010**, *122*, 1266–1272.
50. Naczk, M.; Pegg, R.B.; Zadernowski, R.; Shahidi, F. Radical scavenging activity of canola hull phenolics. *J. Am. Oil Chem. Soc.* **2005**, *82*, 255–260.

Modelling Extraction of White Tea Polyphenols: The Influence of Temperature and Ethanol Concentration

Sara Peiró [1,2,3], Michael H. Gordon [4], Mónica Blanco [5], Francisca Pérez-Llamas [6], Francisco Segovia [3] and María Pilar Almajano [3,*]

[1] Department of Health Microbiology and Parasitology, Faculty of Pharmacy, Barcelona University, Avenue Joan XXIII s/n, 08028 Barcelona, Spain; E-Mail: sara.peiro@yahoo.es

[2] IRIS-Innovació I Recerca Industrial i Sostenible, Avda. Carl Friedrich Gauss nº11, 08860 Barcelona, Spain

[3] Chemical Engineering Department, Technical University of Catalonia, Avda Diagonal 647, 08028 Barcelona, Spain; E-Mail: segoviafj@gmail.com

[4] Department of Food and Nutritional Science, University of Reading, Whiteknights, P.O. Box 226, Reading RG6 6AP, UK; E-Mail: m.h.gordon@reading.ac.uk

[5] Department of Applied Mathematics III, Technical University of Catalonia, ESAB, Campus del Baix Llobregat, Esteve Terradas 8, 08860 Barcelona, Spain; E-Mail: monica.blanco@upc.edu

[6] Department of Physiology and Pharmacology, School of Biology, University of Murcia, Campus de Espinardo, 30100 Murcia, Spain; E-Mail: frapella@um.es

* Author to whom correspondence should be addressed; E-Mail: m.pilar.almajano@upc.edu

External Editors: Maria G. Miguel and João Rocha

Abstract: The optimization of the extraction of natural antioxidants from white tea has fostered intensive research. This study has investigated the effects of ethanol-water mixtures, temperature and time on the extraction of polyphenols and antioxidant components from white tea. The response surface methodology was applied to identify the best extraction conditions. The best conditions to maximize the extraction of total polyphenols were: ethanol, 50%, for 47.5 min. Although the yield of polyphenols was optimal at 65 °C, the maximum antioxidant capacity was achieved with an extraction temperature of 90 °C. This study has identified the optimal conditions for the extraction of tea liquor with the best antioxidant properties. Epigallocatechin gallate, epicatechin gallate, epigallocatechin and epicatechin were extracted from white tea at concentrations up to 29.6 ± 10.6, 5.40 ± 2.09, 5.04 ± 0.20 and 2.48 ± 1.10 mg/100 g.

Keywords: white tea; polyphenols; extraction; antioxidant; RSM; MECK

1. Introduction

Natural antioxidants are increasingly appreciated by consumers due to both their inherent positive effects [1] and to the possibility of using them as a source of natural additives to replace synthetic ones [2–4]. Tea is a natural plant that is a rich source of natural antioxidants and provides a high free radical scavenger activity [5,6].

Tea, the infusion from *Camellia sinensis* (L), is one of the world's most widely consumed beverages [7,8]. Its medicinal properties have been widely explored [9]. In addition, its health promoting properties have been known from the early periods of the Chinese civilization, going back almost 5000 years [10].

Teas vary in properties depending on geographical origin, climatic conditions and processing methods [7], but in general, they can be classified into three types: unfermented (green and white teas), partially fermented (oolong tea) and completely fermented (black tea) [9,11]. White tea is the least processed tea. It is exclusively prepared from very young tea leaves and buds, which are harvested before being fully open and are processed by air drying [10]. This less extended processing confers to white tea its special and highly appreciated odour and flavour characteristics [12,13].

Some *in vitro* studies have reported and characterised the antioxidant activity of white tea [11,14,15]. Other studies have focused on its protective effects on live cells subject to induced oxidative stress [16–20]. Some of them have concluded that white tea extract is a neuroprotector, because it reduces the effect of hydrogen peroxide on cells [6]. This could be relevant to enhance the protection against neurodegenerative diseases, such as Alzheimer's or Parkinson's. Recent studies [21] suggest that the protective action of white tea in oxidative stress *in vitro* is related to the maintenance of the normal redox status of cells when they are susceptible to damage by free radicals.

The antioxidant capacity of white tea extracts, which can be measured *in vitro* by various assays, including oxygen radical absorbance capacity (ORAC), Trolox equivalent antioxidant capacity (TEAC), ferric reducing antioxidant power (FRAP) or diphenylpicrylhydrazyl assay (DPPH), among others [22–24]. This activity is linked to the high content of flavan-3-ols, which are also known as catechins. The major catechins present in tea are: epigallocatechin (EGC), catechin (C), epigallocatechin gallate (EGCG), epicatechin (EC) and epicatechin gallate (ECG) [8,13,25].

These catechins have a high health benefits and industrial interest. They can be used in the pharmaceutical, cosmetic and food industries as a source of additives or as a source of antioxidants for functional foods [4,9]. As a result of these benefits, more effective extracts are a research focus of interest.

The main goal of this contribution is to determine the optimal extraction conditions of the main antioxidant compounds from white tea. In this study, the response surface methodology (RSM) was used for the optimization of extraction variables (time, temperature and % of ethanol) to enhance the yield of polyphenols and antioxidant activity [15,26,27].

2. Experimental Section

2.1. Chemicals, Reagents and Equipment

Methanol, ethanol, acetone, sodium carbonate, Folin-Ciocalteu reagent, sodium tetraborate, sodium phosphate dibasic and sodium hydroxide were of analytical grade from Panreac (Barcelona, Spain). Gallic acid (GA), rutin, 6-hydroxy-2,5,7,8-tetramethylchroman-2-carboxylic acid (Trolox), 2,4,6-tris (1-pyridyl)-5-triazine (TPTZ), phosphate buffered saline, ferric chloride, potassium persulfate, Tween 20, 2,2′-azino-bis(3-ethylbenzothiazoline-6-sulfonic acid) diammonium salt (ABTS), sodium dodecyl sulphate (SDS), (+)-catechin (C), (−)-epicatechin (EC), (−)-epigallocatechin (EGC), (−)-epicatechin gallate (ECG), (−)-epigallocatechin gallate (EGCG) and caffeine were purchased from Sigma-Aldrich Company Ltd. (Gillingham, UK). Ultrapure water obtained from a Milli-Q system from Millipore (Milford, MA, USA) was used throughout. Spectrophotometric measurements were taken on a Perkin Elmer FTIR spectrometer (Perkin Elmer, Paris, France). Fluorometric measurements were taken with a Florestar omega fluorimeter. Micellar electrokinetic chromatography (MECK) was carried out using the Packard 3DCE capillary electrophoresis system equipped with a diode-array detector from Agilent (Agilent Technologies, Santa Clara, CA, USA).

2.2. Tea Samples and Preparation

White tea was purchased from Manantial de Salud (Herbocat SL, Barcelona, Spain) and was stored at room temperature (22 ± 2 °C) in a desiccator. The samples were extracted by infusion.

Infusions were filtered through Whatman paper filters n.2 (GE Healthcare, Amersham Place, UK). All white tea extracts were protected from light and stored at −20 °C until needed.

2.3. Determination of Total Phenolic Content

The total phenol (TP) content of each extract was determined in duplicate by the Folin-Ciocalteu method according to Almajano *et al.* [16]. The mixture was allowed to stand in the dark at room temperature for 1 h and was finally diluted to an appropriate final volume with distilled water. Absorbance was measured at 765 nm against a blank containing distilled water instead of extract. Values were determined from a calibration curve prepared with gallic acid standard (GAE) (2–14 $mg \cdot L^{-1}$ final concentration) and rutin (1.25–15 $mg \cdot L^{-1}$ final concentration). Results are expressed as mg of gallic acid equivalents g^{-1} of dry weight (mg GAE/g DW) or mg of rutin equivalents/g of dry weight (mg rutin equivalents/g DW), respectively.

2.4. Micellar Electrokinetic Capillary Chromatography (MECK)

MECK is a separation mode of capillary electrophoresis (CE), which can be applied in the separation of neutral and charged compounds. The principle of separation is based on the differential migration of the ionic micelles and the bulk running buffer under electrophoresis conditions and on the interaction between the analyte and the micelle. The micelle is prepared by adding a surfactant, generally sodium dodecyl sulphate (SDS), into the running buffer. MECK is suitable for separation and quantification of natural antioxidants, because it is rapid, gives an efficient separation, uses minimum amounts of sample and reagents and is low cost [17,25,27–29].

MECK was carried out with the diode-array detector set at 200 and 260 nm to detect catechins and gallic acid simultaneously. A fused-silica capillary, with extended light path, 50μm i.d., 34 cm total length and 25.5 cm effective length, was used for the separation. The separation voltage was kept at 30 kV with an intensity of 35 μA. The temperature of the capillary was set to 21 °C and controlled with the help of a thermostat. The capillary was conditioned with NaOH (0.1 M, 10 min) followed by washing with an abundant flow of Milli-Q water and then of the buffer solution (155 min). Buffer (pH 7) was prepared daily with 5 mM sodium tetraborate, 60 mM sodium phosphate dibasic and 50 mM SDS.

All standard solutions (C, EC, GC, EG, EGCG) and sample extracts were injected in triplicate. A calibration curve prepared with standards was used to quantify the components (mg/L).

2.5. Determination of Antioxidant Activity

2.5.1. Trolox Equivalent Antioxidant Capacity (TEAC)

The method used was based on Re *et al.* [30]. 2,2′-Azino-bis(3-ethylbenzothiazoline-6-sulfonic acid) diammonium salt (ABTS, 7 mM) and potassium persulfate (2.45 mM, final concentration) were dissolved separately in water, and then, the mixture was made up to volume in a 10 mL volumetric flask. The mixed solution was transferred to an amber bottle, covered with aluminium foil and allowed to stand at room temperature (RT) for 12–16 h in the dark. The ABTS $^{\bullet+}$ solution was diluted with phosphate buffered saline (PBS) (pH 7.4, 1:100) and equilibrated at 30 °C, to an absorbance of 0.7 ± 0.02) at 734 nm, read in a Hewlett Packard 8452A diode array spectrophotometer (WaldBronn, Germany). An appropriate dilution of the extract was added to ABTS^{+} solution in the proportion of 1:100. PBS (pH 7.4) was used as the blank. After mixing, the absorbance at 734 nm was measured immediately and, then, every minute for 5 min. Duplicate determinations were made for triplicate extractions. The percentage inhibition was calculated from the absorbance values at 5 min.

The relative change in sample absorbance was calculated according to the following equation:

$$\Delta A_{sample} = \frac{A_{t=0(sample)} - A_{t=5(sample)}}{A_{t=0(sample)}} - \frac{A_{t=0(solvent)} - A_{t=5(solvent)}}{A_{t=0(solvent)}} \quad (1)$$

The TEAC value was determined from a Trolox calibration curve (ranging from 1 to 10 μM final concentration). Results are expressed as μmol of Trolox/g of DW.

2.5.2. Oxygen Radical Antioxidant Capacity (ORAC)

The ORAC method [28,31] is widely used in food science and biology. A stock solution of fluorescein (FL) was prepared by dissolving 2 mg of FL in 100 mL of phosphate buffer (PBS) 75 mM and pH 7. The stock solution was stored under refrigeration in the dark. The working FL solution (78 nM) was prepared daily by adequate stock dilution in PBS. The 2,2′-Azobis(2-methylpropionamidine) dihydrochloride (AAPH)radical solution (221 mM) was prepared daily by dilution in PBS. The standard used was Trolox solution.

The ORAC values were calculated using a regression equation between the Trolox concentration and the net area of the fluorescence decay curve (area under curve, AUC). ORAC was expressed as μM Trolox equivalents (μM TE) and was calculated by applying the formula in Equation 2:

$$AUC = \left(0.5 + \left(\sum_{i=1}^{i=31} \frac{f_i}{f_1}\right)\right) \cdot CT \qquad (2)$$

where: i, the number of cycles; f, florescence units; CT, time of each cycle in minutes; in this case, CT is 2 min.

2.6. Statistical Analysis

The results obtained were analysed statistically using Minitab 5.1 for Windows (Minitab Inc., State Collage, PA, USA) and expressed as the means ± standard deviations. Any significant difference between solvents and samples was determined by one-way analysis of variance (ANOVA, considering significant differences at $p < 0.05$.

2.7. Response Surface Methodology (RSM)

The data were modelled and analysed by RSM, which is a collection of mathematical and statistical techniques suitable for problems in which a response of interest is influenced by several variables. It uses quantitative data from appropriate experimental designs to model and optimize the combination of factors that yield a desired response near the optimum [26].

In order to determine the optimum conditions for the extraction of tea polyphenols and to evaluated their activity with TEAC and ORAC, assays were performed using low and high levels for the independent variables, EtOH (%), time (min) and temperature (°C), in accordance with a factorial experimental design. The results of the preliminary trials of a two-level, three-variable full factorial design were taken into account, which involved three replicated runs. A two-level, three-factor and central composite design was chosen for this experimental design using the Minitab package for Windows software (Minitab, State Collage, PA, USA). Coded levels for independent variables are presented in Table 1.

Table 1. Variables and ranges used on the experimental design for the study of temperature (T), time (t) and % ethanol.

Variables	Range and Level		
	1	0	1
Temperature (°C)	40	65	90
Time (min)	5	47.5	90
% EtOH	0	50	100

Coefficients of the full model were evaluated by regression analysis and tested for their significance. The non-significant coefficients ($p > 0.05$) were eliminated on the basis of p-values after examining the coefficients, and the models were finally refined. For the first-order model, we considered a 2^3 factorial design augmented by three centre points. For the second-order model, we augmented the design with six points (star design). From the values (displayed in Table 1) and assuming a second order polynomial model, at least 17 experiments must be carried out to solve the matrix and the error evaluation. The resulting factorial central composite design for the two-level and three-factor scheme with 17 treatments in total is described in Table 2. The response surface values are the concentrations of resulting tea polyphenols, which are shown in the results section.

3. Results and Discussion

3.1. Selection of Extraction Solvents

It is reported by other researchers [26,32,33] that ethanol and methanol were effective solvents for extracting phenolic compounds. For this study, only ethanol was used, since it is food-grade and also cheaper than methanol.

3.2. Experimental Design

The experimental design was carried out to evaluate the effects of temperature, solvent concentration and time on antioxidant extraction. The variables and ranges used are shown in Table 1. The p-value for each term analysed in each parameter is shown in Table 2. The initial model and the final reducted model are shown in Table 3. The final reducted model, with all statistically significant terms, has a higher predicted R^2, which means that it is more reliable in estimating a response.

TP, ORAC, TEAC and caffeine show high number interactions with statistical significant terms ($p < 0.05$), and a good adjustment was obtained.

3.3. Total Polyphenol Content

The extraction efficiency of different concentrations of aqueous ethanol, temperature and time for the extraction of total polyphenols from white tea leaves was investigated using a central composite design. The total phenolic content in the white tea extracts ranged from 20.93 to 178.70 mg as GAE/g tea; see Table 2. It was observed that the best yield occurred with 47.5 min of extraction at 65 °C, using 48% ethanol.

All of the linear coefficients and two quadratic coefficients ($EtOH^2$ and t^2) were significant. The final response model (using uncoded units) to predict the yield of tea polyphenols is shown in Equation 3:

$$TP = 24.1036 + 2.3871\ EtOH + 1.7821\ t + 0.6350\ T - 0.0270\ EtOH^2 - 0.0156\ t^2 \quad (3)$$

where TP is the response variable and $EtOH$, t and T are the values of the independent variables, namely the concentration of ethanol, extraction time and temperature, respectively.

Table 2. Experimental design obtained and the experimental values obtained from the determination of total polyphenols, antioxidant capacity of the extracts assessed by TEAC and ORAC and polyphenol content.

	Assay nº	% EtOH	*t* (min)	*T* (°C)	Experimental Values							
					TP	**TEAC**	**ORAC**	**Caffeine**	**EGC**	**EGCG**	**ECG**	**EC**
2^3 factorial design	1	0	5	40	62.4 ± 3.1	596 ± 27	710 ± 83	11.31 ± 6.34	1.19 ± 0.53	3.22 ± 1.97	0.46 ± 0.26	0.55 ± 0.37
	2	96	5	40	20.9 ± 2.2	209 ± 23	437 ± 102	2.05 ± 0.43	1.86 ± 0.05	4.97 ± 0.44	1.08 ± 0.08	0.00 ± 0.00
	3	0	90	40	95.7 ± 4.1	945 ± 27	1396 ± 146	21.08 ± 0.90	4.00 ± 1.24	9.97 ± 3.58	0.97 ± 0.40	1.86 ± 0.53
	4	96	90	40	36.5 ± 8.6	348 ± 92	718 ± 156	3.68 ± 1.42	3.42 ± 1.83	10.50 ± 5.76	2.58 ± 1.36	0.33 ± 0.31
	5	0	5	90	96.1 ± 3.5	1067 ± 60	1053 ± 213	28.16 ± 2.13	0.64 ± 0.68	1.92 ± 1.04	0.30 ± 0.06	0.14 ± 0.05
	6	96	5	90	31.9 ± 2.9	341 ± 32	548 ± 177	3.52 ± 0.28	1.38 ± 0.21	1.21 ± 0.78	0.28 ± 0.17	0.22 ± 0.06
	7	0	90	90	112.6 ± 6.6	1165 ± 132	1355 ± 134	27.84 ± 1.88	1.16 ± 0.61	3.55 ± 2.44	0.64 ± 0.55	0.45 ± 0.45
	8	96	90	90	104.8 ± 2.1	1083 ± 50	1405 ± 45	9.18 ± 0.39	5.04 ± 0.20	10.90 ± 4.97	2.17 ± 0.93	1.29 ± 0.17
star design	9	0	47.5	65	94.8 ± 7.9	1042 ± 82	1244 ± 60	22.72 ± 2.26	0.80 ± 0.68	1.60 ± 0.76	0.18 ± 0.08	0.18 ± 0.20
	10	96	47.5	65	113.3 ± 24.9	529 ± 126	906 ± 285	3.49 ± 0.83	4.79 ± 2.28	18.73 ± 12.90	4.66 ± 3.15	0.68 ± 0.34
	11	48	5	65	148.5 ± 6.8	1777 ± 25	1914 ± 171	27.84 ± 0.82	1.71 ± 1.60	7.81 ± 6.41	1.25 ± 1.07	0.20 ± 0.34
	12	48	90	65	138.1 ± 25.6	1433 ± 221	1749 ± 296	20.36 ± 3.38	1.06 ± 0.32	5.88 ± 1.91	0.90 ± 0.43	0.15 ± 0.12
	13	48	47.5	40	144.4 ± 16.0	1466 ± 124	2387 ± 467	23.08 ± 1.73	2.01 ± 0.52	10.05 ± 3.88	1.71 ± 0.89	0.44 ± 0.19
	14	48	47.5	90	173.4 ± 7.9	1940 ± 117	2174 ± 245	26.66 ± 0.64	4.01 ± 0.20	19.51 ± 2.59	3.20 ± 0.44	1.28 ± 0.42
central design	15	48	47.5	65	163.3 ± 13.8	1674 ± 130	1811 ± 105	25.08 ± 0.11	3.96 ± 0.72	21.25 ± 6.78	3.58 ± 1.02	1.77 ± 0.40
	16	48	47.5	65	164.9 ± 7.9	1745 ± 130	1936 ± 205	26.85 ± 1.72	5.03 ± 1.06	27.45 ± 7.87	4.83 ± 1.26	1.31 ± 0.20
	17	48	47.5	65	178.7 ± 5.2	1820 ± 60	2132 ± 342	27.48 ± 2.37	4.71 ± 1.23	29.60 ± 10.60	5.40 ± 2.09	2.48 ± 1.10

Results are expressed as the mean of three replicates ± standard deviation; TP, total polyphenols (mg gallic acid equivalents (GAE)/g); TEAC, Trolox equivalent antioxidant capacity (μM TE/g); ORAC, oxygen radical antioxidant capacity (μM Trolox equivalents (TE)/g tea); Caffeine, epigallocatechin (EGC), epigallocatechin gallate (EGCG), epicatechin (EC) and epicatechin gallate (ECG) expressed as mg/g.

Table 3. The *p*-values for each of the constants in the equation of the mathematical model.

	Term	*p*-Value							
		Response							
		TP	Caffeine	EGC	EGCG	ECG	EC	ORAC	TEAC
Complete Model	Constant	0.986	0.562	0.230	0.742	0.496	0.150	0.003	0.174
	% EtOH	0.000	0.000	0.000	0.118	0.000	0.690	0.000	0.000
	t (min)	0.008	0.156	0.002	0.007	0.003	0.007	0.170	0.416
	T (°C)	0.115	0.337	0.798	0.752	0.960	0.458	0.053	0.724
	% EtOH × % EtOH	0.000	0.000	0.610	0.045	0.786	0.270	0.000	0.000
	t (min) × *t* (min)	0.000	0.550	0.004	0.002	0.004	0.000	0.013	0.130
	T × *T*	0.157	0.959	0.470	0.717	0.550	0.563	0.049	0.958
	EtOH × *t* (min)	0.204	0.692	0.490	0.590	0.715	0.175	0.740	0.133
	% EtOH × *T* (°C)	0.344	0.004	0.720	0.730	0.209	0.269	0.276	0.538
	t × *T*	0.184	0.269	0.400	0.940	0.385	0.980	0.671	0.220
Reducted Model	Constant	0.038	0.003	0.027	0.344	−0.771	0.069	0.001	0.003
	% EtOH	0.000	0.000	0.017	0.060	0.016	-	0.000	0.000
	t (min)	0.000	0.000	0.096	0.000	0.130	0.041	0.006	0.000
	T (°C)	0.000	0.000	-	-	-	-	0.089	0.000
	% EtOH × % EtOH	0.000	0.000	-	0.018	-	-	0.000	0.000
	t (min) × *t* (min)	-	-	0.001	0.000	0.001	0.000	0.052	-
	T × *T*	0.000	-	-	-	-	-	-	-
	EtOH × *t* (min)	-	-	-	-	-	-	-	-
	% EtOH × *T* (°C)	-	0.003	-	-	-	-	-	-
	t × *T*	-	-	-	-	-	-	-	-

- : Term not applied in the Model

The coefficient of determination in a multiple regression equation measures the strength of the relationship between the independent variables and the (dependent) response. The value of the determination coefficient for the equation for tea polyphenols is $R^2 = 0.873$, which indicates that only 13% of the total variation is not explained by the model.

Focusing on the RSM (represented in Figure 1A), it can deduced that the TP dependency is quadratic, and the optimal results could be obtained by extracting with 45% ETOH for 57 min, at 90 °C.

The results of this study are different from those of other authors; Venditti *et al.* [10] obtained significantly higher values in white tea after steeping in cold water (RT) for two hours. This study found that extraction of polyphenols was poor for water and aqueous ethanol at 40 °C.

Acid pH is reported to improve polyphenol extraction [14,15,17,34]. Zimmerman [34] reported that at pH = 3, ECG extraction increased 20% and suggested that this may be due to two possible mechanisms. At a low pH, either the diffusion of flavonoids from the leaf into the aqueous phase increased or degradation of the leaf structure occurred, permitting better accessibility of the solvent to the leaf components. However, when lemon juice was used to reduce the pH, the antioxidant capacity of the lemon juice bioactive compounds was not considered [35,36].

3.4. Polyphenol and Caffeine Composition

Caffeine, EGC, EGCG, ECG, EC, catechin and gallic acid content were analysed by MECK. Catechin and gallic acid were not detected. All results obtained are shown in Table 2.

Caffeine shows a final response model (Equation 4) with good adjustment ($R^2 = 0.89$) in the model.

$$\text{Caffeine} = 7.93192 + 0.41115EtOH + 0.21984T - 0.00481EtOH^2 - 0.00166EtOH \times T \quad (4)$$

The caffeine extracted was in the range of 2.05–28.3 mg/g dry tea. The minimum values were observed in samples extracted with a high ethanol concentration. This is consistent with the low polarity of the caffeine molecule. There is no literature about the caffeine content of white tea samples.

EC, ECG and EG have a reduce number of terms and interactions with statistical significant p-values (Table 3). They have an equation with low R^2. This fact suggested that the TP model responds to the synergic effect of all catechins contained in the sample.

Despite this bad adjustment, experimental results (Table 2) show that tea ECG content was in the range from 0.64 to 5.04 mg/g GAE. Lopez *et al.* [37], obtained similar ECG values (7.95 mg/g of tea) using an acid extraction method. Rusak *et al.* [13] reported a very high ECG content (42.3 mg/g of tea) after acid extraction and HPLC analysis. The latter author obtained 34.4 mg ECG/g dry tea using 40% ethanol for 30 min and 12.9 mg/g dry tea using 70% ethanol.

The EGC yield was 1.21–29.60 mg/g dry tea. The maximum yield was obtained with ethanol between 48% and 96% with an extraction time of 47.5 min and a temperature range of 56–90 °C. Rusak *et al.* [13] reported a range between 40.2 and 154 mg of EGC/g of dry white tea in bags and a range between 38.9 and 129 mg/g of EGC/g of dry tea in loose leaves analysed by HPLC. Lopez *et al.* [37] reported 11.1 mg EGC/g white tea, which is well within the range that was found by MECK analysis in this study.

The ECG content of the extracts corresponded to values between 0.18 and 5.40 mg ECG as GAE/g tea. Values reported by Lopez *et al.* [37] (3.19 mg/g dry tea) extracted from acidified samples were within this range. However, Rusak *et al.*[13] reported that 36.6 mg ECG as GAE/g dry white tea was extracted using 40% ethanol for a 15–30 min extraction time.

In EC analysis, a maximum extraction of 2.48 mg/g GAE was obtained in this study, which is similar to the value reported by Lopez *et al.* [37] (2.13 mg/g).

3.5. Antioxidant Capacity

The antioxidant activity of the different white tea infusions was assessed by the TEAC and ORAC assays. Some authors have shown significant differences in free radical scavenging activity according to the assay method used [23,38] and have reported that the TEAC assay is simpler and cheaper than the ORAC assay, but may give an underestimate of the antioxidant capacity. Tabart *et al.* [23] proposed that the mean of the values obtained using four different tests should be used.

A comparison between the two methods used in this study is presented, and the results are analysed using RSM.

3.5.1. TEAC

The contour plot (Figure 1B) shows that the TEAC values were highest for EtOH concentrations between 40% to 50%. The extract with the maximum TEAC value was obtained with 43.7% EtOH for 90 min. Surface plots shows that there was a quadratic dependence on %EtOH and a linear dependence on extraction time. All of the linear coefficients were again significant, but just one quadratic coefficient ($EtOH^2$) was included. The final response model (using uncoded units) to predict TEAC is shown in Equation 5:

$$TEAC = 325.540 + 33.821EtOH + 2.314t + 8.126T - 0.384EtOH^2 \qquad (5)$$

where TEAC is the response variable. The value of the determination coefficient for the equation is $R^2 = 0.91$ (only about 9% of the total variation is not explained by the model), so it can be concluded that there is a very good correlation between TEAC and the independent variables.

Samples 14 and 17, which had the highest TEAC values, had the highest TP values (Table 2). The polyphenols present were rich in EGCG, with moderate concentrations of ECG and EGC and a low concentration of EC (see Table 3). Salah *et al.* [39] reported that the relative antioxidant activity of the tea catechins assessed by the TEAC assay was in the order ECG > EGCG > EGC > EC, so it is clear that ECG and EGCG made important contributions to the antioxidant capacity.

3.5.2. ORAC

For assessment of antioxidant capacity by the ORAC assay, only two linear coefficients (EtOH and *t*) and one quadratic coefficient ($EtOH^2$) were significant. The final response model (using uncoded units) to predict ORAC is defined in Equation 6:

$$ORAC = 932.682 + 38.017EtOH + 4.615t - 0.415EtOH^2 \qquad (6)$$

where ORAC is the response variable. The value of the determination coefficient for the equation is $R^2 = 0.791$, which indicates that less than 21% of the total variation was not explained by the model.

3.6. Correlation between TP, TEAC and ORAC

The study of the correlation between experimental data obtained in TP, TEAC and ORAC was also evaluated. Several authors have reported correlations between the results found for radical scavenging activity assessed by the TEAC and ORAC methods [39]. The results of this study show a correlation between TP, TEAC and ORAC values (Figure 2). The best correlation is that between total polyphenols (TP) and TEAC values ($R^2 = 0.90$).

The correlation between TP and the ORAC values is not so good ($R^2 = 0.86$). The most likely reason for this is that the TEAC assay only measures single electron transfer (SET) and can only measure the extent of inhibition by antioxidants. ORAC combines the time and magnitude of the inhibition, so the effects of slow reacting and fast reacting antioxidants differ in this assay, but not in the TP or TEAC assays.

All three variables in the extraction process had a significant effect on the polyphenols and TEAC value. EtOH and time had a much greater effect on the ORAC value, while temperature had less effect.

Figure 1 displays the surface plots of RSM, in which the fitted responses were plotted against changes in the factors, EtOH and time, whereas the temperature was held at three different levels (40 °C, 65 °C and 90 °C). EtOH and time are represented on the *x*-axis, while TP, TEAC and ORAC are shown on the *y*-axis.

Figure 1. Response surface plots (**left**) and contour plots (**right**) showing the effect of ethanol concentration % in white tea extractions and the relations with antioxidant capacity evaluated by TP, TEAC and ORAC. (**A**) Response surface methodology (RSM) for TP and EtOH, (**B**) RSM for TEAC and ETOH and (**C**) RSM for ORAC and EtOH.

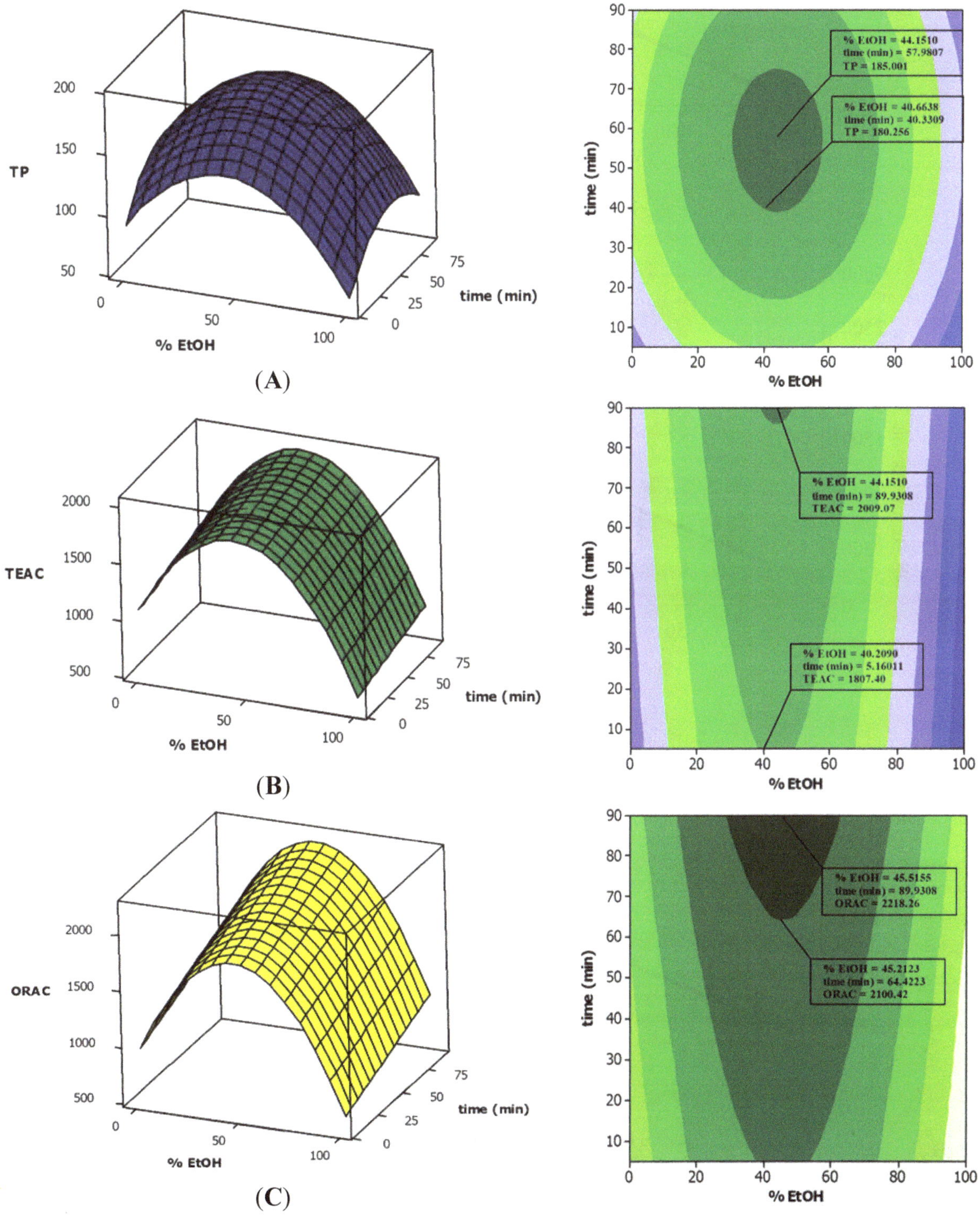

Figure 2. Correlation between TEAC, ORAC and total polyphenols (TP). (**A**) Correlation between TEAC and TP, (**B**) ORAC and TP and (**C**) ORAC and TEAC.

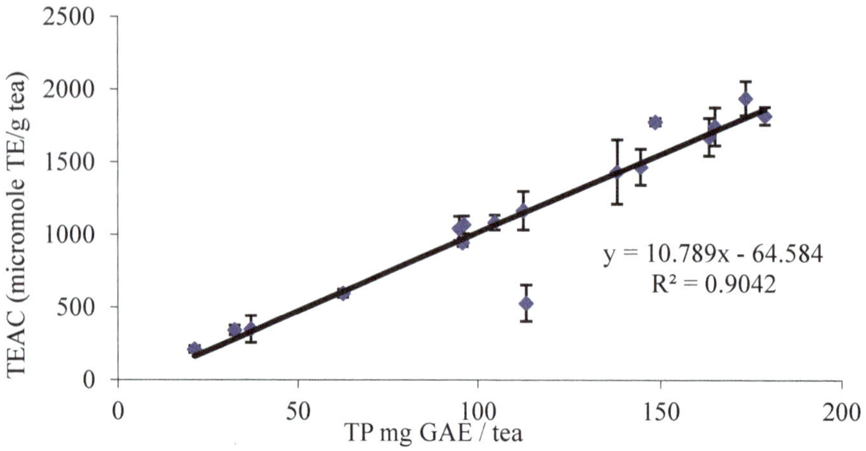

(**A**)

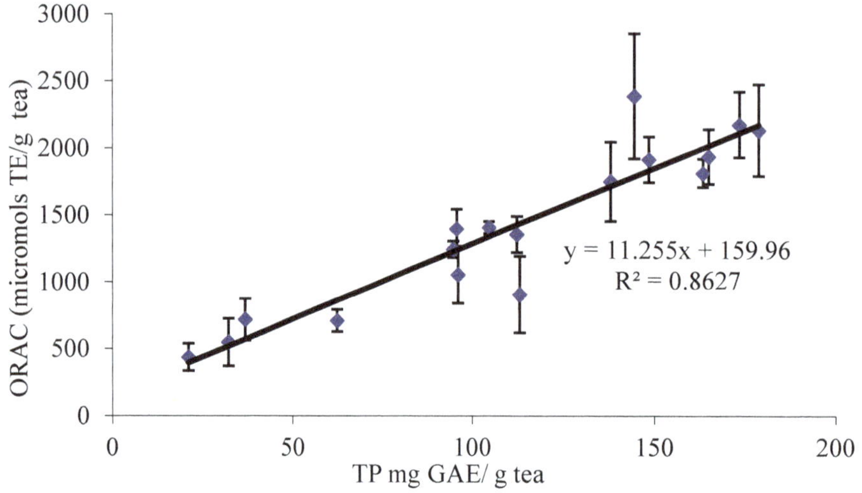

(**B**)

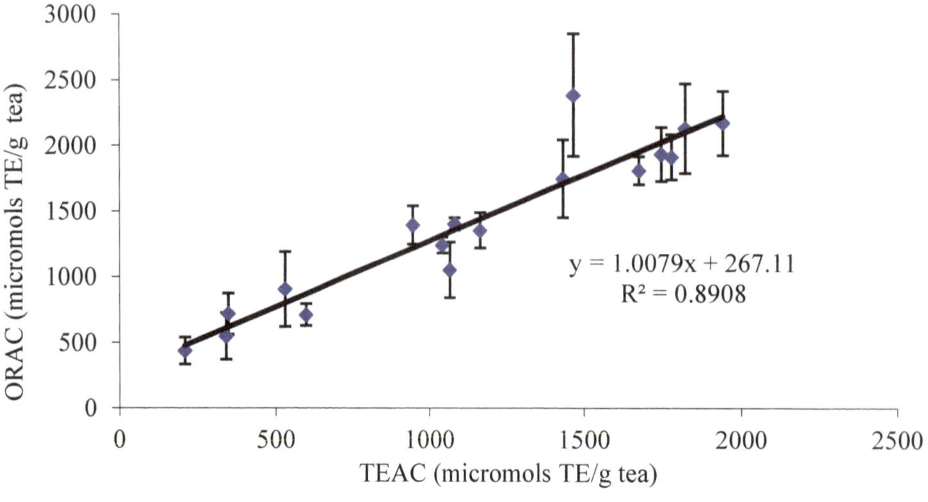

(**C**)

The optimal conditions for the extraction of tea polyphenols predicted by the equation were: EtOH = 45%, t = 57 min and T = 90 °C. Likewise, the conditions for extracts with optimal TEAC values predicted by the equation were: EtOH = 40%, t = 90 min and T = 90 °C. Finally, the conditions for extracts with optimal ORAC values predicted by the equation were: EtOH = 40%–60% and t = 60 min, with no significant effect of temperature.

4. Conclusions

The highest amounts of polyphenols (both individually and as determined by the Folin-Ciocalteu assay) were extracted at intermediate values of the conditions studied (about 48% ethanol, 47.5 min and 65 °C). The TP values correlated with antioxidant capacity determined by both the ORAC and TEAC assays, although the best temperature for the extraction of radical scavenging components assessed by the TEAC assay was 90 °C. Time is the factor that was less important, in the ranges studied, and %EtOH had the greatest influence. This study has identified optimal conditions for the extraction of tea liquor with the best antioxidant properties.

Acknowledgments

This project was supported by the *Agencia d'Ajuts Universitaris i de Recerca (AGAUR)* (Catalonia, Spain). The authors thanks to Technical University from Catalonia for their support in this research.

Author Contributions

María Pilar Almajano and Francisca Pérez-Llamas conceived and designed the study. Mónica Blanco performed the statistical analysis. Sara Peiró and Francisco Segovia did the experimental work, analysed the data and interpreted results. Sara Peiró wrote the paper and Michael H. Gordon helped with English, with concepts and revised it critically. Finally, María Pilar Almajano reviewed all the manuscript and the final version to be submitted.

Conflicts of Interest

The authors declare no conflict of interest.

References

1. Oh, J.; Jo, H.; Cho, A.R.; Kim, S.; Han, J. Antioxidant and antimicrobial activities of various leafy herbal teas. *Food Control* **2013**, *31*, 403–409.
2. Gülçin, I. Antioxidant activity of food constituents: An overview. *Arch. Toxicol.* **2012**, *86*, 345–391.
3. Silva-Weiss, A.; Ihl, M.; Sobral, P.J.A.; Gómez-Guillén, M.C.; Bifani, V. Natural Additives in Bioactive Edible Films and Coatings: Functionality and Applications in Foods. *Food Eng. Rev.* **2013**, *5*, 200–216.
4. Perumalla, A.V.S.; Hettiarachchy, N.S. Green tea and grape seed extracts—Potential applications in food safety and quality. *Food Res. Int.* **2011**, *44*, 827–839.

5. Gramza, A.; Korczak, J. Tea constituents (*Camellia sinensis* L.) as antioxidants in lipid systems. *Trends Food Sci. Technol.* **2005**, *16,* 351–358.
6. Almajano, M.P.; Carbo, R.; Jimenez, J.A.L.; Gordon, M.H. Antioxidant and antimicrobial activities of tea infusions. *Food Chem.* **2008**, *108*, 55–63.
7. Cabrera, C.; Artacho, R.; Giménez, R. Beneficial effects of green tea—A review. *J. Am. Coll. Nutr.* **2006**, *25*, 79–99.
8. Sharangi, A.B. Medicinal and therapeutic potentialities of tea (*Camellia sinensis* L.)—A review. *Food Res. Int.* **2009**, *42*, 529–535.
9. Moderno, P.M.; Carvalho, M.; Silva, B.M. Recent patents on *Camellia sinensis*: Source of health promoting compounds. *Recent Pat. Food Nutr. Agric.* **2009**, *1*, 182–192.
10. Venditti, E.; Bacchetti, T.; Tiano, L.; Carloni, P.; Greci, L.; Damiani, E. Hot *vs.* cold water steeping of different teas: Do they affect antioxidant activity? *Food Chem.* **2010**, *119*, 1597–1604.
11. Hilal, Y.; Engelhardt, U. Characterisation of white tea—Comparison to green and black tea. *J. Verbrauch. Lebensm.* **2007**, *2*, 414–421.
12. Müller, N.; Ellinger, S.; Alteheld, B.; Ulrich-Merzenich, G.; Berthold, H.K.; Vetter, H.; Stehle, P. Bolus ingestion of white and green tea increases the concentration of several flavan-3-ols in plasma, but does not affect markers of oxidative stress in healthy non-smokers. *Mol. Nutr. Food Res.* **2010**, *54*, 1636–1645.
13. Rusak, G.; Komes, D.; Likić, S.; Horžić, D.; Kovač, M. Phenolic content antioxidative capacity of green and white tea extracts depending on extraction conditions and the solvent used. *Food Chem.* **2008**, *110*, 852–858.
14. Unachukwu, U.J.; Ahmed, S.; Kavalier, A.; Lyles, J.T.; Kennelly, E.J. White and Green Teas (*Camellia sinensis* var. sinensis): Variation in Phenolic, Methylxanthine, and Antioxidant Profiles. *J. Food Sci.* **2010**, *75*, C541–C548.
15. Komes, D.; Belščak-Cvitanović, A.; Horžić, D.; Rusak, G.; Likić, S.; Berendika, M. Phenolic composition and antioxidant properties of some traditionally used medicinal plants affected by the extraction time and hydrolysis. *Phytochem. Anal.* **2011**, *22*, 172–180.
16. Almajano, M.P.; Vila, I.; Gines, S. Neuroprotective effects of white tea against oxidative stress-induced toxicity in striatal cells. *Neurotox. Res.* **2011**, *20*, 372–378.
17. Lopez, V.; Isabel Calvo, M. White Tea (*Camellia sinensis* Kuntze) Exerts Neuroprotection against Hydrogen Peroxide-Induced Toxicity in PC12 Cells. *Plant Foods Hum. Nutr.* **2011**, *66*, 22–26.
18. Thring, T.S.A.; Hili, P.; Naughton, D.P. Antioxidant and potential anti-inflammatory activity of extracts and formulations of white tea, rose, and witch hazel on primary human dermal fibroblast cells. *J. Inflamm.* **2011**, *8*, 27.
19. Yen, W.-J.; Chyau, C.-C.; Lee, C.-P.; Chu, H.-L.; Chang, L.-W.; Duh, P.-D. Cytoprotective effect of white tea against H_2O_2-induced oxidative stress *in vitro*. *Food Chem.* **2013**, *14*, 4107–4114.
20. Camouse, M.M.; Domingo, D.S.; Swain, F.R.;Conrad, E.P.; Matsui, M.S.; Maes, D.; Declercq, L.; Cooper, K.D.; Stevens, S.R.; Baron, E. Topical application of green and white tea extracts provides protection from solar-simulated ultraviolet light in human skin. *Exp. dermatol.* **2009**, *18*, 522–526.
21. Dias, R.; Alves, M.G.; Socorro, S.; Silva, B.M.; Oliveira, P.F. White Tea as a Promising Antioxidant Medium Additive for Sperm. *J. Agric. Food Chem.* **2014**, *62*, 608–617.

22. Hengst, C.; Werner, S.; Müller, L.; Fröhlich, K.; Böhm, V. Determination of the antioxidant capacity: Influence of the sample concentration on the measured values. *Eur. Food Res. Technol.* **2009**, *230*, 249–254.
23. Tabart, J.; Kevers, C.; Pincemail, J.; Defraigne, J.-O.; Dommes, J. Comparative antioxidant capacities of phenolic compounds measured by various tests. *Food Chem.* **2009**, *113*, 1226–1233.
24. Rusaczonek, A.; Swiderski, F.; Waszkiewicz-Robak, B. Antioxidant Properties of Tea and Herbal Infusions—A Short Report. *Pol. J. Food Nutr. Sci.* **2010**, *60*, 33–35.
25. Peres, R.G.; Tonin, F.G.; Tavares, M.F.M.; Rodriguez-Amaya, D.B. Determination of catechins in green tea infusions by reduced flow micellar electrokinetic chromatography. *Food Chem.* **2011**, *127*, 651–655.
26. Sun, Y.; Liu, D.; Chen, J.; Ye, X.; Yu, D. Effects of different factors of ultrasound treatment on the extraction yield of the all-*trans*-β-carotene from citrus peels. *Ultrason. Sonochem.* **2011**, *18*, 243–249.
27. Terabe, S. Capillary separation: Micellar electrokinetic chromatography. *Annual Rev. Anal. Chem.* **2009**, *2*, 99–120.
28. Prior, R.L.; Wu, X.L.; Schaich, K. Standardized methods for the determination of antioxidant capacity and phenolics in foods and dietary supplements. *J. Agric. Food Chem.* **2005**, *53*, 4290–4302.
29. Herrero, M.; Ibáñez, E.; Cifuentes, A. Analysis of natural antioxidants by capillary electromigration methods. *J. Sep. Sci.* **2005**, *28*, 883–897.
30. Re, R.; Pellegrini, N.; Proteggente, A.; Pannala, A.; Yang, M.; Rice-Evans, C. Antioxidant activity applying an improved ABTS radical cation decolorization assay. *Free Radic. Biol. Med.* **1999**, *26*, 1231–1237.
31. Ou, B.; Hampsch-Woodill, M.; Prior, R.L. Development and validation of an improved oxygen radical absorbance capacity assay using fluorescein as the fluorescent probe. *J. Agric. Food Chem.* **2001**, *49*, 4619–4626.
32. Ferruzzi, M.G.; Green, R.J. Analysis of catechins from milk-tea beverages by enzyme assisted extraction followed by high performance liquid chromatography. *Food Chem.* **2006**, *99*, 484–491.
33. Mariya John, K.M.; Vijayan, D.; Raj Kumar, R.; Premkumar, R. Factors influencing the efficiency of extraction of polyphenols from young tea leaves. *Asian J. Plant Sci.* **2006**, *5*, 123–126.
34. Zimmermann, B.F.; Gleichenhagen, M. The effect of ascorbic acid, citric acid and low pH on the extraction of green tea: How to get most out of it. *Food Chem.* **2011**, *124*, 1543–1548.
35. Del Río, J.A.; Fuster, M.D.; Gómez, P.; Porras, I.; García-Lidón, A.; Ortuño, A. *Citrus limon*: A source of flavonoids of pharmaceutical interest. *Food Chem.* **2004**, *84*, 457–461.
36. González-Molina, E.; Moreno, D.A.; García-Viguera, C. Genotype and harvest time influence the phytochemical quality of Fino lemon juice (*Citrus limon* (L.) Burm. F.) for industrial use. *J. Agric. Food Chem.* **2008**, *56*, 1669–1675.
37. López, M.D.M.C.; Vilariño, J.M.L.; Rodríguez, M.V.G.; Losada, L.F.B. Development, validation and application of Micellar Electrokinetic Capillary Chromatography method for routine analysis of catechins, quercetin and thymol in natural samples. *Microchem. J.* **2011**, *99*, 461–469.

38. Zulueta, A.; Esteve, M.J.; Frasquet, I.; Frígola, A. Vitamin C, vitamin A, phenolic compounds and total antioxidant capacity of new fruit juice and skim milk mixture beverages marketed in Spain. *Food Chem.* **2007**, *103*, 1365–1374.
39. Salah, N.; Miller, N.J.; Paganga, G.; Tijburg, L.; Bolwell, G.P.; Rice-Evans, C. Polyphenolic flavanols as scavengers of aqueous phase radicals and as chain-breaking antioxidants. *Arch. Biochem. Biophys.* **1995**, *322*, 339–346.

6

Optimization of the Aqueous Extraction of Phenolic Compounds from Olive Leaves

Chloe D. Goldsmith [1,*], Quan V. Vuong [1], Costas E. Stathopoulos [2], Paul D. Roach [1] and Christopher J. Scarlett [1]

[1] School of Environmental & Life Sciences, University of Newcastle, Ourimbah, NSW 2258, Australia; E-Mails: vanquan.vuong@newcastle.edu.au (Q.V.V.); paul.roach@newcastle.edu.au (P.D.R.); c.scarlett@newcastle.edu.au (C.J.S.)

[2] Faculty of Bioscience Engineering, Ghent University Global Campus, Incheon 406-840, South Korea; E-Mail: costas.stathopoulos@ghent.ac.kr

* Author to whom correspondence should be addressed; E-Mail: chloe.d.goldsmith@uon.edu.au

External Editors: Maria G. Miguel and João Rocha

Abstract: Olive leaves are an agricultural waste of the olive-oil industry representing up to 10% of the dry weight arriving at olive mills. Disposal of this waste adds additional expense to farmers. Olive leaves have been shown to have a high concentration of phenolic compounds. In an attempt to utilize this waste product for phenolic compounds, we optimized their extraction using water—a "green" extraction solvent that has not yet been investigated for this purpose. Experiments were carried out according to a Box Behnken design, and the best possible combination of temperature, extraction time and sample-to-solvent ratio for the extraction of phenolic compounds with a high antioxidant activity was obtained using RSM; the optimal conditions for the highest yield of phenolic compounds was 90 °C for 70 min at a sample-to-solvent ratio of 1:100 g/mL; however, at 1:60 g/mL, we retained 80% of the total phenolic compounds and maximized antioxidant capacity. Therefore the sample-to-solvent ratio of 1:60 was chosen as optimal and used for further validation. The validation test fell inside the confidence range indicated by the RSM output; hence, the statistical model was trusted. The proposed method is inexpensive, easily up-scaled to industry and shows potential as an additional source of income for olive growers.

Keywords: olive leaves; phenolic compounds; green extraction solvents; waste valorisation; *Olea europaea*; response surface methodology (RSM)

1. Introduction

Adherence to a Mediterranean-style diet has been associated with a reduced risk for cardiovascular disease and certain types of cancers [1]. These associations have been linked, in part, to the high consumption of olive oil, more specifically, the consumption of the unique phenolic compounds found in olive oil [2–4]. The same compounds believed to be responsible for the health-promoting properties attributed to olive oil consumption have also been identified in olive leaves [5]. Hence, the potential applications for the health promoting compounds extracted from olive leaves are extensive. These include their use as food additives or health supplements, as well as their continued use in future research into potential anti-cancer [6], anti-inflammatory [7] or anti-fungal [8] agents. It is therefore important to optimize the extraction of these compounds. An understanding of the parameters affecting the extraction of phenolic compounds is paramount to establishing the foundations for this future work.

Mediterranean countries account for around 98% of the world's olive cultivation (approximately ten million hectares); they produce about 1.9 million metric tonnes per annum of olive oil and 1.1 million tonnes of table olives [9]. Olive leaves are an agricultural waste of the olive oil and table olive production industry. This waste is produced as a result of pruning olive trees during the growing season, as well as accounting for approximately 10% of the weight of materials received by olive mills. Currently, this by-product is not profitable, given that in many countries, olive leaves are used as animal feed or simply burned with excess branches gathered from pruning [10,11]. Many olive oil producers even charge a fee to the olive farmer for the disposal of olive leaves.

The market for natural additives and ingredients is rapidly growing, with some natural products obtaining high prices. Moreover, the possible toxicity of certain synthetic compounds [5,12] has led to an increased interest in natural product research from the cosmetic, pharmaceutical and food additive industries. This has led to improved extraction, fractionation and purification technologies being developed in the last few years. However, these modern purification and separation technologies can be expensive and sometimes hazardous, rendering it near impossible for farmers to profit directly.

A number of methods have been proposed for the extraction of phenolic compounds from olive leaves, including the use of advanced technologies, such as microwave, pressurized liquid extraction and ultra-sonic extraction methods [13–15]. However, these practices can often have high energy costs and lead to the production of excessive solvent waste, which can be more hazardous to dispose of than the actual agricultural waste itself. Therefore, there is a need for the development of "green" extraction procedures. Water is a cheap, non-hazardous polar extraction solvent. It has been shown to efficiently extract a vast array of phenolic compounds with high antioxidant activities from a number of plant materials [16–18].

Therefore, in the present study, we aimed to optimize the extraction of phenolic compounds from olive leaves using the inexpensive, non-hazardous and easily obtainable solvent, water. The parameters of time, temperature and sample-to-solvent ratio were chosen for optimization, as they are easy for farmers or processors to control. The influence of these extraction parameters on antioxidant activity was also investigated.

2. Experimental Section

2.1. Materials and Reagents

Folin–Ciocalteu's reagent, sodium carbonate, gallic acid, 1,1-diphenyl-2-picrylhydrazyl (DPPH), 6-hydroxy-2,5,7,8-tetramethylchroman-2-carboxylic acid (trolox), 2,4,6-Tris(2-pyridyl)-*s*-triazine (TPTZ), ferric chloride, sodium acetate, acetic acid, copper (II) chloride, ammonium acetate (NH_4Ac), neocuproine methanol and ethanol were purchased from Sigma Aldrich (Castle Hill, Australia). Ultra-pure (type 1) de-ionized (DI) water was prepared by reverse osmosis and filtration using a Mili-Q direct 16 system (Milipore Australia Pty Ltd., North Ryde, Australia).

2.2. Sample Preparation

Corregiola olive leaves were obtained from Houndsfield Estate in the Hunter Valley of NSW Australia. The leaves were dried at 120 °C for 90 min according to [19], ground to a size of 0.1 mm and stored at −20 °C until further analysis.

2.3. Response Surface Methodology (RSM)

The RSM with the Box–Behnken design was then employed to design the experiment to investigate the influence of three independent parameters, temperature, time and sample-to-solvent ratio, on the extraction of total phenolic compounds (TPC) and on the antioxidant activity of the resultant extracts. The optimal ranges of temperature (70–90 °C), time (50–70 min) and sample-to-solvent ratio (1:10–1:100 g/mL) were determined based on preliminary experiments. The independent variables and their code variable levels are shown in Table 1. To express the TPC or antioxidant capacity as a function of the independent variables, a second-order polynomial equation was used as follows and as previously described by Vuong *et al.* [20]: $Y = \beta_0 + \sum_{i=1}^{k} \beta_i X_i + \sum_{i=1}^{k=1} \sum_{j}^{k} \beta_{ij} X_i X_j + \sum_{i=1}^{k} \beta_{ii} X_i^2$, where various X_i values are independent variables affecting the response Y; β_0, β_i, β_{ii}, and β_{ij} are the regression coefficients for the intercept and the linear, quadratic and interaction terms, respectively, and k is the number of variables.

2.4. Total Phenolic Compounds

The TPC was determined according to Thaipong *et al.* [21]. Briefly, the appropriately diluted samples (300 μL) were added to Folin–Ciocalteu's reagent (300 μL) and left to equilibrate for 2 min before adding 2.4 mL of 5% sodium carbonate solution and incubating in the dark for 1 h. Absorbance was then read at 760 nm using a UV spectrophotometer (Varian, Melbourne, Australia). Gallic acid was used as the standard, and results were expressed as mg of gallic acid equivalents per g of sample (mg GAE/g).

Table 1. Values of the independent parameters and their coded forms with their symbols employed in RSM for optimization of olive leaf extraction using water.

Independent Parameters	Symbols of the Parameters	Original Values of the Parameters	Parameter Coded Forms *
Temperature (°C)	X_1	70	−
		80	0
		90	+
Time (min)	X_2	50	−
		60	0
		70	+
Ratio (mg/mL)	X_3	10	−
		55	0
		100	+

* Parameter coded forms −, 0 and + are the minimum point, centre point and maximum point (respectively) for the independent parameters temperature, time and ratio.

2.5. Antioxidant Activity Assays

Three assays were employed in order to assess the antioxidant activity of the olive leaf extracts:

For the ferric reducing antioxidant power (FRAP) assay, the extract was diluted within the appropriate range, and then, their ferric ion reducing capacity was determined according to Thaipong *et al.* [21].

Stock solutions were: (1) 300 mM acetate buffer; (2) 10 mM TPTZ solution in 40 mM HCL; (3) 20 mM $FeCl_3$ solution. The fresh working solution was prepared by mixing 25 mL acetate buffer, 2.5 mL TPTZ solution and 2.5 mL $FeCl_3$ and then warming to 37 °C. Olive leaf extracts, standards and blanks (150 μL) were then added to 2.85 mL of the working FRAP solution and left to incubate in the dark at 37 °C for 30 min. Absorbance was read at 593 nm. Results were expressed as mg trolox equivalents per gram of sample dry weight (mg Trolox Equivalents (TE)/g).

For the cupric reducing antioxidant capacity (CUPRAC) assay, the extracts were diluted within the appropriate range, and their cupric ion reducing capacity was determined as described by Apak *et al.* [22].

The stock solutions were: (1) 10 mM $CuCl_2$ solution; (2) ammonium acetate buffer at pH 7.0; (3) 7.5 mM neocuproine (Nc) solution in 95% ethanol. A working solution of the three reagents (1:1:1 v/v) was prepared, 3 mL of which was added to 1.1 mL of the diluted extracts, standards and blanks and left to react in the dark for 1 hour. Absorbance was read at 450 nm. Results were expressed as mg of trolox equivalents per gram of sample dry weight (mg TE/g).

The DPPH free radical scavenging activity of the extracts was analysed using the 1,1-diphenyl-2-picrylhydrazyl (DPPH) assay, as described by Vuong *et al.* [23]. Briefly, the appropriately diluted samples, standards and blank (150 μL) were added to 2.85 mL of DPPH working solution (made to an absorbance of 1.1 ± 0.01 at 760 nm) and left to react in a dark at room temperature for 3 h. Trolox was used as a standard. The results were expressed as mg of trolox equivalents per g of sample dry weight (mg TE/g).

2.6. Statistical Analysis

The RSM experimental design and analysis was conducted using JMP software (Version 11, SAS, Cary, NC, USA). The software was also used to establish the model equation, graph the 3D plot with 2D contour of the responses and to predict the optimum values for the three response variables in order to obtain the maximum TPC level. All experiments were carried out in triplicate.

3. Results and Discussion

3.1. Fitting the Models for the Prediction of Total Phenolic Compounds and Antioxidant Capacity

The experimental design is presented in Table 1, while Table 2 indicates the effects of temperature, time and the ratio of sample-to-solvent on the extraction of TPC from olive leaves using water. The predicted yield of TPC ranged from 22.36 to 38.25 mg GAE/g depending on the combination of extraction parameters.

Table 2. Analysis of variance for the determination of the fit of the model. TPC, total phenolic compounds; FRAP, ferric reducing antioxidant power; CUPRAC, cupric reducing antioxidant capacity; PRESS, predicted residual sum of squares.

Sources of Variation	TPC	Antioxidant Capacity		
		FRAP	CUPRAC	DPPH
Lack of fit (p-value)	0.1991	0.0168 *	0.1369	0.1377
R^2	0.8	0.95	0.97	0.92
Adjusted R^2	0.44	0.87	0.92	0.78
PRESS	1149.1	1500.72	1097.5	1988.1
F-ratio of model	2.2025	11.54	19.6	6.639
p of model > F	0.1991	0.0075 *	0.0022 *	0.0258 *

* Significant difference with $p < 0.05$.

Table 2 shows the reliability of the RSM mathematical model in predicting optimal variances and accurately representing the real interrelationships between the selected parameters. The results for the analysis of variances of the Box–Behnken design are shown in Table 2. Figure 1 indicates the correlation between the predicted and experimental values.

Figure 1 and Table 2 indicate that there was no significant difference between the actual and predicted values for TPC ($p > 0.05$). Furthermore, the coefficient of determination (R^2) value for the correlation between the predicted and actual values was 0.8, indicating that the model can predict 80% of the actual data for TPC. Table 2 also showed that the "lack of fit" for the model was also not significant ($p = 0.1991$). In addition, the PRESS (predicted residual sum of squares) was 1149.1 and the F-ratio was 2.2025. PRESS is a measure of how well each point fits into the experimental design, further identifying the appropriateness of the model's fit.

It was therefore concluded that the second-order polynomial equation for the following three independent variables could be used: temperature (X_1), time (X_2) and sample-to-solvent ratio (X_3). The predictive equation for the response of total phenolic compounds (Y) was as follows:

$$Y = 26.02 + 1.31\,X_1 + 0.42\,X_2 + 4.88\,X_3 - 0.14\,X_1\,X_2 + 1.42\,X_1\,X_3 + 1.91\,X_2\,X_3 + (0.09\,X_1)^2 + (3.79\,X_2)^2 + (1.23\,X_3)^2 \quad (1)$$

Figure 1. Prediction profiler plots for the effects of the test parameters on the extraction of phenolic compounds from olive leaves.

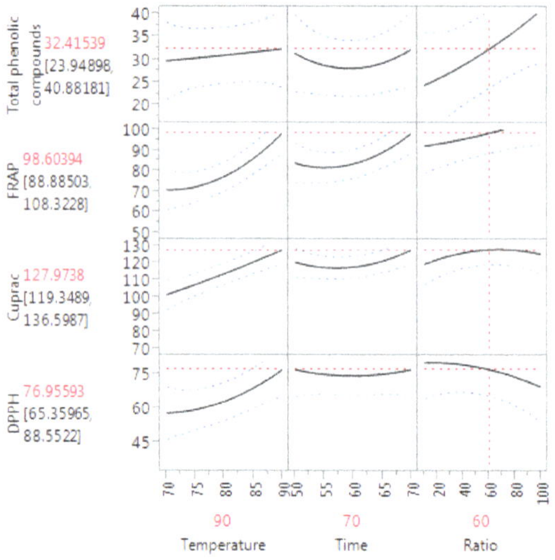

The model fit for the antioxidant activity of the olive leaf extract was also investigated. Figures 2–4 show the relationship between the actual and predicted values, while Table 2 represents the analysis of variance results for the determination of the fit of the model. The *p*-values for the model fit were 0.0168, 0.1369 and 0.1377 for FRAP, CUPRAC and DPPH, respectively. This shows that there was no difference between actual and predicted values for CUPRAC and DPPH. However, there was a significant difference between the actual and predicted values for FRAP.

The coefficients of determination were 0.95, 0.97 and 0.92 for FRAP, CUPRAC and DPPH, respectively. This highlighted the close correlation between the actual and predicted values. This relationship is further supported with the values for PRESS and the F-ratios of the model: 1500.72 and 11.54 for FRAP, 1097.5 and 19.6 for CUPRAC and 1988.1 and 6.639 for DPPH, respectively. This indicated that the mathematical models were reliable predictors of the antioxidant activity of the olive leaf water extracts. Therefore, the following second order polynomials could be used:

FRAP:

$$Y = 64.66 + 10.51\,X_1 + 4.58\,X_2 + 7.45\,X_3 + 3.05\,X_1X_2 + 2.16\,X_1X_3 - 2.66\,X_2X_3 + (7.39\,X_1)^2 + (7.64\,X_2)^2 + (1.4\,X_3)^2 \tag{2}$$

CUPRAC:

$$Y = 104.53 + 11.76\,X_1 + 1.91\,X_2 + 11.31\,X_3 + 2.06\,X_1X_2 - 6.14\,X_1X_3 - 2.27\,X_2X_3 + (1.01\,X_1)^2 + (6.45\,X_2)^2 - (5.33\,X_3)^2 \tag{3}$$

DPPH:

$$Y = 60.08 + 9.29\,X_1 + 0.39\,X_2 + 7.02\,X_3 + 0.68\,X_1X_2 - 3.4\,X_1X_3 - 8.98\,X_2X_3 + (4.43\,X_1)^2 + (2.71\,X_2)^2 - (3.03\,X_3)^2 \tag{4}$$

Figure 2. Correlation between the actual and the predicted values for the total phenolic compounds (TPC) and antioxidant capacity of olive leaf water extract (FRAP, DPPH and CUPRAC).

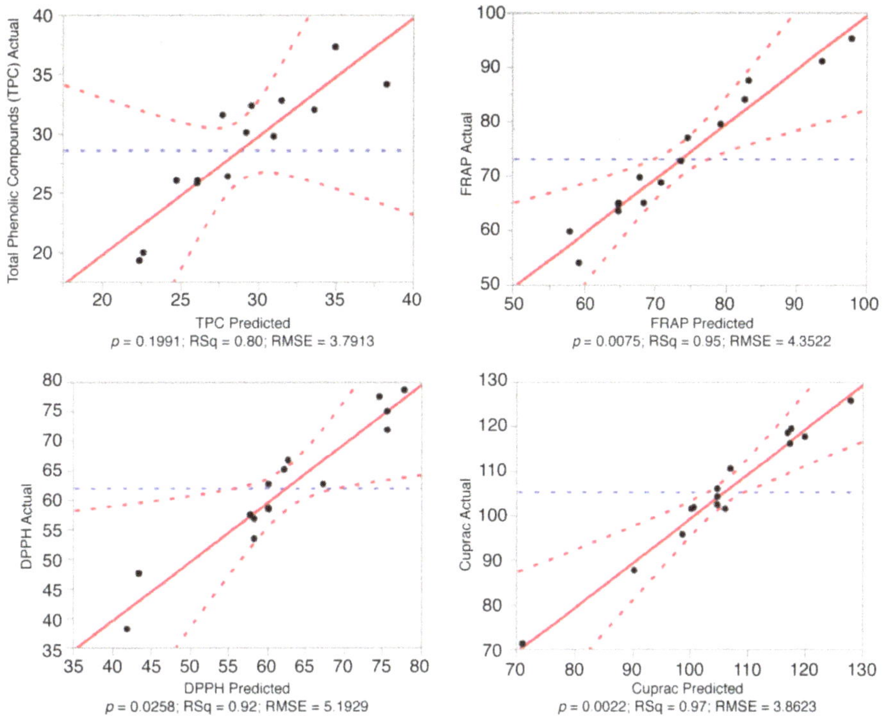

Figure 3. 3D response surface and 2D contour plots for the effects of the test parameters on total phenolic compounds.

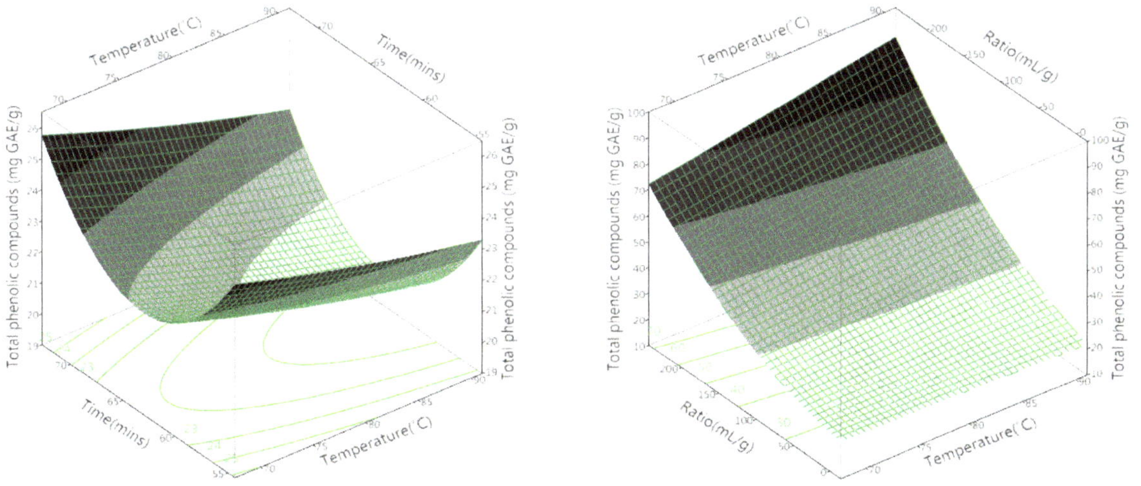

Figure 3. *Cont.*

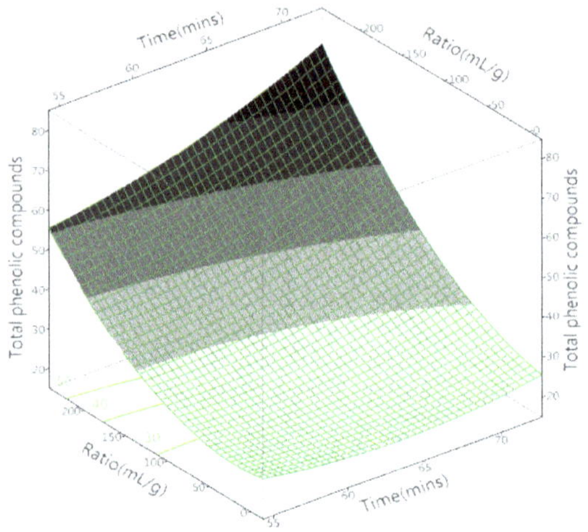

Figure 4. 3D response surface and 2D contour plots for the effects of the test parameters on antioxidant activity.

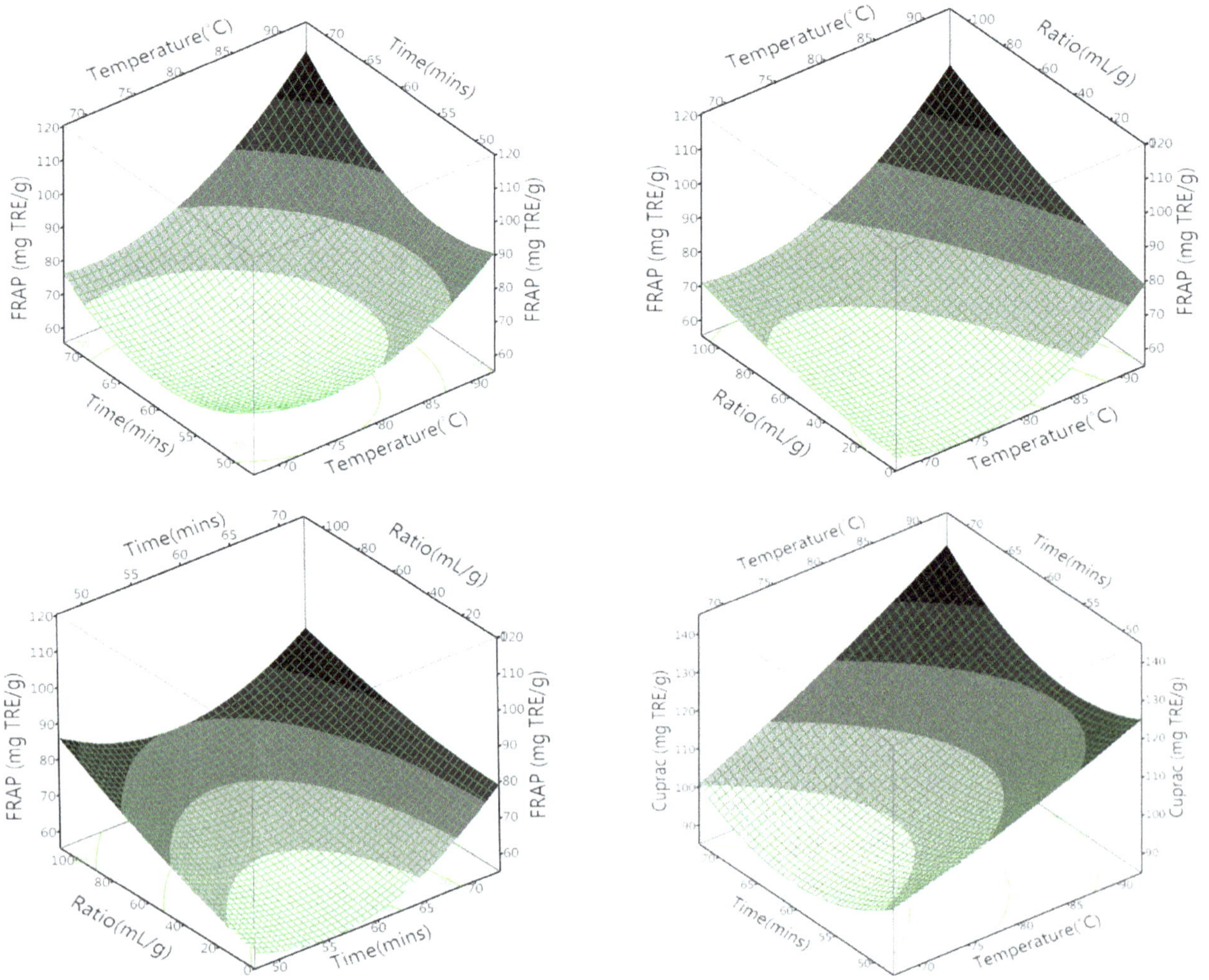

Figure 4. *Cont.*

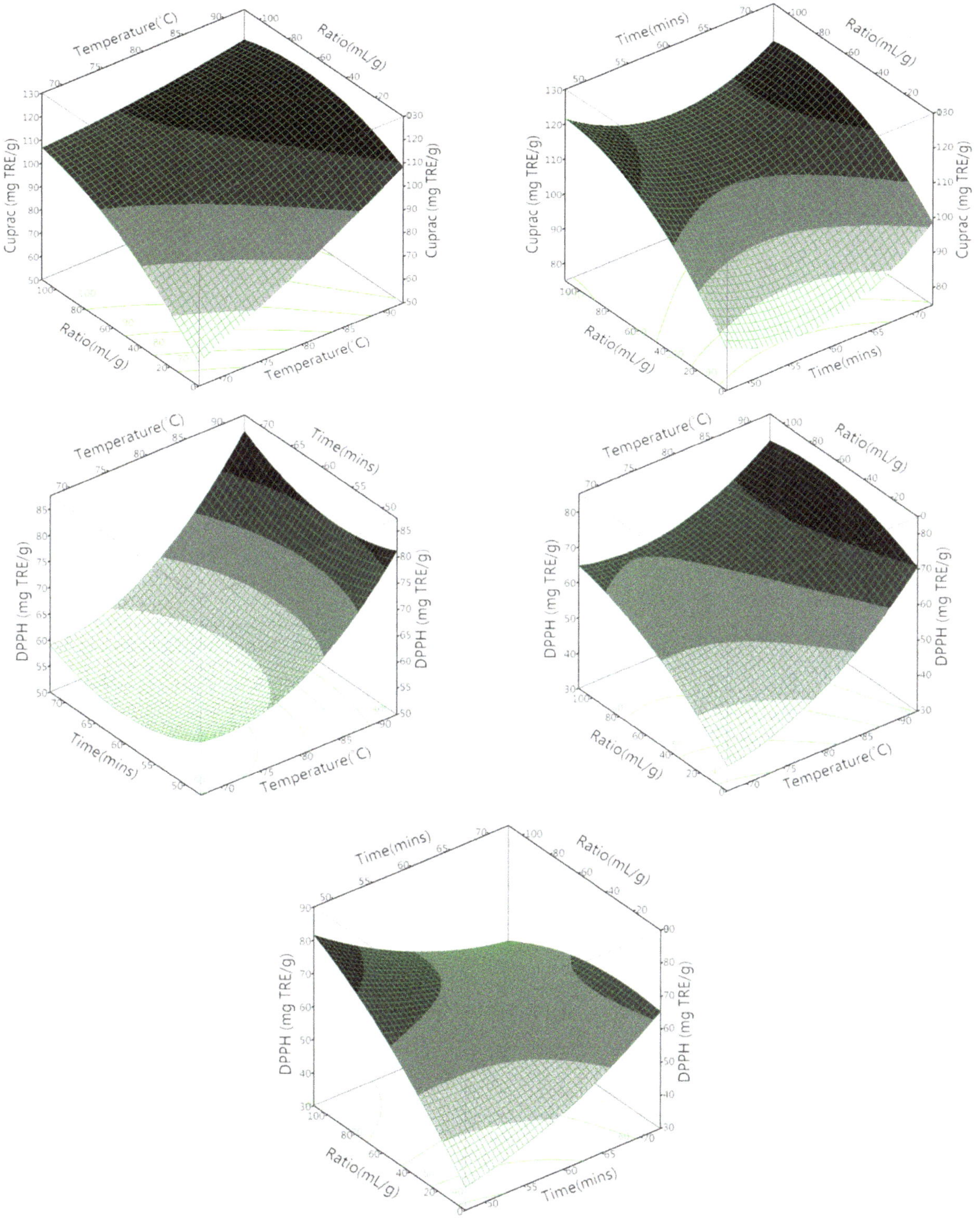

3.2. The Effect of the Different Variables on the Total Phenolic Compounds

Table 3 presents the linear regression coefficients and indicates their statistical significance. Temperature, time and ratio were all shown to have a positive influence on the extraction of TPC. However, the only parameter to significantly affect the extraction efficiency was the sample-to-solvent ratio ($p = 0.01$). Temperature and time had no significant effect on TPC ($p > 0.05$), nor did any of the various combinations of factors (temperature × time, temperature × ratio or time × ratio) (Table 3). This was unexpected, since time has previously been shown to have a significant effect on the extraction of TPC from olive leaves when using ultrasonic assistance [15]. Extraction time has also been identified as a significant extraction parameter for the extraction of natural polyphenols from wheat bran [24]. However, in both of these studies, the use of advanced technologies could account for the observed differences.

Table 3. The analysis of variance for the experimental results.

Parameter	DF	TPC		Antioxidant Capacity					
				Frap		DPPH		CUPRAC	
		F	Prob > F	F	Prob > F	F	Prob > F	F	Prob > F
β_0	1	26.02	<0.0001	64.66	<0.0001	60.08	<0.0001	104.53	<0.0001
β_1	1	1.31	0.37	10.51	0.001 *	9.29	0.004 *	11.76	0.0003 *
β_2	1	0.42	0.77	4.58	0.031 *	0.39	0.84	1.91	0.22
β_3	1	4.88	0.01 *	7.45	0.005 *	7.02	0.01 *	11.31	0.0004 *
β_{12}	1	−0.14	0.94	3.05	0.22	0.68	0.8	2.06	0.34
β_{13}	1	1.42	0.49	2.16	0.37	−3.4	0.25	−6.14	0.02 *
β_{23}	1	1.91	0.36	−2.66	0.28	−8.98	0.02 *	−2.27	0.29
β_{11}	1	0.09	0.96	7.39	0.02 *	4.43	0.16	1.01	0.63
β_{22}	1	3.79	0.11	7.64	0.02 *	2.71	0.36	6.45	0.02 *
β_{33}	1	1.23	0.56	1.4	0.56	−3.03	0.31	−5.33	<0.05 *

* Significantly difference with $p < 0.05$; β_0: intercept; β_1, β_2 and β_3: linear regression coefficients for temperature, time and ratio; β_{12}, β_{13} and β_{23}: regression coefficients for interaction between temperature × time, temperature × ratio and time × ratio; β_{11}, β_{22} and β_{33}: quadratic regression coefficients for temperature × temperature, time × time and ratio × ratio; Prob = probability.

The sample-to-solvent ratio was shown to have a significant effect on the extraction of TPC. This is consistent with mass transfer principles, which outline that the concentration gradient (the driving force) is higher when there is more solvent present, leading to higher diffusion rates.

3.3. The Effect of the Different Variables on Antioxidant Activity

The temperature and ratio were both found to significantly impact the antioxidant activity of the olive leaf extract measured via FRAP, CUPRAC and DPPH (p = 0.001, 0.004, 0.0003, respectively). However, time was only shown to significantly affect the antioxidant capacity measured via FRAP. The temperature × ratio had a negative influence on the DPPH measurements ($p < 0.05$).

3.4. Optimization of Aqueous Extraction Conditions for Maximizing the Total Phenolic Content and Antioxidant Capacity of Olive Leaf Extract

Based on the predictive models shown in Figures 3 and 4 the optimal conditions for the aqueous extraction of phenolic compounds were a temperature of 90 °C for 70 min at a sample-to-solvent ratio of 1:100 g/mL. These conditions were the same for the optimization of antioxidant capacity via FRAP. However, the optimal conditions for CUPRAC and DPPH varied slightly (CUPRAC: temperature 90 °C, time 70 min, sample-to-solvent ratio of 1:60 g/mL, DPPH: temperature 90 °C, time 70 min, sample-to-solvent ratio of 1:20 g/mL). Therefore, the extraction conditions of a temperature at 90 °C for

70 min and at a sample-to-solvent ratio of 1:60 g/mL were chosen for the extraction of phenolic compounds, as the extracts also displayed a high level of antioxidant activity. Furthermore, consuming less extraction solvent is practical from an economic point of view. For this reason, the sample-to-solvent ratio of 1:60 g/mL was used for validation. Increases in antioxidant activity with increasing temperature have previously been linked to the thermal degradation of higher molecular weight compounds into lower molecular weight ones [25,26]. This is one example of the non-specificity of the Folin–Ciocalteu method.

In order to validate the conditions predicted by the models, these extraction conditions (temperature 90 °C, time 70 min, sample-to-solvent ratio of 1:60 g/mL) were tested. The resulting values fell inside of the predicted ranges for TPC and all three antioxidant capacity assays (Table 4). These conditions are therefore proposed as optimal for the aqueous extraction of phenolic compounds with a high antioxidant capacity from olive leaves.

Table 4. Validation of the experimental model. GAE, gallic acid equivalents.

Assay	Values of TPC and Antioxidant Capacity	
	Predicted	**Experimental (n = 3)**
TPC (mg GAE/g)	32.42 ± 8.66	32.4 ± 2.06
FRAP (mg TE/g)	98.6 ± 9.71	91.03 ± 6.13
DPPH (mg TE/g)	76.96 ± 11.56	85.26 ± 3.54
CUPRAC (mg TE/g)	127.97 ± 8.62	121.97 ± 5.45

4. Conclusions

The optimal conditions for the aqueous extraction of phenolic compounds from olive leaves were proposed to be at 90 °C for 70 min at a sample-to-solvent ratio of 1:60 g/mL. Using olive leaves as a starting material for the extraction of phenolic compounds via this simple and inexpensive method constitutes a viable use for this agricultural waste product and may potentially serve as an additional source of income for olive growers/olive oil producers.

Acknowledgments

We acknowledge the following funding support: Ramaciotti Foundation (ES2012/0104); Cancer Australia and Cure Cancer Australia Foundation (1033781); The University of Newcastle; Terrigal Trotters; and special thanks to Christine and Jo Ashcroft for providing all samples.

Author Contributions

Chloe Goldsmith participated in the experimental design and completion, as well as interpretation, manuscript design and preparation. Quan Vuong and Costas Stathopoulos participated in the experimental design and data interpretation. Chris Scarlett and Paul Roach participated in manuscript design and preparation. All authors read and approved the final manuscript.

Conflicts of Interest

The authors declare no conflict of interest.

References

1. Trichopoulou, A.; Lagiou, P.; Kuper, H.; Trichopoulos, D. Cancer and Mediterranean dietary traditions. *Cancer Epidemiol. Biomark. Prev.* **2000**, *9*, 869–873.
2. Covas, M.I. Olive oil and the cardiovascular system. *Pharmacol. Res.* **2007**, *55*, 175–186.
3. Cicerale, S.; Conlan, X.A.; Sinclair, A.J.; Keast, R.S. Chemistry and health of olive oil phenolics. *Crit. Rev. Food Sci. Nutr.* **2009**, *49*, 218–236.
4. Bogani, P.; Galli, C.; Villa, M.; Visiolia, F. Postprandial anti-inflammatory and antioxidant effects of extra virgin olive oil. *Atherosclerosis* **2007**, *190*, 181–186.
5. Goulas, V.; Exarchou, V.; Troganis, A.N.; Psomiadou, E.; Fotsis, T.; Briasoulis, E.; Gerothanassis, I.P. Phytochemicals in olive-leaf extracts and their antiproliferative activity against cancer and endothelial cells. *Mol. Nutr. Food Res.* **2009**, *53*, 600–608.
6. Fabiani, R.; de Bartolomeo, A.; Rosignoli, P.; Servili, M.; Montedoro, G.F.; Morozzi, G. Cancer chemoprevention by hydroxytyrosol isolated from virgin olive oil through G1 cell cycle arrest and apoptosis. *Eur. J. Cancer Prev.* **2002**, *11*, 351–358.
7. Beauchamp, G.K.; Keast, R.S.J.; Morel, D.; Lin, J.; Pika, J.; Han, Q.; Lee, C.-H.; Smith, A.B.; Breslin, P.A.S. Phytochemistry: Ibuprofen-like activity in extra-virgin olive oil. *Nature* **2005**, *437*, 45–46.
8. Bisignano, G.; Tomaino, A.; Lo Cascio, R.; Crisafi, G.; Uccella, N.; Saija, A. On the *in vitro* antimicrobial activity of oleuropein and hydroxytyrosol. *J. Pharm. Pharmacol.* **1999**, *51*, 971–974.
9. Hashim, Y.Z.; Rowland, I.R.; McGlynn, H.; Servili, M.; Selvaggini, R.; Taticchi, A.; Esposto, S.; Montedoro, G.; Kaisalo, L.; Wähälä, K.; *et al.* Inhibitory effects of olive oil phenolics on invasion in human colon adenocarcinoma cells *in vitro*. *Int. J. Cancer* **2008**, *122*, 495–500.
10. Tsatsanis, C.; Androulidaki, A.; Venihaki, M.; Margioris, A.N. Signalling networks regulating cyclooxygenase-2. *Int. J. Biochem. Cell Biol.* **2006**, *38*, 1654–1661.
11. Sebolt-Leopold, J.S.; Herrera, R. Targeting the mitogen-activated protein kinase cascade to treat cancer. *Nat. Rev. Cancer* **2004**, *4*, 937–947.
12. Femia, A.P.; Dolara, P.; Servili, M.; Esposto, S.; Taticchi, A.; Urbani, S.; Giannini, A.; Salvadori, M.; Caderni, G. No effects of olive oils with different phenolic content compared to corn oil on 1,2-dimethylhydrazine-induced colon carcinogenesis in rats. *Eur. J. Nutr.* **2008**, *47*, 329–334.
13. Xynos, N.; Papaefstathioua, G.; Gikasb, E.; Argyropouloua, A.; Aligiannisa, N.; Skaltsounisa, A.-L. Design optimization study of the extraction of olive leaves performed with pressurized liquid extraction using response surface methodology. *Sep. Purif. Technol.* **2014**, *122*, 323–330.
14. Taamalli, A.; Arráez-Román, D.; Ibañez, E.; Zarrouk, M.; Segura-Carretero, A.; Fernández-Gutiérrez, A. Optimization of microwave-assisted extraction for the characterization of olive leaf phenolic compounds by using HPLC-ESI-TOF-MS/IT-MS^2. *J. Agric. Food Chem.* **2012**, *60*, 791–798.
15. Şahin, S.; Şamlı, R. Optimization of olive leaf extract obtained by ultrasound-assisted extraction with response surface methodology. *Ultrason. Sonochem.* **2013**, *20*, 595–602.

16. Robards, K.; Obied, H.K.; Bedgood, D.R., Jr.; Prenzler, P.D. Bioscreening of Australian olive mill waste extracts: Biophenol content, antioxidant, antimicrobial and molluscicidal activities. *Food Chem. Toxicol.* **2007**, *45*, 1238–1248.
17. Busnena, B.A.; Foudah, A.I.; Melancon, T.; El Sayed, K.A. Olive secoiridoids and semisynthetic bioisostere analogues for the control of metastatic breast cancer. *Bioorg. Med. Chem.* **2013**, *21*, 2117–2127.
18. Scotece, M.; Gómez, R.; Conde, J.; Lopez, V.; Gómez-Reino, J.J.; Lago, F.; Smith, A.B., III; Gualillo, O. Oleocanthal inhibits proliferation and MIP-1α expression in human multiple myeloma cells. *Curr. Med. Chem.* **2013**, *20*, 2467–2475.
19. Malik, N.S.; Bradford, J.M. Recovery and stability of oleuropein and other phenolic compounds during extraction and processing of olive (*Olea europaea* L.) leaves. *J. Food Agric. Environ.* **2008**, *6*, 8–13.
20. Vuong, Q.V.; Stathopoulos, C.E.; Golding, J.B.; Nguyen, M.H.; Roach, P.D. Optimum conditions for the water extraction of L-theanine from green tea. *J. Sep. Sci.* **2011**, *34*, 2468–2474.
21. Thaipong, K.; Boonprakob, U.; Crosby, K.; Cisneros-Zevallos, L.; Byrne, D.H. Comparison of ABTS, DPPH, FRAP, and ORAC assays for estimating antioxidant activity from guava fruit extracts. *J. Food Compos. Anal.* **2006**, *19*, 669–675.
22. Apak, R.; Güçlü, K.; Özyürek, M.; Karademir, S.E. Novel total antioxidant capacity index for dietary polyphenols and vitamins C and E, using their cupric ion reducing capability in the presence of neocuproine: CUPRAC method. *J. Agric. Food Chem.* **2004**, *52*, 7970–7981.
23. Vuong, Q.V.; Hiruna, S.; Roach, P.D.; Bowyer, M.C.; Phillips, P.A.; Scarletta, C.J. Effect of extraction conditions on total phenolic compounds and antioxidant activities of *Carica papaya* leaf aqueous extracts. *J. Herb. Med.* **2013**, *3*, 104–111.
24. Wang, J.; Sun, B.; Cao, Y.; Tian, Y; Li, X. Optimisation of ultrasound-assisted extraction of phenolic compounds from wheat bran. *Food Chem.* **2008**, *106*, 804–810.
25. Goldsmith, C.D.; Stathopoulos, C.E.; Golding, J.B.; Roach, P.D. Fate of phenolic compounds during olive oil production with the traditional press method. *I. Food Res. J.* **2014**, *21*, 101–109.
26. Klen, T.J.; Vodopivec, B.M. The fate of olive fruit phenols during commercial olive oil processing: Traditional press versus continuous two- and three-phase centrifuge. *LWT-Food Sci. Technol.* **2012**, *49*, 267–274.

Effect of Addition of Natural Antioxidants on the Shelf-Life of "Chorizo", a Spanish Dry-Cured Sausage

Mirian Pateiro, Roberto Bermúdez, José Manuel Lorenzo and Daniel Franco *

Centro Tecnológico de la Carne de Galicia, Rúa Galicia No. 4, Parque Tecnológico de Galicia, San Cibrao das Viñas, 32900 Ourense, Spain; E-Mails: mirianpateiro@ceteca.net (M.P.); robertobermudez@ceteca.net (R.B.); jmlorenzo@ceteca.net (J.M.L.)

* Author to whom correspondence should be addressed; E-Mail: danielfranco@ceteca.net

Academic Editors: Maria G. Miguel and João Rocha

Abstract: The dose effect of the addition of natural antioxidants (tea, chestnut, grape seed and beer extracts) on physicochemical, microbiological changes and on oxidative stability of dry-cured "chorizo", as well as their effect during the storage under vacuum conditions was evaluated. Color parameters were significantly ($p < 0.05$) affected by the addition of antioxidants so that samples that contained antioxidants were more effective in maintaining color. The improving effects were dose-dependent with highest values with the dose of 50 mg/kg during ripening and depend on the extract during vacuum packaging. Addition of antioxidants decreased ($p < 0.05$) the oxidation, showing thiobarbituric acid reactive substances (TBARS) values below 0.4 mg MDA/kg. Natural antioxidants matched or even improved the results obtained for butylated hydroxytoluene (BHT). Regarding texture profile analysis (TPA) analysis, hardness values significantly ($p < 0.001$) decreased with the addition of antioxidants, obtaining the lower results with the dose of 200 mg/kg both during ripening and vacuum packaging. Antioxidants reduced the counts of total viable counts (TVC), lactic acid bacteria (LAB), mold and yeast. Free fatty acid content during ripening and under vacuum conditions showed a gradual and significant ($p < 0.05$) release as a result of lipolysis. At the end of ripening, the addition of GRA_{1000} protected chorizos from oxidative degradation.

Keywords: dry-cured chorizo; natural antioxidants; physicochemical parameters; microbial counts; lipid oxidation

1. Introduction

Agro-industries such as wineries and brewers have an economic relevance in the global market but also produce high quantities of wastes and by-products that could disrupt the environmental balance. There are many alternatives for reusing these materials, and their food utility has gained increasing interest. Their use as "natural" antioxidants could be one of the most efficient uses for these products. Their high content of phenolic compounds and their known antimicrobial power could lead to their use as substitutes for synthetic antioxidants.

For many years, the functional characteristics of many plant extracts have been evaluated because of their antioxidant and antimicrobial activity and their potential to replace synthetic antioxidants [1]. Grape (*Vitis vinifera*), green tea (*Camellia sinensis*) and chestnut (*Castanea sativa*) are of special interest due to their high content of phenolic compounds. Previous works reflect that grape seed extracts have antioxidant and antimicrobial activities in meat [2]; green tea was used to increase the shelf life of meat patties and pig liver pâté [3,4]; and the antioxidant activity of chestnut extract has also been investigated [3,5].

"Chorizo" is a typical dry fermented sausage from Spain. During the manufacturing process of chorizo, microbiological, chemical and physicochemical changes take place. In particular dehydration, fermentation of carbohydrates and acidification, development of color, lipolysis and fat autooxidation and proteolysis takes place [6]. Therefore, the use of antioxidants during processing aims to delay oxidation [3], allowing increase the shelf life of the product. Industries generally used synthetic antioxidants to control this process, such as butylated hydroxyanisole (BHA), butylated hydroxytoluene (BHT) and *tert*-butylhydroquinone (THBQ). However, the use of these synthetic compounds has been linked to health risks (carcinogenic potential) and current research tends for their replacement by natural antioxidants [5]. Thereby, increasing interest in natural antioxidants and a search for naturally occurring compounds with antioxidant activity has increased dramatically [7].

For the conservation and to extent the shelf life of the product, vacuum-packaging under refrigeration together with the use of natural antioxidants could be used to prevent major changes during storage, especially removing oxygen, which is the main cause of food oxidation [8]. To our knowledge, not many studies regarding the effect of natural antioxidants on the oxidation stability of dry ripened sausage "chorizo" were found in the related literature [5]. In addition, not much data about the dose to be used of natural antioxidants is available, only on rosemary and tea extracts in sausages and patties, respectively [9,10]. Therefore, the aim of this study was to evaluate the dose effect of the addition of natural antioxidants (tea, chestnut, grape seed and beer extracts) on physicochemical, microbiological changes and on oxidative stability of dry-cured "chorizo", comparing their effect with a synthetic antioxidant (BHT), as well as knowing the effect of these natural extracts during the storage under vacuum conditions.

2. Experimental Section

2.1. Extraction of Natural Antioxidants

Grape seed extract (GRA) and chestnut extract (CHE) were prepared as previously was indicated in Lorenzo *et al.* [5], while the extraction of green tea extract (TEA) was carried out as described in Lorenzo *et al.* [3]. Beer residue was provided by Hijos de Rivera S.A. (A Coruña, Spain). This residue was used as source of polyphenolic compounds. This suspension residue comes from process of boiling of the must, where the temperature is maintained at 102 °C for 90 min. The objective of boiling is to obtain the necessary density, evaporating the spare water; the sterilizing the must and extracting and dissolving the wanted elements of hops. Lots of 4 L of this residue was transferred to XAD-16 amberlite column (Sigma-Aldrich, Spain). A glass column (7 cm Ø in × 40 cm height) filled with XAD-16 amberlite was equilibrating with distillate water to separate polyphenolic compounds. Four liters of distillate water was poured on the column to remove impurities; later, three liters of ethanol was used to elute polyphenols. This volume was evaporated until 200 mL (or until ethanol was completely removed) remained. Subsequently, the residue was lyophilized using a freeze-dryer (Kinetics EZ-Dryer, Stone Ridge, NY, USA). This lyophilized extract (raw extract of beer by-product) rendered 9.41 ± 34.0 g/L. This extract was subsequently used for the evaluation of the antioxidant capacity.

2.2. Determination of Antioxidant Capacity

2.2.1. Determination of Total Phenolic Content

The total phenolic content was determined using the Folin-Ciocalteu Reagent (FCR) with gallic acid as a standard. Readings were performed at 765 nm and were compared with a standard curve of gallic acid, being the total phenolic content expressed as mg of gallic acid equivalent per g of freeze dried solid (mg GAE/g). Analyses were performed in triplicate.

2.2.2. Trolox Equivalent Antioxidant Capacity (TEAC)

This assay is based on the scavenging of ABTS radical (2,2-azinobis-(3-ethyl-benzothiazoline-6-sulphonate)), observed as a decolorization of blue-green color at 734 nm. The radical scavenging capacity was compared with that of Trolox and results were expressed as g of Trolox equivalent per g of freeze dried solid.

2.2.3. β-Carotene Bleaching Assay

The β-carotene (βC) bleaching assay described by Marco [11] was modified for use with microplates. Absorbance readings (470 nm) were taken at regular intervals in a ThermoFisher Scientific microplate reader until β-carotene was decolored (about 3 h). The antioxidant activity coefficient (AAC) and EC_{50} value (g/L) were calculated as previously described in Lorenzo *et al.* [3] for each antioxidant extract.

2.2.4. α,α-Diphenyl-β-Picrylhydrazyl (DPPH) Radical Scavenging Activity

The antioxidant activity was determined with DPPH as a free radical, using microplate. Antioxidant solutions (10 μL) were added in triplicate to 200 μL of a 60 μM solution of DPPH in 70% ethanol. The decrease in absorbance was followed at 515 nm every 5 min until the reaction reached a plateau (about 2 h). The EC_{50} and BHT equivalent activity were calculated as explained above.

2.3. Manufacture of Dry-Cured Sausages

Four batches (20 units per batch, 3 per ripening time) of dry-cured sausage "chorizo": Control (CON), BHT, grape seed (GRA), chestnut (CHE), green tea (TEA) and beer extracts (BER) were manufactured in the pilot plant of the Meat Technology Center of Galicia. Sausages were manufactured using the primal cuts of shoulder (85%) and pork back fat (15%) from Celta pig breed. The lean and the pork back fat were ground through a 6 mm diameter mincing plate in a refrigerated mincer machine (La Minerva, Bologna, Italy). Mixture was vacuum minced in a vacuum mincer machine (Fuerpla, Valencia, Spain) for 3 min with 5 g/kg of NaCl, 20 g/kg of sweet paprika, 3 g/kg of spicy paprika, 0.5 g/kg of garlic and 200 mg/kg of BHT for BHT batch, 0.05–0.2–1 g/kg of natural extracts. No starter culture was added. The meat mixture was maintained at 3–5 °C for 24 h and then was stuffed into pig gut (diameter 32–34 mm) to obtain an average final sausage weight of 150 g. After stuffing, the sausages were conditioned for two days at 7 °C and 85% of relative humidity. The sausages were transferred to a drying-ripening chamber where they were kept for 48 days at 12 °C and 75%–80% of relative humidity. Below the samples were packed under vacuum conditions five months at 4 °C. Analyses were carried out at 0, 4, 19 and 48 days of ripening time and at 2, 4 and 7 months of vacuum-packaged. The studied parameters were determined in duplicate for every sampling point.

2.4. Determination of pH, Moisture Content, Water Activity and Color Parameters

The pH of samples was measured using a pH-meter (HI 99163, Hanna Instruments, Eibar, Spain) equipped with a glass probe for penetration. Moisture percentage was determined by oven drying (Memmert UFP 600, Schwabach, Germany) at 105 °C until constant weight [12], and calculated as sample (5 g) weight loss. Water activity was determined using a Fast-lab (Gbx, Romans sur Isére, Cédex, France) water activity meter, previously calibrated with sodium chloride. A portable colorimeter (Konica Minolta CM-600d, Osaka, Japan) with pulsed xenon arc lamp, 0° viewing angle geometry and 8 mm aperture size, was used to estimate meat color in the CIELAB space: lightness, (L^*); redness, (a^*); yellowness, (b^*). Each sausage piece was cut (2 cm) and the color of the slices was measured three times for each analytical point.

2.5. Determination of Lipid Oxidation

Lipid stability was evaluated through TBARS index according to the method proposed by Targladis *et al.* [13]. Briefly, the dry-cured sausage sample (10 g) was dispersed in distilled water (50 mL) and homogenized in an Ultra-Turrax (Ika T25 basic, Staufen, Germany) for 2 min. The homogenate was carried to a distillation system with HCl 4N (2.5 mL) and distilled water (47.5 mL) until recover 50 mL of distilled. The filtrate (5 mL) was reacted with a 0.02 M thiobarbituric acid (TBA) solution (5 mL) and incubated in a water bath at 96 °C for 40 min. The absorbance was measured at 538 nm. Thiobarbituric acid reactive substances (TBARS) values were calculated from a standard curve of malonaldehyde with 1,1,3,3-tetraethoxipropane (TEP) and expressed as mg MDA/kg sample.

2.6. Determination of Texture Profile Analysis

Texture profile analysis (TPA) was measured by compressing to 50% with a compression probe of 19.85 cm^2 of surface contact in seven dry-cured sausage slices of 2 cm using a texture analyzer (TA.XTplus, Stable Micro Systems, Vienna Court, UK). Force-time curves were recorded at a crosshead speed of 1 mm/s. Hardness (kg), cohesiveness, springiness (mm), gumminess (kg) and chewiness (kg × mm) were obtained. These parameters were obtained using the available computer software (Texture Exponent 32 (version 1.0.0.68), Stable Micro Systems, Vienna Court, UK).

2.7. Analysis of Free Fatty Acid Content

Total intramuscular lipids were extracted from 5 g of each minced sausage sample, according to Folch *et al.* [14] procedure. Free fatty acids were separated from fifty milligrams of the extracted lipids using aminopropyl (NH_2) mini-columns as described by García-Regueiro *et al.* [15]. This fraction was transesterified with a solution of boron trifluoride (14%) in methanol, according to Carreau and Dubacq [16] and the FAMEs were stored at −80 °C until chromatographic analysis. Separation and quantification of FAMEs was determined following Lorenzo and Franco [17].

2.8. Microbial Analysis

For microbiological analysis, a 10 g sample of dry-cured sausage was aseptically weighted in a sterile plastic bag. Subsequently samples were homogenized with 90 mL of a sterile solution of 0.1% (w/v) peptone water (Oxoid, Unipath, Basingstoke, UK), containing 0.85% NaCl and 1% Tween 80 as emulsifier, for 2 min at 20–25 °C in a Masticator blender (IUL Instruments, Barcelona, Spain), thus making a 1/10 dilution. Serial 10-fold dilutions were prepared by mixing 1 mL of the previous dilution with 9 mL of 0.1% (w/v) sterile peptone water. Total viable counts (TVC) were enumerated in Plate Count Agar (PCA; Oxoid, Unipath Ltd., Basingstoke, UK) and incubated at 30 °C for 48 h; lactic acid bacteria (BAL) were determined on the Man Rogosa Sharpe medium Agar (Oxoid, Unipath Ltd., Basingstoke, UK) (pH 5.6) after an incubation at 30 °C for 5 days. After incubation, plates with 30–300 colonies were counted. The microbiological data were transformed into logarithms of the number of colony forming units (CFU/g).

2.9. Statistical Analysis

For the statistical analysis of the results, an analysis of variance (ANOVA) of one way using SPSS package (SPSS 19.0, Chicago, IL, USA) was performed for all variables considered in the study. The least squares mean (LSM) were separated using Duncan's *t*-test. All statistical test of LSM were performed for a significance level $p < 0.05$.

3. Results and Discussion

3.1. Antioxidant Activity of the Extracts

GRA and TEA extracts showed the highest polyphenol content, mainly flavonoids and flavan-3-ols, which antioxidant activity has been demonstrated [2,18]. The major compounds found in TEA extracts was catechin, epicatechin, cinnamic acids and sugar-linked flavonols [19], while GRA extracts contained benzoic acids, monomer flavan-3-ols and oligomeric procyanidins [18]. The higher activity found in GRA extracts could be associated to its resveratrol content [20]. Regarding polyphenols in CHE and BER extracts, their concentration were significantly lower than the aforementioned natural extracts (28.9 and 89.0 *vs.* 373.0 and 390.0 mg GAE/g extract for BER and CHE *vs.* GRA and TEA extracts, respectively).

TEAC, DPPH and β-carotene were used to assess *in vitro* antioxidant activity of the natural extracts. These methods were directly related to polyphenol contents [4]. Therefore, GRA and TEA extracts showed the highest activities in these methods. In the case of TEAC, the aforementioned extracts displayed values 10 and 15-fold higher Trolox equivalent antioxidant capacity than CHE extract (0.27 *vs.* 2.93 and 4.06 g Trolox/g extract for CHE, GRA and TEA, respectively) and 20 and 40-fold higher than BER extract (0.09 g Trolox/g extract).

The scavenging activity found on DPPH radical showed the higher antioxidant power of BHT standard, followed by GRA and TEA extracts (1.80 and 2.18 g equivalent BHT/g extract, respectively). The values provided by BER and CHE were almost 4 and 8-fold lower than the aforementioned extracts (0.25 and 0.48 g equivalent BHT/g extract, respectively). The EC_{50} values obtained showed the same behavior, the powerful antioxidant activity of TEA and GRA (0.12 and 0.16 g extract/L, respectively) *vs.* BER and CHE extracts (data not shown).

β-carotene bleaching assay of the natural extracts showed similar activity values for CHE and TEA (0.53 and 0.69 g equivalent BHT/g extract, respectively), although GRA were the most active (1.28 g equivalent BHT/g extract) and BER the least active (0.25 g equivalent BHT/g extract). The EC_{50} values obtained displayed rather similar activity values for all the extracts (less than 0.10 g extract/L).

3.2. Effect of Antioxidants on Physicochemical Parameters during the Manufacturing Process and Vacuum Packaging

Changes occurred in pH, moisture content and water activity (a_w) during the manufacturing process and vacuum-packing are given in Figure 1. Ripening time had a significant effect ($p < 0.01$) on pH values. During the first 19 days of ripening, pH values decreased from 5.62 to approximately 5.43 due to the production of lactic acid as a result of carbohydrate breakdown during fermentation [21] and the following increase can be produced by the liberation of peptides, amino acids and ammonia from proteolytic reactions [17]. Although at the end of ripening pH values of sausages were not affected ($p > 0.05$) by the addition of antioxidants, the highest pH values were observed in CHE_{50} and GRA_{50}, followed by CON and BER_{50}. These pH values were similar to those found in other varieties of sausages [22,23]. Regarding the evolution during vacuum packaging, there are not many studies that evaluate the influence of antioxidants on physicochemical parameters of dry-cured sausages. The trend is to continue growing slightly until 120 days to decreasing until the end of storage. The exception to this behavior is found in CON and TEA samples, which pH values continued increasing until the end of storage. Regarding dose effect, only TEA and GRA extracts presented significant ($p < 0.05$) differences on pH during ripening time and vacuum storage.

Moisture content was significantly ($p < 0.01$) affected by ripening time and addition of antioxidants, decreasing during the drying period as a result of moisture loss at high ripening temperature and low percentage of relative humidity (Figure 1). CON was the sample that presented the lowest value at the end of this stage with mean values of 20.6%, while samples manufactured with antioxidants presented higher values, in all cases above 22.0%. The moisture content and water activity followed similar behaviors because are variables that are intrinsically linked. As occurred with water content, the trend of water activity was to decrease over the time, obtaining significant differences during ripening time ($p \leq 0.001$). As can be observed in Figure 1, two steps could be distinguished in its evolution. The first one represented a sharp decline in the values during ripening stage and the second one stabilization during the vacuum packaging until the end of the storage. No significant differences ($p > 0.05$) were observed in moisture and water activity among extracts depending on dose effect during ripening time and vacuum packaging.

Color parameters of the chorizo were significantly ($p < 0.05$) affected by ripening time and addition of antioxidants (Table 1). Regarding redness (a*), the trend was to decrease over time. Thereby, a significant ($p < 0.01$) color loss was observed during ripening time with values ranged between 27.6 and 21.2. This behavior could be due to the partial or total denaturation of nitrosomyoglobin caused by the production of lactic acid. Furthermore, a significant effect ($p < 0.05$) was also observed on redness with the use of antioxidants. Antioxidants were more effective in maintaining color, with higher values of redness in the samples manufactured with antioxidants. In this sense, samples manufactured with GRA_{200}, CHE_{50}, CHE_{200} and BHT presented higher a* values compared to CON batch; so that the addition of natural extracts improved the color stability, showing even better results than those showed by BHT.

During vacuum packaging, a slightly increase was observed in redness values. A similar behavior during packaging under vacuum was reported by Liaros *et al.* [24]. As happened during ripening, significant differences ($p < 0.001$) were also found with the addition of antioxidants. Regarding dose effect, significant differences ($p < 0.01$) were found for this color parameter in samples that contained TEA, CHE and GRA in their composition. During ripening, the highest values were obtained in the samples that containing a dose of 50 mg/kg in their composition, while vacuum packaging the dose more effective depends on the used extract. In this regards, Jayawardana *et al.* [25] showed the capacity of natural extracts to hold the color on pork sausages. As in the present study, the improving effects of extracts were concentration-dependent [1]. In this case, with the exception of samples that contain BER extract, a decrease of redness occurs at increasing levels of antioxidants.

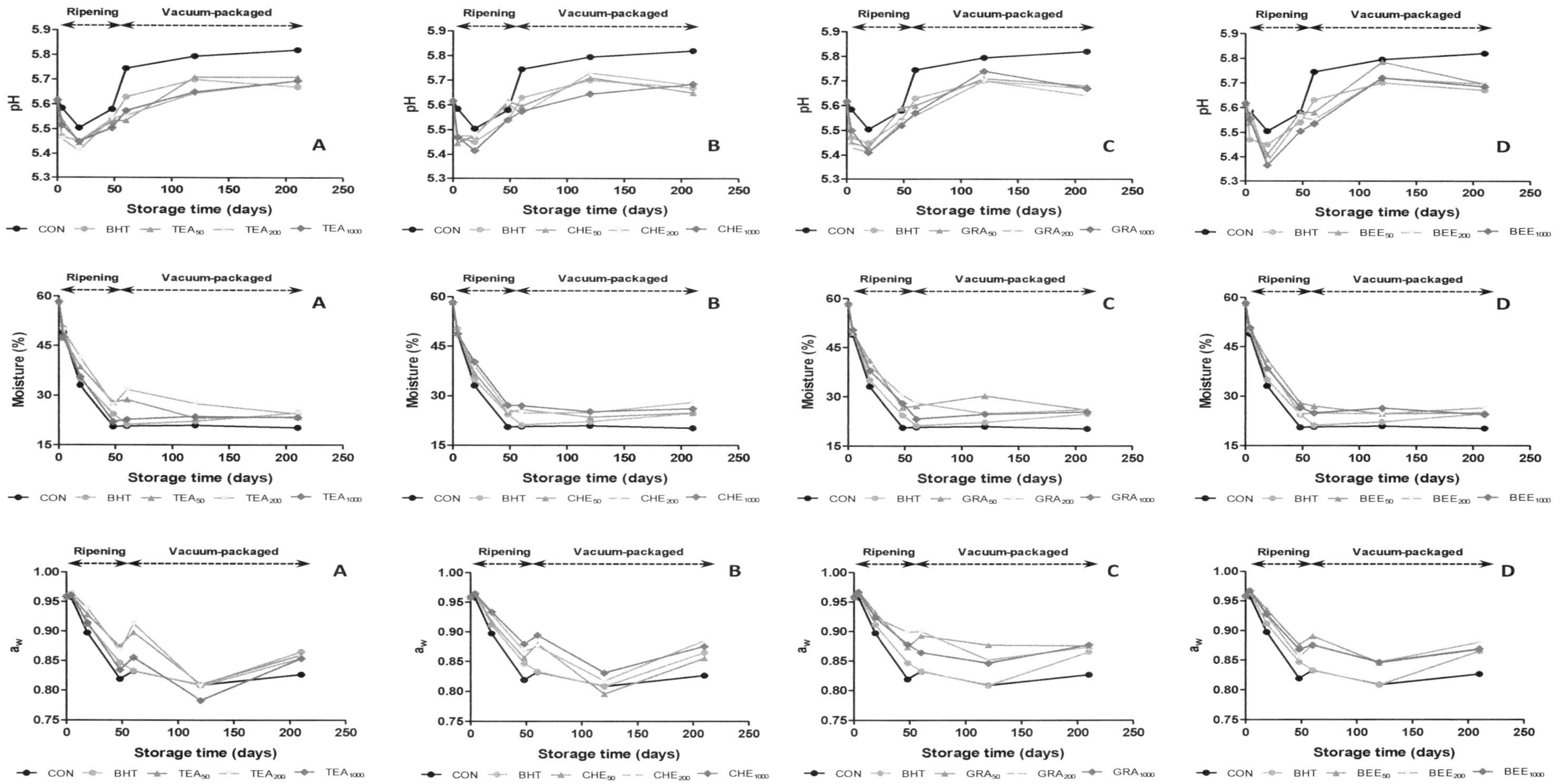

Figure 1. Evolution of pH values, moisture content and water activity in dry-cures sausages treated with butylated hydroxytoluene (BHT) and natural antioxidants during ripening and vacuum-packaged storage.

Table 1. Evolution of color parameters of “chorizo” treated with BHT and natural antioxidants during ripening and vacuum-packaging.

	Days	CON	BHT	TEA			CHE			GRA			BER			p-Value	SEM
			200	50	200	1000	50	200	1000	50	200	1000	50	200	1000		
	4	43.30 ^{c}	44.86 ^{c}	43.79 ^{d}	43.04 ^{c}	42.52 ^{c}	43.94 ^{b}	42.51 ^{b}	43.22 ^{c}	44.61 ^{d}	45.09 ^{d}	42.55 ^{c}	42.34 ^{c}	42.33 ^{c}	43.40 ^{d}	0.186	0.23
	19	34.30 ^{d,1}	40.05 ^{d,3}	41.88 ^{d,3,4}	43.19 ^{c,4}	41.16 ^{c,3,4}	42.49 ^{b,4}	42.60 ^{b,4}	41.62 ^{c,3,4}	42.58 ^{c,d,4}	41.85 ^{d,3,4}	35.70 ^{b,1,2}	36.49 ^{b,2}	37.11 ^{b,2}	37.21 ^{c,2}	0.000	0.58
	48	30.46 ^{c,1}	35.34 $^{c,2-4}$	38.98 ^{c,4}	36.54 ^{b,3,4}	32.03 ^{a,b,1,2}	35.51 $^{a,2-4}$	34.88 ^{a,2,3}	36.24 ^{b,3,4}	36.96 ^{a,b,3,4}	37.36 ^{c,34}	34.70 ^{b,2,3}	33.47 $^{a,1-3}$	32.13 ^{a,1,2}	34.56 ^{b,2,3}	0.008	0.49
	60	29.00 ^{b,c,1,2}	31.68 ^{a,3,4}	38.32 ^{c,8}	39.23 ^{b,c,8}	33.59 ^{b,4,5}	34.48 ^{a,5,6}	33.57 ^{a,4,5}	34.60 ^{a,b,5,6}	37.55 ^{a,b,7,8}	35.94 ^{b,c,6,7}	28.39 ^{a,1}	31.81 ^{a,3,4}	31.08 ^{a,2,3}	29.24 ^{a,1,2}	0.000	0.66
L*	120	28.63 ^{b,1}	34.49 $^{b,c,3-5}$	35.58 ^{b,5,6}	38.40 ^{b,6,7}	32.93 $^{b,2-5}$	33.54 $^{a,2-5}$	35.06 ^{a,4,5}	35.01 ^{a,b,4,5}	39.68 ^{b,c,7}	33.29 $^{a,b,2-5}$	30.81 ^{a,1,2}	31.09 $^{a,1-3}$	31.64 $^{a,1-4}$	31.89 $^{a,1-4}$	0.000	0.59
	210	26.80 ^{a,1}	32.84 ^{a,b,2,3}	31.85 ^{a,2,3}	31.41 ^{a,2}	30.38 ^{a,2}	31.78 ^{a,2,3}	34.86 ^{a,3}	32.34 ^{a,2,3}	34.71 ^{a,3}	30.52 ^{a,2}	30.24 ^{a,2}	31.08 ^{a,2}	32.04 ^{a,2,3}	30.04 ^{a,2}	0.003	0.41
	p-value	0.000	0.000	0.000	0.005	0.000	0.001	0.005	0.001	0.001	0.001	0.000	0.000	0.000	0.000		
	SEM	1.82	1.39	1.20	1.27	1.41	1.43	1.20	1.22	1.05	1.52	1.43	1.24	1.24	1.48		
	4	26.17 $^{c,d,2-5}$	27.30 ^{a,b,5}	26.57 $^{a,3-5}$	25.82 $^{a,1-5}$	24.90 ^{b,c,1,2}	26.07 ^{a,2345}	25.44 $^{a,1-4}$	24.43 ^{a,b,1}	26.60 $^{a,3-5}$	26.94 ^{a,4,5}	24.53 ^{1}	26.24 $^{2-5}$	25.26 $^{1-3}$	26.17 $^{2-5}$	0.007	0.18
	19	23.58 $^{a-c,1}$	28.43 ^{b,c,3,4}	29.43 ^{b,c,4,5}	29.21 ^{b,4,5}	26.90 ^{c,2,3}	29.15 ^{b,4,5}	29.09 ^{b,c,4,5}	27.02 ^{b,c,2,3}	30.51 ^{b,c,5}	29.80 ^{b,4,5}	24.37 ^{1}	26.97 ^{2,3}	26.80 ^{2,3}	26.69 ^{2}	0.000	0.39
	48	21.20 ^{a,1}	25.44 $^{a,b,2-6}$	28.18 ^{a,b,6}	24.82 $^{a,2-6}$	22.23 ^{a,b,1,2}	25.93 $^{a,3-6}$	25.46 $^{a,2-6}$	24.27 $^{a,b,1-5}$	27.67 ^{a,b,5,6}	27.28 $^{a,4-6}$	23.71 $^{1-4}$	24.87 $^{2-6}$	22.73 $^{1-3}$	24.44 $^{1-5}$	0.010	0.43
	60	22.61 ^{a,b,1}	26.57 ^{a,b,3}	31.88 ^{c,5}	32.25 ^{c,5}	26.06 ^{c,2,3}	26.77 ^{a,b,3}	26.79 $^{a-c,3}$	25.54 $^{a-c,2,3}$	29.59 $^{a-c,4}$	29.53 ^{b,4}	23.02 ^{1}	25.89 ^{2,3}	26.97 ^{3}	24.22 ^{1,2}	0.000	0.56
a*	120	23.94 $^{a-c,1}$	30.42 ^{c,4,5}	30.73 ^{b,c,4,5}	32.33 ^{c,5}	26.07 $^{c,1-3}$	29.16 $^{b,3-5}$	29.90 ^{c,4,5}	28.07 $^{c,2-4}$	31.91 ^{c,5}	26.06 $^{a,1-3}$	23.11 ^{1}	24.60 ^{1}	24.88 ^{1,2}	25.23 ^{1,2}	0.000	0.61
	210	24.93 $^{b-d}$	24.93 ^{a}	26.17 ^{a}	25.07 ^{a}	20.85 ^{a}	25.00 ^{a}	25.85 ^{a,b}	22.86 ^{a}	27.83 ^{a,b}	26.23 ^{a}	22.80	23.59	25.12	25.25	0.066	0.39
	p-value	0.007	0.030	0.010	0.001	0.016	0.033	0.051	0.046	0.026	0.005	0.805	0.229	0.126	0.117		
	SEM	0.59	0.62	0.67	0.99	0.72	0.54	0.60	0.59	0.60	0.47	0.39	0.43	0.51	0.32		

Table 1. *Cont.*

	Days	CON	BHT	TEA			CHE			GRA			BER			*p*-Value	SEM
			200	50	200	1000	50	200	1000	50	200	1000	50	200	1000		
	4	32.50 ^{c}	35.10 ^{c}	35.28 ^{c}	33.95 ^{b}	33.93 ^{c}	34.11 ^{c}	34.22 ^{b}	32.84 ^{c}	35.85 ^{c}	37.01 ^{d}	33.15 ^{c}	34.04 ^{d}	33.13 ^{c}	33.92 ^{d}	0.249	0.30
	19	21.59 ^{b,1}	30.22 ^{d,3}	30.91 ^{b,c,3}	33.06 ^{b,3}	31.13 ^{c,3}	32.40 ^{c,3}	33.77 ^{b,3}	31.89 ^{c,3}	33.43 ^{c,3}	32.99 ^{c,3}	23.36 ^{b,1,2}	25.22 ^{c,2}	26.13 ^{b,2}	25.06 ^{c,2}	0.000	0.80
	48	19.62 ^{a,b,1}	23.71 $^{c,1-4}$	30.73 ^{b,c,5}	23.01 $^{a,1-4}$	20.55 ^{b,1,2}	25.15 $^{b,1-5}$	26.28 $^{a,2-5}$	25.73 $^{b,2-5}$	28.50 ^{b,4,5}	27.52 $^{b,3-5}$	24.02 $^{b,1-4}$	22.31 $^{b,c,1-3}$	21.86 $^{a,1-3}$	23.30 $^{b,c,1-4}$	0.014	0.66
	60	17.09 ^{a,1,2}	20.17 $^{a,b,2-4}$	27.84 ^{b,7,8}	30.03 ^{b,8}	21.57 $^{b,3-5}$	22.32 $^{a,b,4-6}$	21.64 $^{a,3-5}$	21.88 $^{a,b,3-5}$	25.28 ^{a,b,6,7}	24.10 ^{a,5,6}	16.69 ^{a,1}	18.81 $^{a,b,1-3}$	19.97 $^{a,2-4}$	16.68 ^{a,1}	0.000	0.77
b*	120	17.82 ^{a,b,1}	23.45 ^{b,c,2,3}	26.36 ^{b,3,4}	28.91 ^{b,4}	22.37 $^{b,1-3}$	23.58 ^{b,2,3}	26.30 ^{a,3,4}	25.45 ^{b,3,4}	28.97 ^{b,4}	21.45 $^{a,1-3}$	17.73 ^{a,1}	19.28 ^{a,b,1,2}	20.23 ^{a,1,2}	20.19 ^{a,b,1,2}	0.001	0.76
	210	17.81 ^{a,b}	17.11 ^{a}	19.75 ^{a}	19.87 ^{a}	15.43 ^{a}	18.98 ^{a}	20.62 ^{a}	17.76 ^{a}	21.48 ^{a}	21.73 ^{a}	17.21 ^{a}	16.76 ^{a}	19.38 ^{a}	18.73 ^{a}	0.072	0.42
	p-*value*	0.000	0.000	0.003	0.002	0.000	0.000	0.013	0.003	0.002	0.000	0.000	0.000	0.001	0.000		
	SEM	1.55	1.84	1.51	1.60	1.96	1.66	1.72	1.65	1.49	1.77	1.77	1.76	1.51	1.73		

[a–e] Mean values in the same column (same antioxidant in different weeks) with different letter presented significant differences; [1–8] Mean values in the same row (different antioxidant in the same week) with different number presented significant differences; SEM: standard error of mean; Batches: CON: control; BHT: tert-butyl-4-hydroxytoluene; TEA: tea; CHE: chestnut; GRA: grape seed and BER: beer extracts.

The values of lightness (L*) decreased significantly ($p < 0.01$) until the end of the storage. This behavior could be due as a consequence of moisture losses [17]. Therefore, a significant correlation was found with moisture ($r = 0.87, p < 0.01$) and a_w ($r = 0.80, p < 0.01$). These results were lower than those found by other authors [6,24]. During ripening time the values ranged from 43.8 to 30.5, being CON the samples that showed the lowest values. As in previous studies conducted with these extracts [5], the samples that contained extracts in their composition showed higher L* values. At the end of the ripening process, the samples that contain TEA_{50}, TEA_{200}, GRA_{50} and GRA_{200} in their formulation were those showed the highest values, even improving the results obtaining for BHT (38.98, 36.54, 36.96, and 37.36 *vs.* 35.34, respectively). The same behavior was observed during vacuum packaging, the decline in the values continued, reaching values between 34.9 and 26.8, and the values were higher in samples with added extracts, being CHE_{200} the samples that showed the highest values. Regarding dose effect, significant differences ($p < 0.05$) were found for this color parameter in samples that contained GRA in their composition.

3.3. Effect of Antioxidants on TPA Analysis during the Manufacturing Process and Vacuum Packaging

The evolutions of TPA parameters (hardness, springiness, chewiness, gumminess and cohesiveness) during ripening and vacuum-packaged are shown in Table 2. The major changes take place during fermentation when the pH declines and the myofibrillar proteins aggregate leading to gel formation. Thereby, a significant increase ($p < 0.05$) in hardness, chewiness and gumminess was observed during the first 48 days of ripening, to remain stable until the end of storage. In the case of springiness, the values decreased during ripening to remain constant during vacuum-packaged. Drying is a major factor affecting binding and rheological properties [26]. In fact, a significant correlation ($p < 0.01$) were found between moisture and water activity and TPA parameters. In the case of hardness, chewiness, gumminess and cohesiveness these correlations with moisture ($r = -0.67$, $r = -0.48$, $r = -0.75$, $r = -0.26$, respectively) and water activity were negative ($r = -0.59$, $r = -0.38$, $r = -0.65$, $r = -0.31$, respectively), so an increase in these parameters during ripening time was due to a decrease in moisture and water activity. In addition, other research have shown that polyphenolic compounds are able to react with thiol groups in meat protein to form covalent thiol-quinone adducts [27]. Specifically, it has been hypothesized that polyphenolic compounds from green tea extract can alter the textural properties of Bologna type sausages [28], so especially at elevated concentrations phenolics compounds could interact with the protein thiols to modify water holding capacity and other textural parameters.

The addition of antioxidants significantly ($p < 0.001$) decreased hardness values. The lower results were found in the samples treated with TEA_{200}, CHE_{200} and GRA_{50}, followed by the samples treated with BHT and BER extracts. The values obtained for CON samples at the end of ripening were similar to those obtained in other studies [6] and lower to the results reported by González-Fernández *et al.* [26] in Galician chorizos. On the other hand, significant differences ($p < 0.05$) were found in TEA, CHE and GRA depending on the dose. The lower results of hardness were obtained with the dose of 200 mg/kg both during ripening and vacuum-packaging.

Table 2. Evolution of textural properties of "chorizo" treated with BHT and natural antioxidants during ripening and vacuum-packaging.

	Days	CON	BHT	TEA			CHE			GRA			BER			*p*-Value	SEM
			200	50	200	1000	50	200	1000	50	200	1000	50	200	1000		
Hardness (kg)	4	2.59 $^{a,2-4}$	2.10 $^{a,1-3}$	2.47 ^{a,2,3}	1.49 ^{a,1}	2.01 $^{a,1-3}$	1.78 ^{a,1,2}	2.10 $^{a,1-3}$	1.96 $^{a,1-3}$	2.70 $^{a,3-5}$	2.27 $^{a,1-3}$	3.98 ^{a,6}	3.36 $^{a,4-6}$	3.46 ^{a,,56}	3.69 ^{a,6}	0.000	0.15
	19	8.66 ^{b,7}	7.77 ^{b,6,7}	4.27 $^{b,1-4}$	5.61 $^{b,3-6}$	5.35 $^{b,2-5}$	3.97 $^{b,1-3}$	2.76 ^{a,1}	2.49 ^{a,1}	3.12 ^{a,1,2}	3.51 $^{a,1-3}$	8.73 ^{b,7}	6.29 $^{a,4-6}$	5.08 $^{a,2-5}$	7.20 $^{b,5-7}$	0.000	0.42
	48	17.50 ^{d,7}	9.84 $^{b,3-5}$	6.89 ^{c,1}	6.56 ^{b,1}	8.60 ^{c,2,3}	7.01 ^{c,1}	6.97 ^{b,1}	7.59 ^{b,1,2}	7.47 ^{b,1,2}	8.91 $^{b,c,2-4}$	10.30 ^{c,4,5}	10.30 ^{b,4,5}	12.00 ^{b,c,6}	10.52 ^{b,c,5}	0.000	0.55
	60	15.77 ^{c,5}	9.98 ^{b,3}	7.78 ^{c,1,2}	6.82 ^{b,1}	8.52 ^{c,2}	7.37 ^{c,1,2}	7.60 ^{b,1,2}	7.68 ^{b,1,2}	7.78 ^{b,1,2}	7.89 ^{b,1,2}	11.17 ^{c,3}	10.91 ^{b,3}	13.93 ^{c,4}	10.24 ^{b,3}	0.000	0.50
	120	16.07 ^{c,d,7}	9.36 $^{b,3-6}$	10.70 ^{d,5,6}	6.30 ^{b,1}	8.74 $^{c,2-5}$	7.45 $^{c,1-3}$	8.21 $^{b,1-4}$	7.30 $^{b,1-3}$	7.12 ^{b,1,2}	9.81 $^{c,4-6}$	11.23 ^{c,6}	11.29 ^{b,6}	10.22 $^{b,4-6}$	9.78 $^{b,4-6}$	0.000	0.47
	210	16.07 ^{c,d,5}	9.01 ^{b,12}	11.27 ^{d,2,3}	8.58 ^{c,1,2}	9.57 ^{c,1,2}	9.22 ^{d,1,2}	7.78 ^{b,1}	8.24 ^{b,1}	9.55 ^{c,1,2}	9.66 ^{c,1,2}	12.83 ^{d,3,4}	14.44 ^{c,4,5}	13.64 $^{c,3-5}$	13.84 $^{c,3-5}$	0.000	0.52
	p-value	0.000	0.001	0.000	0.000	0.005	0.000	0.000	0.000	0.000	0.000	0.000	0.001	0.000	0.005		
	SEM	1.73	0.85	0.96	0.67	0.84	0.75	0.77	0.79	0.76	0.91	0.86	1.12	1.24	0.99		
Springiness (mm)	4	0.42 ^{b,4}	0.38 ^{b,3,4}	0.34 $^{1-3}$	0.36 ^{2,3}	0.29 ^{1}	0.33 ^{1,2}	0.36 ^{d,2,3}	0.31 ^{1,2}	0.31 ^{1,2}	0.30 ^{1,2}	0.30 ^{1}	0.30 ^{1,2}	0.31 ^{1,2}	0.29 ^{1}	0.001	0.01
	19	0.31 ^{a}	0.35 ^{b}	0.28	0.31	0.30	0.33	0.32 ^{b,c}	0.33	0.32	0.31	0.35	0.32	0.32	0.36	0.390	0.01
	48	0.31 ^{a}	0.28 ^{a}	0.31	0.29	0.27	0.28	0.29 ^{a,b}	0.30	0.29	0.34	0.30	0.35	0.33	0.32	0.246	0.01
	60	0.30 $^{a,1-4}$	0.27 ^{a,1,2}	0.32 $^{2-5}$	0.31 $^{2-5}$	0.26 ^{1}	0.28 $^{1-3}$	0.28 $^{a,1-3}$	0.34 ^{4,5}	0.30 $^{1-4}$	0.31 $^{2-5}$	0.29 ^{1234}	0.31 $^{2-5}$	0.36 ^{5}	0.33 $^{3-5}$	0.016	0.01
	120	0.31 ^{a}	0.29 ^{a}	0.31	0.30	0.30	0.27	0.31 ^{b,c}	0.30	0.30	0.30	0.30	0.29	0.29	0.28	0.666	0.01
	210	0.31 $^{a,1-3}$	0.29 ^{a,1}	0.29 ^{1}	0.29 ^{1}	0.30 ^{1,2}	0.32 $^{1-4}$	0.33 $^{c,d,1-4}$	0.31 $^{1-3}$	0.32 $^{1-3}$	0.31 $^{1-3}$	0.33 $^{1-4}$	0.35 $^{2-4}$	0.37 ^{4}	0.36 ^{3,4}	0.035	0.01
	p-value	0.000	0.001	0.074	0.318	0.190	0.376	0.007	0.410	0.625	0.649	0.218	0.424	0.130	0.069		
	SEM	0.01	0.01	0.01	0.01	0.01	0.01	0.01	0.01	0.01	0.01	0.01	0.01	0.01	0.01		
Chewiness (kg × mm)	4	0.25 ^{a,3}	0.14 ^{a,1,2}	0.14 ^{a,1,2}	0.18 ^{a,1}	0.10 ^{a,1}	0.10 ^{a,1}	0.15 ^{a,1,2}	0.09 ^{a,1}	0.12 ^{a,1,2}	0.09 ^{a,1}	0.20 ^{a,2,3}	0.16 ^{a,1,2}	0.19 ^{a,2,3}	0.17 $^{a,1-3}$	0.009	0.01
	19	0.97 ^{b,4}	0.62 ^{b,3}	0.19 ^{a,1}	0.17 ^{a,b,1,2}	0.31 ^{a,b,1,2}	0.22 ^{a,1,2}	0.21 ^{a,1,2}	0.18 ^{a,1}	0.20 ^{a,1,2}	0.24 ^{a,1,2}	0.86 ^{b,4}	0.43 ^{b,2,3}	0.33 ^{a,b,1,2}	0.63 ^{b,3}	0.000	0.05
	48	1.29 ^{c,6}	0.74 ^{b,2,3}	0.65 ^{b,1,2}	0.27 ^{b,c,1}	0.64 ^{b,c,1,2}	0.53 ^{b,1}	0.55 ^{b,1}	0.66 ^{b,1,2}	0.63 ^{b,1,2}	0.97 ^{b,4,5}	0.88 ^{b,3,4}	1.08 ^{d,5}	1.04 ^{c,d,4,5}	0.89 $^{b,3-5}$	0.000	0.05
	60	1.11 ^{b,c,5}	0.65 ^{b,1,2}	0.75 $^{b,c,1-3}$	0.30 ^{c,,12}	0.56 $^{a-c,1}$	0.61 ^{b,1,2}	0.58 ^{b,c,1}	0.85 $^{c,2-4}$	0.69 ^{b,1,2}	0.79 $^{b,1-4}$	0.85 $^{b,2-4}$	1.02 ^{d,4,5}	1.41 ^{d,e,6}	0.97 $^{b,3-5}$	0.000	0.05
	120	1.29 ^{c,4}	0.75 $^{b,1-3}$	1.01 ^{d,3}	0.32 ^{c,1,2}	0.80 $^{c,1-3}$	0.51 ^{b,1}	0.76 $^{b,c,1-3}$	0.62 ^{b,1,2}	0.66 ^{b,1,2}	0.83 ^{b,2,3}	0.86 ^{b,2,3}	0.83 ^{c,2,3}	0.81 $^{b,c,1-3}$	0.77 $^{b,1-3}$	0.003	0.04
	210	1.16 $^{b,c,1-3}$	0.72 ^{b,1}	0.84 ^{c,1,2}	0.27 ^{c,1}	0.84 ^{c,1,2}	0.87 ^{c,1,2}	0.82 ^{c,1,2}	0.73 ^{b,1}	0.98 $^{c,1-3}$	0.87 ^{b,1,2}	1.33 $^{c,2-4}$	1.48 ^{e,3,4}	1.69 ^{e,4}	1.47 ^{c,3,4}	0.005	0.07
	p-value	0.000	0.004	0.000	0.018	0.037	0.001	0.002	0.000	0.002	0.006	0.000	0.000	0.004	0.001		
	SEM	0.13	0.07	0.10	0.02	0.09	0.08	0.08	0.09	0.09	0.11	0.10	0.13	0.17	0.12		

Table 2. *Cont.*

	Days	CON	BHT	TEA			CHE			GRA			BER			*p*-value	SEM
			200	50	200	1000	50	200	1000	50	200	1000	50	200	1000		
Gumminess (kg)	4	$0.58^{a,3-5}$	$0.35^{a,1-3}$	$0.39^{a,1-4}$	$0.27^{a,1}$	$0.35^{a,1-3}$	$0.30^{a,1,2}$	$0.43^{a,1-5}$	$0.28^{a,1}$	$0.41^{a,1-4}$	$0.30^{a,1,2}$	$0.67^{a,5}$	$0.54^{a,2-5}$	$0.61^{a,4,5}$	$0.59^{a,3-5}$	0.013	0.03
	19	$3.00^{b,5}$	$1.80^{b,3,4}$	$0.66^{a,1,2}$	$0.77^{a,1,2}$	$1.04^{a,b,1,2}$	$0.69^{a,1,2}$	$0.68^{a,1,2}$	$0.54^{a,1}$	$0.58^{a,1}$	$0.78^{a,1,2}$	$2.41^{b,4,5}$	$1.33^{a,2,3}$	$1.05^{a,1,2}$	$1.85^{b,3,4}$	0.000	0.15
	48	$4.26^{c,6}$	$2.62^{c,3-5}$	$2.07^{b,1-3}$	$1.68^{b,1}$	$2.34^{c,2-4}$	$1.87^{b,1,2}$	$1.89^{b,1,2}$	$2.19^{b,c,1-3}$	$2.16^{b,1-3}$	$2.86^{b,4,5}$	$2.93^{c,5}$	$3.14^{b,5}$	$3.15^{b,c,5}$	$2.84^{b,c,4,5}$	0.000	0.13
	60	$3.79^{c,5}$	$2.44^{c,1,2}$	$2.35^{b,1}$	$2.05^{b,1}$	$2.21^{b,c,1}$	$2.12^{b,1}$	$2.11^{b,1}$	$2.52^{d,1-3}$	$2.30^{b,1}$	$2.48^{b,1,2}$	$2.93^{c,2-4}$	$3.26^{b,4}$	$3.93^{c,d,5}$	$3.01^{c,3,4}$	0.000	0.12
	120	$4.23^{c,6}$	$2.60^{c,1-5}$	$3.30^{d,5}$	$2.01^{b,1,2}$	$2.69^{c,2-5}$	$1.93^{b,1}$	$2.46^{b,1-4}$	$2.08^{b,1,2}$	$2.20^{b,1-3}$	$2.83^{b,3-5}$	$2.94^{c,4,5}$	$2.95^{b,4,5}$	$2.82^{b,3-5}$	$2.81^{b,c,3-5}$	0.000	0.12
	210	$3.76^{c,3-5}$	$2.47^{c,1}$	$2.92^{c,1-3}$	$2.25^{b,1}$	$2.70^{c,1,2}$	$2.66^{c,1,2}$	$2.48^{b,1}$	$2.36^{c,d,1}$	$3.09^{c,1-4}$	$2.74^{b,1,2}$	$3.98^{d,3-5}$	$4.28^{c,5}$	$4.52^{d,5}$	$4.14^{d,4,5}$	0.001	0.16
	p-value	0.000	0.000	0.000	0.002	0.012	0.000	0.001	0.000	0.000	0.000	0.000	0.000	0.000	0.002		
	SEM	0.44	0.25	0.33	0.23	0.29	0.25	0.25	0.27	0.30	0.32	0.30	0.39	0.44	0.34		
Cohesiveness	4	$0.23^{a,b}$	0.17^{a}	0.16^{a}	0.10^{a}	0.18^{a}	0.17^{a}	0.21^{a}	0.16^{a}	0.15^{a}	0.14^{a}	0.17^{a}	0.16^{a}	0.17^{a}	0.16^{a}	0.256	0.01
	19	$0.34^{c,5}$	$0.23^{b,2-4}$	$0.18^{a,1,2}$	$0.23^{a,1}$	$0.20^{a,1,2}$	$0.18^{a,1,2}$	$0.22^{a,1-}$	$0.22^{b,1-4}$	$0.19^{a,1,2}$	$0.20^{a,1,2}$	$0.27^{b,4}$	$0.22^{b,1-4}$	$0.21^{a,1-3}$	$0.26^{b,3,4}$	0.000	0.01
	48	0.24^{b}	0.27^{b}	0.30^{c}	0.48^{b}	0.28^{b}	0.27^{b}	$0.28^{a,b}$	0.30^{c}	0.29^{b}	0.32^{b}	0.29^{b}	0.30^{c}	0.26^{b}	0.27^{b}	0.260	0.01
	60	$0.24^{b,1}$	$0.25^{b,1}$	$0.30^{c,3-5}$	$0.64^{b,3-5}$	$0.26^{b,1,2}$	$0.29^{b,2-4}$	$0.28^{a,b,1-3}$	$0.33^{d,5}$	$0.30^{b,2-5}$	$0.32^{b,4,5}$	$0.26^{b,1,2}$	$0.30^{c,3-5}$	$0.29^{b,2-4}$	$0.30^{b,2-5}$	0.001	0.01
	120	0.27^{b}	0.28^{b}	0.31^{c}	0.61^{b}	0.31^{b}	0.26^{b}	0.30^{b}	0.29^{c}	0.31^{b}	0.29^{b}	0.26^{b}	0.27^{c}	0.28^{b}	0.29^{b}	0.102	0.01
	210	$0.23^{a,b,1}$	$0.28^{b,1-4}$	$0.26^{b,1,2}$	$0.66^{b,1-3}$	$0.28^{b,1-4}$	$0.29^{b,2-5}$	$0.32^{b,4,5}$	$0.29^{c,2-5}$	$0.33^{b,5}$	$0.29^{b,2-5}$	$0.31^{b,3-5}$	$0.30^{c,2-5}$	$0.33^{c,5}$	$0.30^{b,2-5}$	0.006	0.01
	p-value	0.002	0.027	0.000	0.003	0.003	0.002	0.030	0.000	0.000	0.006	0.010	0.001	0.001	0.002		
	SEM	0.01	0.01	0.02	0.07	0.02	0.02	0.01	0.02	0.02	0.02	0.02	0.02	0.02	0.02		

$^{a-e}$ Mean values in the same column (same antioxidant in different weeks) with different letter presented significant differences; $^{1-7}$ Mean values in the same row (different antioxidant in the same week) with different number presented significant differences.

Gumminess and chewiness were significantly ($p < 0.05$) increased during ripening time, reaching mean values of 4.26 kg and 1.29 kg × mm, respectively. This increase indicated that gumminess changed from short to pasty gummy through ripening. The highest values were observed in CON samples, while the lowest were found in the samples treated with TEA and GRA extracts. Within the dose effect, significant differences ($p < 0.05$) were found among samples treated with different concentrations of TEA, CHE and GRA extracts. In the case of TEA, the samples treated with 200 mg/kg were that showed the lowest values, while a dose of 50 mg/kg were enough in CHE and GRA samples. Chewiness values indicated that sausages became tougher during ripening period. As occurred with gumminess, the lowest values were found in the samples treated with TEA extract. Significant differences ($p < 0.05$) were found among batches during ripening time. However, regarding dose effect, only significant differences were found in samples treated with TEA extract (0.65 *vs.* 0.27 *vs.* 0.64 kg × mm in TEA_{50}, TEA_{200} and TEA_{1000}, respectively).

Springiness values have been related to the elastic properties of sausages [29]. The values decreased during ripening to remain constant during vacuum-packaged, but only in CON and samples treated with BHT and CHE_{200} this decreased was significant ($p < 0.01$). This result could be also due to water removal during the ripening period. No significant differences ($p > 0.05$) were observed among treatments and the end of ripening, but these differences were significant ($p < 0.05$) at the end of the vacuum-packaging (Table 2).

3.4. Effect of Antioxidants on Oxidative Stability during the Manufacturing Process and Vacuum Packaging

The influence of antioxidants on oxidative stability during the manufacturing process and vacuum packaging was evaluated as TBARS index (Figure 2). Significant changes ($p < 0.05$) were detected in TBARS values among samples during storage time. According to other authors [5,10], CON batches showed more intense lipid oxidation. Thereby, samples with contained antioxidants in their composition showed values below 0.4 mg MDA/kg, showing that these extracts reduced lipid oxidation of the dry-cured sausage. The maximum TBARS values were observed at the end of ripening time (values between 0.23 to 0.78 mg MDA/kg at 0 and 48 days, respectively) followed by a decline until day 60, to remain constant to the end of vacuum packaging. The values found during ripening and vacuum packaging were higher than those found by other authors [6,30] and similar to those found in other dry-cured sausages [17]. The levels obtained during storage period were significantly ($p < 0.05$) lower than the limit (2.0 mg MDA/kg) which is accepted as deterioration level [31].

Regarding the effect of the addition of antioxidants, the results obtained were equal or even better to those found with BHT. Thereby, the samples treated with CHE and GRA reached mean values of 0.17 mg MDA/kg at the end of vacuum packaging, while the samples that contained BHT showed values of 0.24 mg MDA/kg. These results are in agreement with previously published studies [4], which reported higher effectiveness of natural products compared to synthetic antioxidants and suggesting the possibility of using these extracts as replacers of commercial compounds. Within dose effect, no significant differences ($p > 0.05$) were found among samples on the same extract, so that the lower concentration of natural extract would be sufficient to improve the results obtained in CON samples.

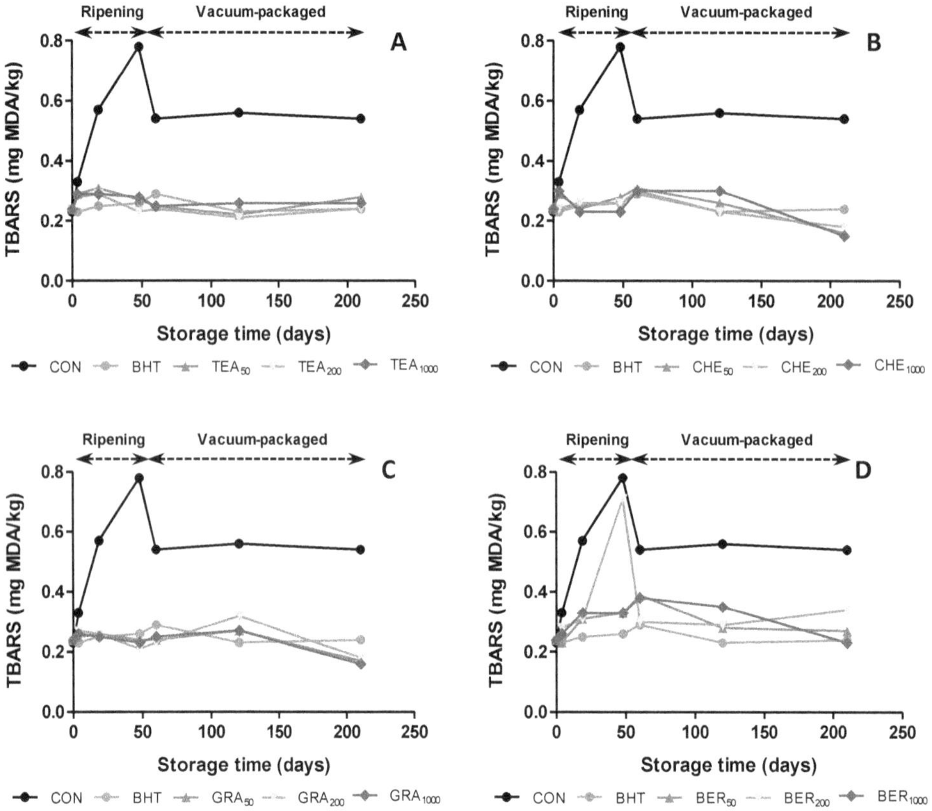

Figure 2. Evolution of thiobarbituric acid reactive substances (TBARS) in dry-cures sausages treated with BHT and natural antioxidants during ripening and vacuum-packaged storage.

3.5. Effect of Antioxidants on Microbial Counts during the Manufacturing Process and Vacuum Packaging

Changes in the microbial populations, TVC, LAB and mold/yeasts, during the manufacturing process and vacuum packaging of dry-cured sausage "chorizo" are shown in Table 3. The initial TVC, LAB and mold/yeasts counts ranged from 10^3 to 10^5 CFU/g (data not shown). Significant differences ($p < 0.05$) in microbial counts were detected among batches and during ripening and storage period. TVC counts increased from 5.17 to 8.62 $\log_{10}$ CFU/g ($p < 0.001$) during the first 19 days of ripening, remaining stable until the end of ripening (reaching values of 8.72 $\log_{10}$ CFU/g). TVC counts continued to increase up to day 60, to proceed decreasing gradually during vacuum storage. Among batches, significant differences ($p < 0.001$) were found with the addition of antioxidants. At the end of ripening process, samples that contained extracts in their composition showed lower TVC counts than CON. Excluding CHE_{200}, natural extracts showed lower results, getting to improve the results of BHT (8.12 $\log_{10}$ CFU/g *vs.* counts below 8 $\log_{10}$ CFU/g in TEA_{200}, CHE_{200}, GRA_{1000}, BER_{200} and BER_{1000}). Regarding dose effect, significant differences ($p < 0.001$) were found for all batches studied. Increase the level of natural antioxidant usually decreased the TVC counts. In fact, samples manufactured with GRA_{1000} showed lower values than GRA_{50}, GRA_{200} and CON (7.91 $\log_{10}$ CFU/g *vs.* 8.14, 8.15 and 8.55 $\log_{10}$ CFU/g, respectively). In the other extracts, the dose of 200 mg/kg showed the lowest values (7.97, 7.88 and 7.95 $\log_{10}$ CFU/g for TEA, CHE and BER, respectively).

Table 3. Evolution of TVC, BAL and mold/yeast of "chorizo" treated with BHT and natural antioxidants during ripening and vacuum-packaging.

	Days	CON	BHT	TEA			CHE			GRA			BER			*p*-Value	SEM
			200	50	200	1000	50	200	1000	50	200	1000	50	200	1000		
TVC	4	6.35 ^{a,8}	6.06 ^{a,4,5}	6.34 ^{a,8}	5.88 ^{a,2}	6.12 ^{a,5,6}	6.11 ^{a,5,6}	5.97 ^{a,3}	5.94 ^{a,3}	6.14 ^{a,6,7}	6.09 $^{a,4–6}$	6.18 ^{a,7}	6.12 ^{a,5,6}	5.72 ^{a,1}	6.04 ^{a,4}	0.000	0.03
	19	8.62 ^{e,8}	8.38 ^{e,4}	8.42 ^{d,4,5}	8.59 ^{c,7,8}	8.51 ^{d,6}	8.55 ^{f,6,7}	8.46 ^{d,5}	8.22 ^{d,2}	8.19 ^{c,1,2}	8.32 ^{e,3}	8.31 ^{e,3}	8.21 ^{d,2}	8.16 ^{c,1}	8.58 ^{e,7,8}	0.000	0.03
	48	8.55 ^{d,8}	8.12 $^{c,5–7}$	8.06 ^{b,4}	7.97 ^{b,3}	8.08 ^{c,4,5}	8.10 $^{d,4–6}$	7.88 ^{c,1}	8.72 ^{f,9}	8.14 ^{c,6,7}	8.15 ^{d,7}	7.91 ^{c,1,2}	8.10 $^{c,4–6}$	7.95 ^{b,2,3}	7.99 ^{c,3}	0.000	0.04
	60	9.07 ^{f,8}	8.24 ^{d,2}	8.30 ^{c,2}	8.70 ^{d,6}	8.55 ^{d,4}	8.28 ^{e,2}	8.84 ^{e,7}	8.62 ^{e,5}	8.43 ^{d,3}	8.65 ^{f,5,6}	8.27 ^{e,2}	8.48 ^{e,3}	8.46 ^{d,3}	8.15 ^{d,1}	0.000	0.05
	120	7.61 ^{c,1}	7.92 ^{b,3}	8.00 ^{b,4}	7.90 ^{b,3}	7.99 ^{c,4}	7.99 ^{c,4}	7.87 ^{c,3}	7.80 ^{b,2}	7.91 ^{b,3}	7.61 ^{b,1}	8.12 ^{d,6}	8.06 ^{c,5}	8.13 ^{c,6}	7.92 ^{b,3}	0.000	0.03
	210	7.47 ^{b,1}	8.15 ^{c,5}	9.92 ^{e,6}	7.91 ^{b,3,4}	7.88 ^{b,3,4}	7.92 ^{b,3,4}	7.73 ^{b,2}	8.15 ^{c,5}	7.93 ^{b,4}	7.81 ^{c,2,3}	7.85 ^{b,3,4}	7.86 ^{b,3,4}	7.93 ^{b,4}	7.93 ^{b,4}	0.000	0.11
	p-value	0.000	0.000	0.000	0.000	0.000	0.000	0.000	0.000	0.000	0.000	0.000	0.000	0.000	0.000		
	SEM	0.37	0.24	0.32	0.28	0.25	0.24	0.27	0.28	0.23	0.25	0.22	0.23	0.28	0.24		
LAB	4	6.62 ^{a,2}	6.91 ^{a,6}	6.87 ^{a,6}	6.62 ^{a,2,3}	7.07 ^{a,7}	7.11 ^{a,7}	6.54 ^{a,1}	6.71 ^{a,4}	6.78 ^{a,5}	6.89 ^{a,6}	6.88 ^{a,6}	6.55 ^{a,1}	6.67 ^{a,3,4}	6.78 ^{a,5}	0.000	0.03
	19	7.97 ^{c,6,7}	7.90 ^{b,c,3,4}	7.95 ^{c,5,6}	7.93 ^{b,c,4,5}	8.00 $^{b,7–9}$	8.00 $^{b,7–9}$	7.95 ^{d,5,6}	7.84 ^{b,2}	8.01 ^{b,c,8,9}	7.90 ^{c,3,4}	7.86 ^{a,b,2,3}	7.97 $^{b,6–8}$	7.76 ^{b,1}	8.03 ^{c,9}	0.000	0.04
	48	8.12 ^{d,6,7}	8.07 ^{d,5,6}	8.01 ^{c,4,5}	8.02 ^{c,d,45}	8.18 ^{c,7}	7.97 ^{b,3,4}	7.80 ^{c,d,2}	8.57 ^{c,8}	7.93 ^{b,3}	7.96 ^{c,3,4}	7.70 ^{a,b,1}	8.16 ^{b,7}	8.03 ^{b,4,5}	7.84 ^{b,2}	0.000	0.04
	60	11.08 ^{f,8}	7.96 ^{c,d,1,2}	7.91 ^{c,1}	8.06 ^{d,3}	8.00 ^{b,2,3}	8.20 ^{b,4,5}	8.51 ^{e,6}	8.72 ^{c,7}	8.23 ^{c,5}	8.49 ^{d,6}	8.14 ^{b,4}	8.05 ^{b,3}	8.44 ^{c,6}	8.18 ^{d,4,5}	0.000	0.15
	120	8.32 $^{e,2–4}$	8.32 $^{e,2–4}$	7.95 $^{c,1–4}$	7.85 $^{b,1–3}$	8.51 ^{d,3,4}	8.50 ^{b,3,4}	7.34 ^{b,1}	8.00 $^{b,1–4}$	9.71 ^{d,6}	7.58 ^{b,c,1,2}	8.66 ^{b,4,5}	8.03 $^{b,1–4}$	8.00 $^{b,1–4}$	9.28 ^{e,5,6}	0.000	0.13
	210	7.45 ^{b}	7.75 ^{b}	7.54 ^{b}	7.90 ^{b}	8.22 ^{c}	7.87 ^{b}	7.73 ^{c}	7.94 ^{b}	8.09 ^{b,c}	7.32 ^{a,b}	7.68 ^{a,b}	7.69 ^{b}	8.11 ^{b,c}	7.95 ^{b,c}	0.104	0.06
	p-value	0.000	0.000	0.001	0.000	0.000	0.038	0.000	0.000	0.000	0.002	0.042	0.003	0.000	0.000		
	SEM	0.50	0.13	0.13	0.15	0.14	0.14	0.18	0.20	0.26	0.16	0.18	0.17	0.17	0.22		
Mold/Yeast	4	6.09 ^{d,6}	5.78 ^{a,3}	5.86 ^{d,3,4}	5.67 ^{b,2}	5.80 ^{b,3}	6.00 ^{c,5}	6.44 ^{c,7}	5.56 ^{c,1}	5.89 ^{d,4}	5.89 ^{c,4}	6.48 ^{e,7}	5.78 ^{b,3}	5.94 ^{c,4,5}	6.13 ^{d,6}	0.000	0.05
	19	7.74 ^{e,4}	7.10 ^{b,1,2}	7.69 ^{e,3,4}	7.52 $^{c,2–4}$	7.16 ^{c,1,2}	7.10 ^{d,1,2}	7.60 ^{d,3,4}	7.28 $^{d,1–3}$	7.31 $^{e,1–4}$	7.07 ^{f,1}	7.38 $^{f,1–4}$	7.05 ^{c,1}	7.75 ^{e,4}	7.67 ^{e,3,4}	0.006	0.06
	48	6.17 ^{d,6}	5.61 ^{a,1}	5.91 ^{d,4}	5.71 ^{b,2}	5.96 ^{b,4}	5.81 ^{c,3}	5.91 ^{b,c,4}	5.73 ^{c,2}	5.79 ^{d,3}	5.61 ^{c,1}	6.19 ^{d,6}	6.12 ^{b,5}	6.15 ^{d,5,6}	6.24 ^{d,7}	0.000	0.04
	60	5.92 ^{c,9}	5.58 ^{a,6}	5.20 ^{c,3}	5.78 ^{b,7,8}	5.51 ^{b,5}	5.28 ^{a,b,4}	5.23 ^{b,3}	5.31 ^{b,4}	5.08 ^{b,2}	4.93 ^{a,1}	5.78 ^{c,7,8}	5.89 ^{b,9}	5.74 ^{b,7}	5.82 ^{c,8}	0.000	0.06
	120	5.58 ^{b,5}	5.21 $^{a,3–5}$	4.53 ^{b,1,2}	5.05 $^{a,b,2–5}$	5.08 $^{b,2–5}$	5.50 ^{b,4,5}	4.37 ^{a,1}	4.60 $^{a,1–3}$	4.89 $^{a,1–4}$	5.15 $^{b,2–5}$	5.43 ^{a,4,5}	5.23 $^{a,3–5}$	5.58 ^{b,5}	5.53 ^{b,4,5}	0.004	0.08
	210	4.49 ^{a,2}	5.12 ^{a,2,3}	2.79 ^{a,1}	4.66 ^{a,2,3}	3.25 ^{a,1}	5.04 ^{a,2,3}	4.48 ^{a,2}	4.74 ^{a,2,3}	5.32 ^{c,2,3}	5.48 ^{c,2,3}	5.57 ^{b,3}	5.23 ^{a,2,3}	4.67 ^{a,2,3}	5.23 ^{a,2,3}	0.000	0.16
	p-value	0.000	0.043	0.000	0.003	0.000	0.000	0.000	0.000	0.000	0.000	0.000	0.001	0.000	0.000		
	SEM	0.36	0.22	0.45	0.28	0.36	0.20	0.35	0.27	0.24	0.21	0.20	0.19	0.28	0.24		

$^{a–f}$ Mean values in the same column (same antioxidant in different weeks) with different letter presented significant differences; $^{1–9}$ Mean values in the same row (different antioxidant in the same week) with different number presented significant differences; SEM: standard error of mean; Batches: CON: control; BHT: tert-butyl-4-hydroxytoluene; TEA: tea; CHE: chestnut; GRA: grape seed and BER: beer extracts.

The lactic acid bacteria (LAB) counts showed significant ($p < 0.05$) differences during ripening and storage. A rapid increase in the LAB population was observed during the first 19 days of fermentation, increasing counts from 4.7 to 8.0 log_{10} CFU/g. Until the end of fermentation and ripening process the counts remained constant, reaching values of between 8.57 and 7.70 log_{10} CFU/g. The samples than contained CHE_{1000} in their formulation showed the highest values, followed by CON, TEA_{1000} and BER_{50}, that presented similar values around 8.15 log_{10} CFU/g. Except for CON samples, the trend during vacuum packaging was to decreased slightly (33% *vs.* mean values of 5.1%).

Dose effect showed significant ($p < 0.05$) differences among batches during ripening process on LAB counts. GRA and BER extracts showed a decrease in the population of LAB with the increase of the concentration, obtaining lower values in GRA_{1000} (7.70 *vs.* 7.93 and 7.96 for GRA_{1000}, GRA_{50} and GRA_{200}, respectively) and BER_{1000} (7.84 *vs.* 8.16 and 8.03 for BER_{1000}, BER_{50} and BER_{200}, respectively). In contrast, samples treated with TEA and CHE showed lower counts for the minor dose (Table 3). As happened with the aforementioned microbial groups, mold and yeasts counts increased rapidly during the first 19 days of ripening, from 3.4 to 7.7 log_{10} CFU/g, to decrease slightly until the end of ripening process with values ranged between 6.2 and 5.61 log_{10} CFU/g.

3.6. Effect of Antioxidants on Free Fatty Acid Content during the Manufacturing Process and Vacuum Packaging

The free fatty acid (FFA) content of the different batches expressed as mg of fatty acid/g of fat is shown in Tables 4 and 5. The predominated fatty acids both the end of ripening time and the end of storage at vacuum-packaging were monounsaturated fatty acids (MUFA), followed by saturated fatty acids (SFA) and polyunsaturated fatty acids (PUFA). These results are in agreement with other studies conducted in dry-ripened "chorizo" [6], being oleic, linoleic, palmitic, stearic and palmitoleic acids the predominated ones in the stages of ripening.

The free fatty acid content during ripening and vacuum packaging showed a gradual and significant release of these compounds as a result of lipolysis. Significant differences ($p < 0.05$) were detected among samples at the end of ripening time. In all cases, CON samples showed lower release values than those obtained for the samples treated with antioxidants. MUFAs were the FFA that showed the highest increases, greater than 70%. Oleic acid was the predominant fatty acid presented in all the batches, with values that ranged between 49% and 55%. These values reached the highest levels between days 4 and 19 of ripening, being the samples treated with BHT, TEA_{1000}, CHE_{200} and BER_{1000} that reached the maximum release values. Unlike other authors [32], the percentages of oleic acid continued to increase until the end of ripening and during vacuum storage.

Table 4. Evolution of free fatty acid composition (mg/100g) of "chorizo" treated with BHT and natural antioxidants during ripening and vacuum-packaging.

	Days	CON	BHT	TEA			CHE			GRA			BER			p-Value	SEM
			200	50	200	1000	50	200	1000	50	200	1000	50	200	1000		
C16:0	4	274.05 ^{a,b}	119.86 ^{a}	280.66 ^{a}	233.36 ^{a}	219.68 ^{a}	207.05 ^{a}	247.90 ^{a}	159.44 ^{a}	250.42 ^{a}	215.51 ^{a}	223.47 ^{a}	244.31	194.55	167.68	0.711	13.10
	19	166.96 ^{a}	393.41 ^{b}	350.09 ^{a}	289.73 ^{a}	281.74 ^{a}	324.98 ^{a,b}	414.83 ^{a}	227.25 ^{a}	227.42 ^{a}	210.42 ^{a}	303.13 ^{a}	271.75	288.36	297.27	0.227	16.63
	48	169.33 ^{a,1}	353.28 ^{b,3,4}	375.84 ^{a,4}	322.48 $^{a,2–4}$	507.68 ^{b,5}	331.06 $^{a,b,2–4}$	341.90 $^{a,2–4}$	396.63 ^{b,4}	317.90 $^{a,2–4}$	373.45 ^{a,4}	270.06 ^{a,2,3}	330.28 $^{2–4}$	255.06 ^{2}	314.01 $^{2–4}$	0.000	15.12
	210	426.16 ^{b}	700.61 ^{c}	743.92 ^{b}	1227.20 ^{b}	689.07 ^{c}	801.26 ^{b}	936.76 ^{b}	820.64 ^{c}	749.68 ^{b}	1006.20 ^{b}	691.26 ^{b}	600.02	594.27	673.13	0.082	44.54
	p-value	0.052	0.002	0.056	0.004	0.002	0.083	0.056	0.000	0.006	0.009	0.012	0.090	0.060	0.079		
	SEM	43.88	79.57	75.17	158.85	71.85	97.41	111.39	97.54	82.71	128.05	73.48	60.72	64.69	80.04		
C16:1	4	29.32 ^{a}	1.64 ^{a}	27.99 ^{a}	23.72 ^{a}	14.56 ^{a}	26.28 ^{a}	26.27 ^{a}	11.72 ^{a}	26.66 ^{a}	13.88 ^{a}	16.66 ^{a}	18.07 ^{a}	17.05 ^{a}	8.47 ^{a}	0.438	2.17
	19	14.16 ^{a}	53.91 ^{b}	39.91 ^{a}	34.66 ^{a}	32.77 ^{a}	35.07 ^{a}	50.31 ^{a}	23.83 ^{a}	26.46 ^{a}	21.55 ^{a}	35.22 ^{a}	29.97 ^{a}	34.88 ^{a}	37.55 ^{a}	0.081	2.39
	48	19.26 ^{a,1}	49.30 ^{b,2,3}	54.19 ^{a,3}	43.30 ^{a,2,3}	89.21 ^{b,4}	48.80 ^{a,2,3}	43.91 ^{a,2,3}	55.45 ^{b,3}	46.49 ^{a,2,3}	57.41 ^{b,3}	38.35 ^{a,2}	47.80 ^{a,2,3}	36.01 ^{a,2}	43.39 ^{a,2,3}	0.000	2.95
	210	76.10 ^{b,1}	103.16 ^{c,1,2}	106.19 ^{b,1,2}	167.12 ^{b,3}	112.31 ^{b,1,2}	102.71 ^{b,1,2}	123.62 $^{b,1–3}$	109.10 ^{c,1,2}	116.46 ^{b,1,2}	149.79 ^{c,2,3}	97.67 ^{b,1}	91.46 ^{b,1}	97.90 ^{b,1}	103.37 ^{b,1,2}	0.038	5.09
	p-value	0.009	0.003	0.010	0.001	0.001	0.006	0.030	0.000	0.013	0.000	0.009	0.019	0.032	0.026		
	SEM	9.60	13.85	11.72	22.18	15.27	11.56	15.07	14.33	14.60	20.50	11.92	11.11	12.45	13.90		
C18:0	4	143.67	89.44 ^{a}	143.01	114.11 ^{a}	104.97 ^{a}	107.01	130.30	91.68 ^{a}	121.10 ^{a}	129.85 ^{a}	124.49 ^{a,b}	135.63	111.36	102.55	0.900	5.87
	19	92.34	179.57 ^{b}	166.29	139.56 ^{a}	142.81 ^{a,b}	166.92	203.95	130.09 ^{a}	124.16 ^{a}	122.16 ^{a}	154.38 ^{a,b}	142.90	157.77	167.49	0.269	6.87
	48	69.20 ^{1}	116.22 ^{a,1,2}	123.21 ^{2}	96.85 ^{a,1,2}	176.92 ^{b,3}	101.52 ^{1,2}	106.23 ^{1,2}	133.60 ^{a,2,3}	115.99 ^{a,1,2}	144.12 ^{a,2,3}	104.71 ^{a,1,2}	137.68 ^{2,3}	99.33 ^{1,2}	126.71 ^{2}	0.021	5.54
	210	149.17	223.33 ^{b}	259.89	461.58 ^{b}	240.33 ^{c}	345.60	351.84	280.00 ^{b}	232.18 ^{b}	395.92 ^{b}	245.44 ^{b}	215.10	211.30	321.35	0.552	22.98
	p-value	0.262	0.006	0.254	0.055	0.014	0.296	0.226	0.003	0.015	0.067	0.109	0.269	0.131	0.256		
	SEM	16.60	20.44	25.51	62.53	19.71	49.61	45.73	27.68	19.22	48.29	23.56	16.33	19.62	41.52		

Table 4. *Cont.*

	Days	CON	BHT	TEA			CHE			GRA			BER			*p*-value	SEM
			200	50	200	1000	50	200	1000	50	200	1000	50	200	1000		
	4	529.15 ^{a}	213.02 ^{a}	468.86 ^{a}	443.31 ^{a}	338.88 ^{a}	438.33	459.15	339.35 ^{a}	511.66 ^{a}	409.07 ^{a}	452.25 ^{a}	491.65 ^{a}	400.20 ^{a}	344.88 ^{a}	0.465	22.37
	19	432.85 ^{a,1}	973.25 ^{b,3,4}	853.91 $^{a,2-4}$	733.01 $^{a,1-4}$	688.11 $^{b,1-3}$	799.28 $^{2-4}$	1043.61 ^{4}	624.51 ^{a,b,1,2}	645.37 ^{a,b,1,2}	551.32 ^{a,b,1,2}	671.73 $^{a,1-3}$	682.09 $^{a,1-3}$	696.65 $^{b,1-3}$	784.79 $^{b,2-4}$	0.029	34.50
C18:	48	405.56 ^{a,1}	1030.46 ^{b,4}	855.82 $^{a,2-4}$	875.07 $^{a,2-4}$	1278.56 ^{c,5}	858.61 $^{2-4}$	784.46 ^{2,3}	991.82 ^{b,3,4}	916.95 $^{b,2-4}$	1072.72 ^{b,4}	803.00 ^{a,b,2,3}	952.68 $^{b,2-4}$	732.70 ^{b,2}	894.76 $^{b,2-4}$	0.000	38.41
1n9c	210	899.38 ^{b}	1350.71 ^{c}	1523.51 ^{b}	2341.58 ^{b}	1577.73 ^{c}	1580.00	1791.42	1480.70 ^{c}	1381.90 ^{c}	2130.29 ^{c}	1368.17 ^{b}	1222.02 ^{c}	1306.35 ^{c}	1378.62 ^{c}	0.126	85.05
	p-value	0.050	0.002	0.033	0.007	0.002	0.087	0.144	0.007	0.009	0.004	0.066	0.004	0.004	0.007		
	SEM	81.83	160.15	154.33	286.77	186.32	177.77	220.87	166.56	130.48	261.70	142.51	106.83	126.83	143.54		
	4	305.74 ^{b,5}	150.08 ^{a,1}	270.91 $^{3-5}$	230.20 $^{a,1-5}$	186.19 $^{a,1-3}$	256.88 $^{3-5}$	273.12 ^{4,5}	162.99 ^{a,1,2}	226.89 $^{a,1-5}$	187.06 $^{a,1-3}$	246.44 $^{2-5}$	243.37 $^{a,2-5}$	194.95 $^{a,1-4}$	185.99 $^{a,1-3}$	0.015	9.79
	19	257.29 ^{a,b,1}	397.66 ^{b,3,4}	363.86 $^{2-4}$	351.91 $^{b,1-4}$	286.12 ^{a,b,1,2}	340.63 $^{1-4}$	421.13 ^{4}	301.49 $^{b,1-3}$	326.34 $^{b,1-4}$	261.83 ^{a,b,1}	366.47 $^{2-4}$	373.61 $^{b,2-4}$	330.43 $^{b,1-4}$	372.16 $^{b,2-4}$	0.038	10.72
C18:	48	174.46 ^{a,1}	311.35 ^{b,3,4}	250.16 ^{2,3}	236.18 $^{a,1-3}$	427.53 ^{b,c,5}	252.14 ^{2,3}	205.00 ^{1,2}	344.28 ^{b,c,4}	302.38 ^{a,b,3,4}	339.05 ^{b,4}	311.37 ^{3,4}	352.47 ^{b,4}	308.50 ^{b,3,4}	335.79 ^{b,4}	0.000	13.06
2n6c	210	338.14 ^{b}	335.41 ^{b}	401.34	544.78 ^{c}	438.75 ^{c}	355.56	424.95	418.13 ^{c}	367.91 ^{b}	543.81 ^{c}	364.37	404.98 ^{b}	426.23 ^{c}	464.93 ^{b}	0.299	16.58
	p-value	0.056	0.005	0.176	0.003	0.022	0.351	0.256	0.004	0.048	0.001	0.240	0.042	0.010	0.019		
	SEM	25.70	35.50	28.96	49.16	41.94	24.62	46.38	35.96	21.22	50.99	23.61	24.99	32.28	40.08		
	4	15.81 ^{b,3}	0.00 ^{a,1}	10.18 ^{a,b,2}	10.15 ^{2}	0.00 ^{a,1}	0.00 ^{1}	9.51 ^{2}	0.00 ^{a,1}	3.80 ^{a,1}	2.73 ^{a,1}	0.00 ^{1}	0.00 ^{a,1}	0.00 ^{a,1}	1.72 ^{a,1}	0.000	1.03
	19	6.51 ^{a}	17.06 ^{b}	19.17 ^{b}	13.94	11.99 ^{a,b}	21.82	22.62	11.58 ^{b}	11.91 ^{b}	9.77 ^{a,b}	23.90	14.21 ^{b,c}	11.21 ^{a,b}	16.73 ^{b,c}	0.125	1.21
C18:	48	0.00 ^{a,1}	10.98 ^{a,b,2}	5.52 ^{a,1,2}	4.22 ^{1,2}	20.54 ^{b,3}	7.43 ^{1,2}	6.13 ^{1,2}	12.84 ^{b,2,3}	8.32 ^{a,b,1,2}	11.40 ^{b,2}	8.07 ^{1,2}	10.55 ^{b,2}	6.10 ^{a,1,2}	10.01 ^{a,b,2}	0.024	1.03
3n3	210	19.08 ^{b}	22.13 ^{b}	21.78 ^{c}	30.41	22.83 ^{b}	20.52	20.43	33.65 ^{c}	22.22 ^{c}	28.63 ^{c}	22.37	21.06 ^{c}	22.97 ^{b}	22.60 ^{c}	0.806	1.29
	p-value	0.009	0.055	0.184	0.001	0.054	0.161	0.217	0.001	0.009	0.002	0.092	0.006	0.026	0.022		
	SEM	2.96	3.44	3.05	3.72	3.73	4.16	3.32	4.64	2.66	3.66	4.30	2.96	3.40	3.11		
	4	9.54 $^{a,b,1-3}$	0.00 ^{a,1}	17.79 ^{3}	0.00 ^{1}	8.36 $^{a,1-3}$	0.95 ^{1}	16.32 ^{2,3}	0.00 ^{a,1}	0.00 ^{a,1}	2.31 ^{1}	0.00 ^{a,1}	4.01 ^{1,2}	0.00 ^{a,1}	0.00 ^{a,1}	0.035	1.38
	19	12.65 ^{a,b}	20.26 ^{b}	15.85	14.24	9.99 ^{a,b}	15.12	19.53	8.93 ^{a}	14.12 ^{b}	8.81	23.98 ^{c}	14.91	24.90 ^{c}	18.77 ^{b,c}	0.247	1.25
C20:	48	0.00 ^{a}	9.75 ^{ab}	16.90	7.01	29.31 ^{c}	9.28	9.52	6.23 ^{a}	9.85 ^{b}	8.98	12.27 ^{b}	13.65	10.56 ^{b}	11.27 ^{a,b}	0.081	1.48
4n6	_210	21.43 ^{b}	10.64 ^{a,b}	13.57	15.63	19.21 ^{b}	7.98	7.71	23.62 ^{b}	14.31 ^{b}	17.65	12.72 ^{b}	19.12	23.29 ^{c}	28.56 ^{c}	0.062	1.38
	p-value	0.167	0.084	0.989	0.068	0.012	0.126	0.254	0.026	0.003	0.232	0.019	0.082	0.000	0.035		
	SEM	3.50	3.07	3.63	2.64	3.30	2.23	2.35	3.49	2.24	2.61	3.39	2.36	3.85	4.26		

$^{a-c}$ Mean values in the same column (same antioxidant in different weeks) with different letter presented significant differences; $^{1-5}$ Mean values in the same row (different antioxidant in the same week) with different number presented significant differences; SEM: standard error of mean; Batches: CON: control; BHT: tert-butyl-4-hydroxytoluene; TEA: tea; CHE: chestnut; GRA: grape seed and BER: beer extracts.

Table 5. Evolution of main nutritional index (mg/100g) of "chorizo" treated with BHT and natural antioxidants during ripening and vacuum-packaging.

	Days	CON	BHT	TEA			CHE			GRA			BER			p-value	SEM
			200	50	200	1000	50	200	1000	50	200	1000	50	200	1000		
SFA	4	417.72	209.30 ^{a}	423.68	347.47 ^{a}	324.64 ^{a}	314.06	378.20	251.12 ^{a}	371.51 ^{a}	345.36 ^{a}	347.95 ^{a}	379.94	305.91	270.24	0.780	18.54
	19	259.29	572.99 ^{b}	516.38	429.29 ^{a}	424.54 ^{a}	491.91	618.78	357.33 ^{b}	351.57 ^{a}	332.58 ^{a}	457.51 ^{a}	414.64	446.13	464.77	0.241	23.34
	48	238.52 ^{1}	469.49 $^{b,2-4}$	499.04 ^{3,4}	419.32 $^{a,2-4}$	684.60 ^{b,5}	432.58 $^{2-4}$	448.11 $^{2-4}$	530.23 ^{c,4}	433.89 $^{a,2-4}$	517.57 ^{a,4}	374.78 ^{a,2,3}	467.96 $^{2-4}$	354.39 ^{1,2}	440.72 $^{2-4}$	0.001	20.17
	210	605.07	957.29 ^{c}	1039.82	1754.88 ^{b}	962.22 ^{c}	1195.06	1339.90	1152.74 ^{d}	988.29 ^{b}	1452.09 ^{b}	974.31 ^{b}	853.55	835.24	1032.75	0.189	68.56
	p-value	0.079	0.003	0.072	0.008	0.003	0.123	0.076	0.000	0.007	0.015	0.023	0.130	0.083	0.131		
	SEM	62.73	103.68	103.74	229.84	95.32	152.43	162.25	132.27	102.49	183.00	102.29	84.44	89.03	127.86		
MUFA	4	560.24 ^{a}	214.66 ^{a}	496.85 ^{a}	467.03 ^{a}	353.43 ^{a}	464.60	485.42	351.06 ^{a}	538.32 ^{a}	422.95 ^{a}	468.91 ^{a}	509.72 ^{a}	417.25 ^{a}	353.35 ^{a}	0.432	24.10
	19	447.01 ^{a,1}	1027.17 ^{b,3,4}	893.82 $^{a,2-4}$	767.66 $^{a,1-4}$	720.87 $^{b,1-3}$	834.35 $^{2-4}$	1093.92 ^{4}	648.35 ^{a,b,1,2}	671.83 ^{a,b,1,2}	572.87 ^{a,b,1,2}	706.95 $^{a,1-3}$	712.06 $^{a,1-3}$	731.53 $^{a,b,1-3}$	822.34 $^{b,2-4}$	0.031	36.77
	48	424.81 ^{a,1}	1079.76 ^{b,4,}	910.00 $^{a,2-4}$	918.37 $^{a,2-4}$	1368.99 ^{c,5}	907.41 $^{2-4}$	828.37 ^{2,3}	1047.27 ^{b,3,4}	963.44 $^{b,2-4}$	1130.13 ^{b,4}	841.36 ^{a,b,2,3}	1000.48 $^{b,2-4}$	768.70 ^{b,2}	938.15 $^{b,2-4}$	0.000	41.11
	210	990.17 ^{b}	1469.62 ^{c}	1650.85 ^{b}	2546.22 ^{b}	1712.08 ^{c}	1709.73	1941.17	1608.25 ^{c}	1522.29 ^{c}	2313.74 ^{c}	1481.89 ^{b}	1327.59 ^{c}	1419.73 ^{c}	1498.93 ^{c}	0.112	91.21
	p-value	0.039	0.001	0.028	0.006	0.001	0.077	0.133	0.005	0.008	0.003	0.057	0.004	0.005	0.008		
	SEM	93.32	174.94	168.48	314.60	204.09	193.11	239.88	183.01	148.03	287.04	156.57	119.33	141.34	159.35		
PUFA	4	331.08 ^{b,5}	150.08 ^{a,1}	298.87 ^{4,5}	240.35 $^{a,1-5}$	194.55 $^{a,1-3}$	257.83 $^{3-5}$	298.93 ^{4,5}	162.99 ^{a,1,2}	230.68 $^{a,1-4}$	192.10 $^{a,1-3}$	246.44 $^{2-5}$	247.38 $^{a,2-5}$	194.95 $^{a,1-3}$	187.72 $^{a,1-3}$	0.008	11.29
	19	276.43 ^{a,b,1}	434.97 ^{c,4,5}	398.87 $^{1-5}$	380.08 $^{b,1-5}$	308.09 $^{a,1-3}$	377.57 $^{1-5}$	463.27 ^{5}	321.99 $^{b,1-4}$	352.37 $^{b,1-5}$	280.42 ^{b,1,2}	414.35 $^{3-5}$	402.73 $^{b,2-5}$	366.53 $^{b,1-5}$	407.65 $^{b,c,3-5}$	0.050	12.55
	48	174.46 ^{a,1}	332.07 ^{b,3,4}	272.57 ^{2,3}	247.41 $^{a,1-3}$	477.38 ^{b,5}	268.84 ^{2,3}	220.64 ^{1,2}	363.35 ^{b,4}	320.54 ^{a,b,3,4}	359.42 ^{b,4}	331.71 ^{3,4}	376.66 ^{b,4}	325.15 ^{b,3,4}	357.07 ^{b,4}	0.000	14.81
	210	386.20 ^{b}	372.15 ^{b,c}	444.05	601.60 ^{c}	490.66 ^{b}	394.38	460.58	480.62 ^{c}	414.98 ^{b}	601.65 ^{c}	403.06	451.71 ^{b}	479.32 ^{c}	522.59 ^{c}	0.348	18.41
	p-value	0.050	0.005	0.226	0.002	0.019	0.284	0.256	0.002	0.030	0.001	0.169	0.030	0.008	0.018		
	SEM	32.42	41.13	33.59	56.28	49.13	30.73	51.10	43.67	26.91	58.12	30.73	30.60	39.74	48.02		

Table 5. *Cont.*

	Days	CON	BHT	TEA			CHE			GRA			BER			p-value	SEM
			200	50	200	1000	50	200	1000	50	200	1000	50	200	1000		
P/S	4	0.88	0.72 [b]	0.74	0.72 [b,c]	0.62	0.86	0.83 [b]	0.65 [b]	0.68	0.56	0.72 [a,b]	0.66 [a]	0.66	0.72	0.961	0.03
	19	1.07	0.77 [b]	0.83	0.90 [c]	0.73	0.78	0.76 [b]	0.91 [c]	1.02	0.91	0.91 [b]	0.98 [b]	0.83	0.91	0.597	0.03
	48	0.78 [2–5]	0.72 [b,1–5]	0.55 [1,2]	0.60 [a,b,1–3]	0.70 [1–5]	0.62 [1–3]	0.50 [a,b,1]	0.69 [b,1–4]	0.74 [2–5]	0.71 [1–5]	0.89 [b,4,5]	0.81 [a,b,3–5]	0.92 [5]	0.82 [3–5]	0.015	0.03
	210	0.64 [5]	0.39 [a,1–3]	0.43 [1–4]	0.35 [a,1,2]	0.51 [1–5]	0.35 [1,2]	0.34 [a,1]	0.42 [a,1–4]	0.42 [1–4]	0.43 [1–4]	0.42 [a,1–4]	0.56 [a,3–5]	0.60 [4,5]	0.54 [2–5]	0.036	0.02
	p-value	0.489	0.004	0.140	0.016	0.158	0.203	0.052	0.005	0.084	0.164	0.044	0.056	0.163	0.346		
	SEM	0.09	0.06	0.07	0.08	0.04	0.09	0.08	0.07	0.09	0.08	0.08	0.07	0.06	0.07		
n6/n3	4	20.25 [b,2]	0.00 [a,1]	30.45 [2,3]	22.72 [b,2]	0.00 [a,1]	0.00 [1]	37.77 [2,3]	0.00 [a,1]	59.71 [b,4]	46.83 [b,3,4]	0.00 [a,1]	0.00 [a,1]	0.00 [a,1]	0.00 [a,1]	0.000	4.00
	19	41.49 [c,4]	24.72 [b,1–3]	21.80 [1–3]	26.23 [c,1–3]	25.29 [b,1–3]	19.39 [1,2]	19.59 [1,2]	26.85 [c,1–3]	28.58 [a,2,3]	29.09 [a,2,3]	16.96 [b,1]	27.36 [c,1–3]	31.85 [b,3,4]	23.66 [b,1–3]	0.011	1.29
	48	0.00 [a]	29.71 [b]	48.95	41.03 [d]	24.48 [b]	38.01	45.20	29.04 [c]	43.73 [a,b]	31.36 [a]	40.16 [c]	34.75 [d]	53.04 [c]	35.76 [b]	0.072	2.92
	210	18.85 [b]	17.29 [b]	19.50	18.46 [a]	20.21 [b]	19.06	22.21	13.16 [b]	17.23 [a]	19.61 [a]	19.28 [b]	20.50 [b]	20.56 [b]	22.05 [b]	0.725	0.67
	p-value	0.000	0.014	0.105	0.000	0.018	0.066	0.578	0.007	0.048	0.027	0.005	0.000	0.001	0.009		
	SEM	5.56	4.46	5.05	3.22	4.10	5.66	6.73	4.56	6.61	3.94	5.53	4.92	7.34	5.06		

[a–c] Mean values in the same column (same antioxidant in different weeks) with different letter presented significant differences; [1–5] Mean values in the same row (different antioxidant in the same week) with different number presented significant differences; SEM: standard error of mean; Batches: CON: control; BHT: tert-butyl-4-hydroxytoluene; TEA: tea; CHE: chestnut; GRA: grape seed and BER: beer extracts.

The increases were lower in PUFA and SFA, with values between 3%–145% and 8%–111%, respectively. Regarding PUFA, linoleic and arachidonic were the fatty acids that showed higher percentages of release. The samples treated with CHE_{200} showed a decrease (26%) during ripening that could be associated with the oxidation of PUFA and the decrease in the proportion of the long chain PUFAs such as arachidonic acid (42%) [33]. Within SFA, palmitic and stearic acid were the most abundant, with levels between 17.4%–22.8% and 6.2%–8.1%, respectively. Unlike what happened with oleic acid, stearic decreased toward the end of ripening in CON and in the samples treated with low dose of natural extracts (mean decreases of 52% *vs.* 12%), for increased again until the end of vacuum packaged.

The amount of PUFA can be used as a measurement of the oxidative deterioration of meats, due to containing double bonds in the hydrocarbon chain being preferred substrates in oxidative reactions. Regarding oxidative stability, a significant correlation was found between PUFA and TBARS ($r = -0.22$, $p < 0.05$). As can be seen in Figure 3, the oxidative degradation of PUFA mainly occurred after day 19. At the end of ripening, we observed that the addition of GRA_{1000} and BER_{200} extracts protect chorizos from oxidative degradation since, higher amount of PUFA were observed in these treated samples than in CON (21.5% and 22.4% *vs.* 21.1%, respectively).

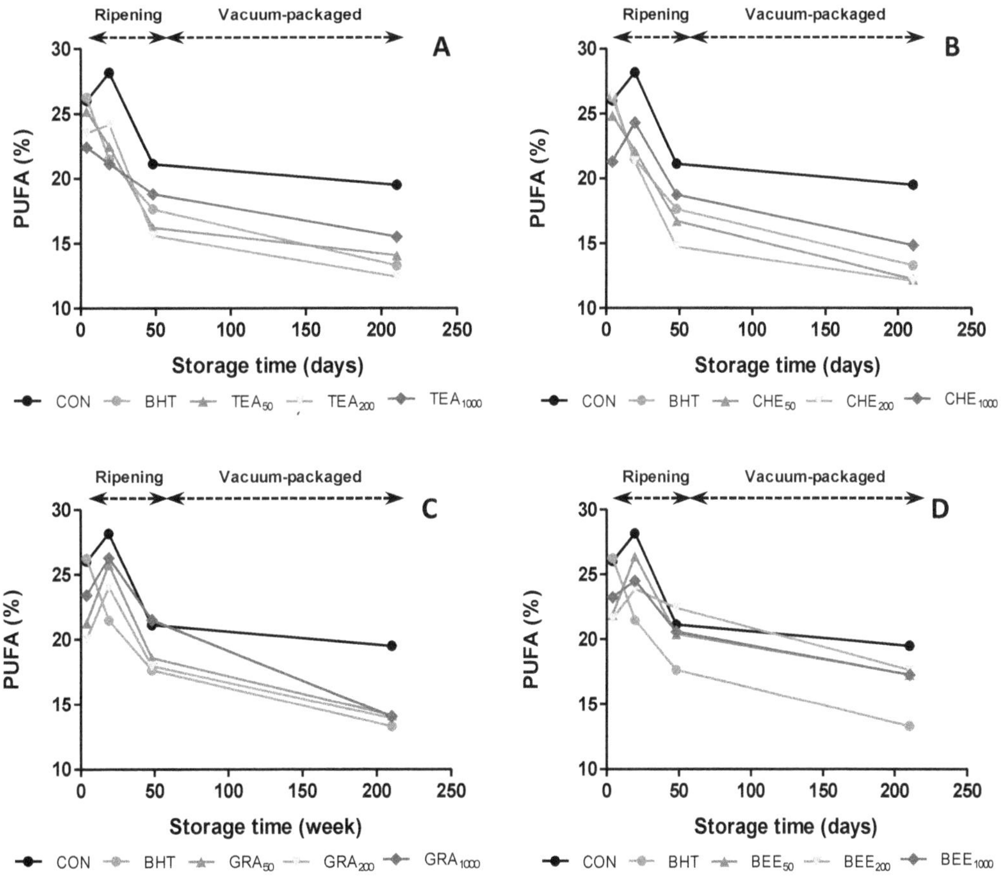

Figure 3. Evolution of PUFA content in dry-cures sausages treated with BHT and natural antioxidants during ripening and vacuum-packaged storage.

To assess the nutritional properties of IMF, the ratios PUFA/SFA and *n*-6/*n*-3 were determined (Table 5). The PUFA/SFA ratio showed mean values of 0.72, being CON and the samples treated with

BER extract which showed the highest values. Significant differences ($p < 0.05$) were found among samples at the end of ripening and within storage time in samples treated with BHT and with higher doses of natural antioxidants (TEA_{200}, CHE_{200}, CHE_{1000} and GRA_{1000}). In general, at the end of ripening the obtained values were within the typical values (0.5–0.7) of the Mediterranean diet [34] and lower than the FAO recommendations [35] for human diet (0.85), while at 210 days the values were lower (mean values lower than 0.46).

4. Conclusions

The addition of natural antioxidants changed the physicochemical properties of dry-cured sausages. The presence of antioxidants and the use of low concentrations improved color maintenance during ripening and under vacuum conditions. The results obtained for TBARS values showed that natural antioxidants matched or even improved the results obtained for BHT, with higher effectiveness for grape and chestnut extracts. The values of hardness decreased significant with the addition of antioxidants, obtaining the lowest values with the intermediate dose. Microbial counts were affected by the addition of antioxidants since lower counts were observed in sausages prepared with natural extracts. Free fatty acid content during ripening and vacuum packaging showed a gradual and significant release. The addition of grape and beer extracts protected sausages from oxidative degradation. Further analysis of how to affect the addition of natural extracts on sensory properties and volatile compounds of chorizo will be addressed.

Acknowledgment

Authors are grateful to Galician government (Xunta de Galicia 09TAL006CT) for the financial support.

Author Contributions

Mirian Pateiro and Roberto Bermúdez have performed the measurements and analyzed the data. Jose Manuel Lorenzo and Daniel Franco have designed and supervised the research. All authors have contributed to the writing of the manuscript and have approved the final paper.

Conflicts of Interest

The authors declare no conflict of interest.

References

1. O'Grady, M.N.; Carpenter, R.; Lynch, P.B.; O'Brien, N.M.; Kerry, J.P. Addition of grape seed extract and bearberry to porcine diets: Influence on quality attributes of raw and cooked pork. *Meat Sci.* **2008**, *78*, 438–446.
2. Jayaprakasha, G.K.; Selvi, T.; Sakariah, K.K. Antibacterial and antioxidant activities of grape (*Vitis vinifera*) seed extracts. *Food Res. Int.* **2003**, *36*, 117–122.
3. Lorenzo, J.M.; Sineiro, J.; Amado, I.R.; Franco, D. Influence of natural extracts on the shelf life of modified atmosphere-packaged pork patties. *Meat Sci.* **2014**, *96*, 526–534.

4. Pateiro, M.; Lorenzo, J.M.; Amado, I.R.; Franco, D. Effect of addition of green tea, chestnut and grape extract on the shelf-life of pig liver pâté. *Food Chem.* **2014**, *147*, 386–394.
5. Lorenzo, J.M.; González-Rodríguez, R.M.; Sánchez, M.; Amado, I.R.; Franco, D. Effects of natural (grape seed and chestnut extract) and synthetic antioxidants (buthylatedhydroxytoluene, BHT) on the physical, chemical, microbiological and sensory characteristics of dry cured sausage "chorizo". *Food Res. Int.* **2013**, *54*, 611–620.
6. Gómez, M.; Lorenzo, J.M. Effect of fat level on physicochemical, volatile compounds and sensory characteristics of dry-ripened "chorizo" from Celta pig breed. *Meat Sci.* **2013**, *95*, 658–666.
7. Moure, A.; Cruz, J.M.; Franco, D.; Domínguez, J.M.; Sineiro, J.; Domínguez, H.; Núñez, M.J.; Parajó, J.C. Natural antioxidants from residual sources. *Food Chem.* **2001**, *72*, 145–171.
8. Summo, C.; Caponio, F.; Paradiso, V.M.; Pasqualone, A.; Gomes, T. Vacuum-packed ripened sausages: Evolution of oxidative and hydrolytic degradation of lipid fraction during long-term storage and influence on the sensory properties. *Meat Sci.* **2010**, *84*, 147–151.
9. Nassu, R.T.; Gonçalves, L.A.G.; da Silva, M.A.A.P.; Beserra, F.J. Oxidative stability of fermented goat meat sausage with different levels of natural antioxidant. *Meat Sci.* **2003**, *63*, 43–49.
10. Tang, S.Z.; Ou, S.Y.; Huang, X.S.; Li, W.; Kerry, J.P.; Buckley, D.J. Effects of added tea catechins on colour stability and lipid oxidation in minced beef patties held under aerobic and modified atmospheric packaging conditions. *J. Food Eng.* **2006**, *77*, 248–253.
11. Marco, G.J. A rapid method for evaluation of antioxidants. *J. Am. Oil Chem. Soc.* **1968**, *45*, 594–598.
12. ISO (International Organization for Standardization). Determination of moisture content, ISO 1442:1997 standard. In *International Standards of Meat and Meat Products*; ISO: Genève, Switzerland, 1997.
13. Targladis, B.G.; Watts, B.M.; Younathan, M.T.; Duggan, L.R. A distillation method for the quantitative determination of malonaldehyde in rancid foods. *J. Am. Oil Chem. Soc.* **1960**, *37*, 44–48.
14. Folch, J.; Lees, M.; Sloane-Stanley, G.H. A simple method for the isolation and purification of total lipids from animal tissues. *J. Biol. Chem.* **1957**, *226*, 497–509.
15. García-Regueiro, J.A.; Gilbert, J.; Díaz, I. Determination of neutral lipids from subcutaneous fat cured ham by capillary gas chromatography and liquid chromatography. *J. Chromatogr. A* **1994**, *667*, 225–233.
16. Carreau, J.P.; Dubacq, J.P. Adaptation of a macro-scale method to the micro-scale for fatty acid methyl transesterification of biological lipid extracts. *J. Chromatogr. A* **1978**, *151*, 384–390.
17. Lorenzo, J.M.; Franco, D. Fat effect on physico-chemical, microbial and textural changes through the manufactured of dry-cured foal sausage. Lipolysis, proteolysis and sensory properties. *Meat Sci.* **2012**, *92*, 704–714.
18. Rubilar, M.; Pinelo, M.; Shene, C.; Sineiro, J.; Nunez, M.J. Separation and HPLC-MS identification of phenolic antioxidants from agricultural residues: Almond hulls and grape pomace. *J. Agric. Food Chem.* **2007**, *55*, 10101–10109.

19. Van der Hooft, J.J.; Akermi, M.; Unlu, F.M.; Mihaleva, V.; Gomez-Roldan, V.; Bino, R.J.; de Vos, R.C.H.; Vervoort, J. Structural annotation and elucidation of conjugated phenolic compounds in black, green, and white tea extracts. *J. Agric. Food Chem.* **2012**, *60*, 8841–8850.
20. Fontecave, M.; Lepoivre, M.; Elleingand, E.; Gerez, C.; Guittet, O. Resveratrol, a remarkable inhibitor of ribonucleotide reductase. *FEBS Lett.* **1998**, *421*, 277–279.
21. Lorenzo, J.M.; Gómez, M.; Fonseca, S. Effect of commercial starter cultures on physicochemical characteristics, microbial counts and free fatty acid composition of dry-cured foal sausage. *Food Control* **2014**, *46*, 382–389.
22. Lorenzo, J.M.; Michinel, M.; López, M.; Carballo, J. Biochemical characteristics of two Spanish traditional dry-cured sausage varieties: Androlla and Botillo. *J. Food Compos. Anal.* **2000**, *13*, 809–817.
23. Lorenzo, J.M.; Temperán, S.; Bermúdez, R.; Cobas, N.; Purriños, L. Changes in physico-chemical, microbiological, textural and sensory attributes during ripening of dry-cured foal salchichón. *Meat Sci.* **2012**, *90*, 194–198.
24. Liaros, N.G.; Katsanidis, E.; Bloukas, J.G. Effect of the ripening time under vacuum and packaging film permeability on processing and quality characteristics of low-fat fermented sausages. *Meat Sci.* **2009**, *83*, 589–598.
25. Jayawardana, B.C.; Hirano, T.; Han, K.; Ishii, H.; Okada, T.; Shibayama, S.; Fukushima, M.; Sekikawa, M.; Shimada, K. Utilization of adzuki bean extract as a natural antioxidante in cured and uncured cooked pork sausages. *Meat Sci.* **2011**, *89*, 150–153.
26. González-Fernández, C.; Santos, E.M.; Rovira, J.; Jaime, I. The effect of sugar concentration and starter culture on instrumental and sensory textural properties of "chorizo", Spanish dry-cured sausage. *Meat Sci.* **2006**, *74*, 467–475.
27. Jongberg, S.; Gislason, N.E.; Lund, M.N.; Skibsted, L.H.; Waterhouse, A.L. Thiol-quinone adduct formation in myofibrillar proteins detected by LC-MS. *J. Agric. Food Chem.* **2011**, *59*, 6900–6905.
28. Jongberg, S.; Tørngren, M.A.; Gunvig, A.; Skibsted, L.H.; Lund, M.N. Effect of green tea or rosemary extract on protein oxidation in Bologna type sausages prepared from oxidatively stressed pork. *Meat Sci.* **2013**, *93*, 538–546.
29. Bozkurt, H.; Bayram, M. Colour and textural attributes of sucuk during ripening. *Meat Sci.* **2006**, *73*, 344–350.
30. Coşkuner, O.; Ertaş, A.H.; Soyer, A. The effect of processing method and storage time on constituents of Turkish sausages (sucuk). *J. Food Process. Preserv.* **2010**, *34*, 125–135.
31. Greene, B.E.; Cumuze, T.H. Relationship between TBA numbers and inexperienced panelist's assessments of oxidized-avor in cooked beef. *J. Food Sci.* **1982**, *47*, 52–58.
32. Hierro, E.; de la Hoz, L.; Ordóñez, J.A. Contribution of microbial and meat endogenous enzymes to the lipolysis of dry fermented sausages. *J. Agric. Food Chem.* **1997**, *45*, 2989–2995.
33. Qiu, C.; Zhao, M.; Sun, W.; Zhou, F.; Cui, C. Changes in lipid composition, fatty acid profile and lipid oxidative stability during Cantonese sausage processing. *Meat Sci.* **2013**, *93*, 525–532.
34. Ulbricht, T.L.V.; Southgate, D.A.T. Coronary heart disease: Seven dietary factors. *Lancet* **1991**, *338*, 985–992.

35. FAO (Food and Agriculture Organization of the United Nations). Fat and fatty acid requirements for adults. In *Fats and Fatty Acids in Human Nutrition*; FAO: Rome, Italy, 2010; pp. 55–62.

Oxidation Stability of Pig Liver Pâté with Increasing Levels of Natural Antioxidants (Grape and Tea)

Mirian Pateiro [1], José M. Lorenzo [1], José A. Vázquez [2] and Daniel Franco [1,*]

[1] Centro Tecnológico de la Carne de Galicia, Parque Tecnológico de Galicia, San Cibrao das Viñas, 32900 Ourense, Spain; E-Mails: mirianpateiro@ceteca.net (M.P.); jmlorenzo@ceteca.net (J.M.L.)

[2] Grupo de Reciclado y Valorización de Residuos (REVAL), Instituto de Investigaciones Marinas (IIM-CSIC), Eduardo Cabello, 6, 36208 Vigo, Spain; E-Mail: jvazquez@iim.csic.es

* Author to whom correspondence should be addressed; E-Mail: danielfranco@ceteca.net

Academic Editors: Maria G. Miguel and João Rocha

Abstract: The present study investigated the effect of the addition of increasing levels of the natural antioxidants tea (TEA) and grape seed extracts (GRA) on the physiochemical and oxidative stability of refrigerated stored pig pâtés. In addition, a synthetic antioxidant and a control batch were used, thus a total of eight batches of liver pâté were prepared: CON, BHT, TEA (TEA_{50}, TEA_{200} and TEA_{1000}) and GRA (GRA_{50}, GRA_{200} and GRA_{1000}). Pâté samples were analyzed following 0, 4, 8 and 24 weeks of storage. Color parameters were affected by storage period and level of antioxidant extract. Samples with TEA_{200} and GRA_{1000} levels of extracts showed lower total color difference between 0 and 24 weeks. At the end of storage period, the lower TBARs values were obtained in samples with the highest concentration on natural extract. Overall, the evolution of volatile compounds showed an increase in those ones that arise from the lipid oxidation and samples with TEA_{1000} extract showed the lowest values.

Keywords: antioxidant level; storage time; oxidative stability; volatile compounds

---

1. Introduction

The prevention of lipid oxidation along the lines of processing and storage of meat products is important with regards to its quality and healthiness [1]. Nitrites have been widely used in cooked meat products alone or joined to natural antioxidants [2]. However, the potential health risks related to residual nitrite levels and the formation of harmful *N*-nitrosamines in meat products has led the meat industry to reduce the use of sodium nitrite, significantly. On the other hand, due to the fact that synthetic antioxidant may constitute a potential health hazard for consumers, interest in natural antioxidants and search of naturally occurring compounds with antioxidant activity has increased dramatically [3,4].

In the last years, many researchers have evaluated the antioxidant capacity of extracts from different plants derived foods; specifically studies with grape and tea extracts have been developed in meat patties [1,5], in cooked products [6] or even in films with positive results [7]. Concerning liver pâtés and to our knowledge, in only one study has been employed this type of extracts with levels of 1000 ppm to prove the extract efficacy [8]. This level is necessary due to chemical composition of pâtés (high amounts of fat and non-heme iron, and low content of natural antioxidants) and their manufacturing process (high temperature) this product is highly susceptible to lipid oxidation [2,9]. However, the disadvantages of the use of natural antioxidants are that they are usually more expensive than synthetic antioxidants and may impart color or taste to the product [3] so studies that investigate the dose effect in "real" (*i.e.*, food) matrix are necessary to obtain a better extract characterization. In addition, many of them are by-products of agro-industries (e.g., grape extract) and their use could represent a significant step towards an effective and economically valuable valorization of such waste [10].

The aim of this work was therefore to evaluate the effect of the addition of increasing amounts of natural antioxidants (tea (TEA) and grape seed (GRA) extracts) on the physiochemical and oxidative stability of refrigerated stored pig pâtés, and to compare the effects observed with those of a synthetic antioxidant (butylated hydroxytoluene BHT) and a control (CON).

2. Experimental Section

2.1. Grape Seed Extract (GRA) and Green Tea Extract (TEA)

The procedures for the preparation of the TEA and GRA extracts were carried out as previously described [1,10].

2.2. Determination of Antioxidant Activity

The antioxidant activity of these extracts was previously described by Pateiro *et al.* [8].

2.3. Manufacture of Liver Pâté

For this study, eight batches of liver pâté were prepared: CON (control), BHT (BHT, 200 mg/kg), TEA (tea extract; 50, 200 and 1000 mg/kg; hereafter TEA_{50}, TEA_{200} and TEA_{1000}, respectively) and GRA (grape seed extract; 50, 200 and 1000 mg/kg of liver pate; hereafter GRA_{50}, GRA_{200} and GRA_{1000}, respectively). The pâtés were prepared in the pilot plant of the Meat Technology Center of Galicia. An identical formula was used for all batches, except for the addition of the different antioxidants. The ingredients (%) were as follows: subcutaneous fat (40%), lean meat (15%), liver (18%), cold water (23%), sodium chloride (2%), and sodium caseinate (2%). Fat, meat and liver were from the Celta pig breed. First, fat and liver were chopped in to small cubes and scalded at 65 °C for 10 min. The cooked fat and liver, after being allowed to cool at room temperature, were mixed with the remaining ingredients in a Talsa bowl chopper (Talsabell, S.A., Valencia, Spain). After that, the total mass was divided in five batches of 3 kg each. Antioxidants (BHT, tea, and grape seed extract) were added in the corresponding batch (BHT, TEA and GRA, respectively) and the mass was mixed with a beater. Finally, the mixture was packed in glass containers (150 g) and cooked by immersion in a hot water bath at 80 °C for 30 min. After the meat samples were allowed to cool at room temperature, they were stored in the dark at 4 °C for 24 weeks. Batches were made in triplicate. Two units of pâté from each batch were taken at 0, 4, 8 and 24 weeks to determine the following parameters: pH, color, thiobarbituric acid reactive substances (TBARs), and fatty acid composition. The volatile compounds profile of the manufactured liver pâté was determined at the beginning and at the end of the storage period.

2.4. Analytical Methods

Analytical procedures (color parameters, TBARS index, fatty acid and volatile profile) were carried out as previously described by Pateiro *et al.* [8].

2.4.1. Physical Analysis

The pH of samples was measured using a pH-meter (HI 99163, Hanna Instruments, Eibar, Spain) equipped with a glass probe for penetration.

A portable colorimeter (Konica Minolta CM-600d, Osaka, Japan) with the next settings machine (pulsed xenon arc lamp, 0° viewing angle geometry, standard illuminant D65 and aperture size of 8 mm) was used to measure the pâté color in the CIELAB space. Results were expressed as lightness (L*), redness (a*) and yellowness (b*).

The total color difference (ΔE) between pâtés at day 0 and week 24 of storage was calculated by the next formula [11].

$$\Delta E_{0-24} = \left[(L_{24} - L_0)^2 + (a_{24} - a_0)^2 + (b_{24} - b_0)^2\right]^{1/2}$$

The relative content of metmyoglobin (METOX) on the surface of the samples is based on measurements of reflex attenuance of incident light at the isobestic points 572, 525, 473 and 730 nm [12].

2.4.2. Lipid Oxidation

Lipid stability was evaluated using the method proposed by Vyncke [13]. Thiobarbituric acid reactive substances (TBARs) were calculated from a standard curve of malonaldehyde (MDA) produced from with 1,1-3,3 tetraethoxypropane (TEP) and expressed as mg MDA/kg sample.

2.4.3. Analysis of Fatty Acid Methyl Esters

The fat was extracted using chloroform/metanol (2/1; v/v) and stored at −80 °C until analysis. Lipids were trans-esterified with a solution of boron trifluoride (14%) in methanol. For total fatty acids analysis, 50 mg of the extracted lipids were esterified to form fatty acid methyl esters (FAMEs) which were stored at −80 °C until chromatographic analysis. Separation and quantification of the FAMEs were carried out using a gas chromatograph (GC-Agilent 6890N; Agilent Technologies Spain, S.L., Madrid, Spain) equipped with a flame ionization detector and an automatic sample injector HP 7683, and using a Supelco SPTM-2560 fused silica capillary column (100 m, 0.25 mm i.d., 0.2 μm film thickness, Sigma-Aldrich, Spain). The chromatographic conditions were as follows: the initial column temperature (120 °C) was maintained for 5 min, then programmed to increase at a rate of 5 °C/min up to 200 °C maintaining this temperature for 2 min, then at 1 °C/min up to 230 °C maintaining this temperature for 3 min, and then increasing again at 2 °C/min up to a final temperature of 235 °C which is then held for 10 min. The injector and detector were maintained at 260 and 280 °C, respectively. Helium was used as carrier gas at a constant flow-rate of 1.1 mL/min, with the column head pressure set at 37.73 psi. The split ratio was 1:50 and 1 μL of solution was injected. Nonanoic acid methyl ester (C9:0 ME) at 0.3 mg/mL was used as internal standard and added to the samples prior to injection. Individual FAMEs were identified by comparing their retention times with those of authenticated standards (Supelco 37 component FAME Mix).

Data regarding FAME composition were expressed as percentage of total area of FAMEs. The proportion of polyunsaturated (PUFA), monounsaturated (MUFA), and saturated (SFA) fatty acid contents and the ratios PUFA/SFA, *n*-6/*n*-3 and nutritional value (NV) = ∑(C14:0 + C16:0)/∑(C18:1 + C18:2*n*6c) were calculated.

2.4.4. Analysis of Volatile Compounds

The volatile compounds profile was studied at the beginning and the end of the refrigerated storage. The extraction of the volatile compounds was performed using solid-phase microextraction (SPME). For headspace SPME (HS-SPME) extraction, 2 g of each ground sample were weighed in a 40 mL vial and screw-capped with a laminated Teflon-rubber disk. The fiber, previously conditioned by heating in a gas chromatograph injection port at 270 °C for 60 min, was inserted into the sample vial through the septum and then exposed to headspace. The extractions were carried out in an oven at 35 °C for 30 min, after equilibration of the samples for 15 min at the extraction temperature, ensuring a homogeneous temperature for sample and headspace. Once sampling was finished, the fiber was withdrawn into the needle and transferred to the injection port of the gas chromatograph−mass spectrometer (GC−MS) system.

A gas chromatograph 6890N (Agilent Technologies, Santa Clara, CA, USA) equipped with a mass selective detector 5973N (Agilent Technologies) was used with a DB-624 capillary column of 30 m × 0.25 mm i.d., 1.4 μm film thickness (J & W Scientific, Folsom, CA, USA). The SPME fiber was desorbed and maintained in the injection port at 260 °C during 5 min. The sample was injected in splitless mode. Helium was used as a carrier gas with a linear velocity of 40 cm/s. The temperature program was isothermal for 10 min at 40 °C, raised to 200 °C at a rate of 5 °C/min, and then raised to

250 °C at a rate of 20 °C/min and held for 5 min. Injector and detector temperatures were both set at 260 °C. The mass spectra were obtained using a mass selective detector at 70 eV, with a multiplier voltage of 1953 V and collected data at a rate of 6.34 scans/s over a mass range of m/z 40–300. Compounds were identified comparing their mass spectra with those contained in the NIST05 (National Institute of Standards and Technology, Gaithersburg, MD, USA) library and/or by calculation of their retention index relative to a series of standard alkanes (C5–C14) for calculating the Kovats indexes and matching them with data reported in literature. Results for each volatile compound were the mean value of three replicates and they were expressed as AU (area units) $\times 10^6$.

2.5. Statistical Analysis

For the statistical analysis of the results, one way analysis of variance (ANOVA) using SPSS package (SPSS 19.0, Chicago, IL, USA) was performed for all variables considered in the study. Least-squares means were compared among treatments using the Duncan's post hoc test for a significance level $p < 0.05$. Correlations between variables were determined using correlation analyses using Pearson's correlation coefficient (r) with the above mentioned statistical software package.

3. Results and Discussion

3.1. Antioxidant Activity of the Extracts

GRA and TEA extracts showed a high polyphenol content (373.0 *vs.* 390.0 mg GAE/g extract, respectively), mainly flavonoids and flavan-3-ols, which antioxidant activity has been demonstrated [14,15]. The major compounds found in TEA extracts were catechin, epicatechin, cinnamic acids and sugar-linked flavonols [16], while GRA extracts contained benzoic acids, monomer flavan-3-ols and oligomeric procyanidins [15]. The higher activity found in GRA extracts could be associated to its resveratrol content [17].

TEAC, DPPH and β-carotene were used to assess *in vitro* antioxidant activity of the natural extracts. These methods were directly related to polyphenol contents [8]. In the case of TEAC, the aforementioned extracts displayed values of 2.93 and 4.06 g Trolox/g extract for GRA and TEAextracts, respectively.

The scavenging activity found on DPPH radical showed the higher antioxidant power of BHT standard, followed by GRA and TEA extracts (1.80 and 2.18 g equivalent BHT/g extract, respectively). The EC_{50} values obtained showed values of 0.12 and 0.16 g extract/L for TEA and GRA extracts, respectively.

β-carotene bleaching assay of the natural extracts showed that GRA was the most active (1.28 g equivalent BHT/g extract), while TEA displayed values of 0.69 g equivalent BHT/g extract. The EC_{50} values obtained displayed rather similar activity values for both extracts (less than 0.10 g extract/L).

3.2. Color Properties of Pig Liver Pâtés during Storage Time

Except for week 0, no significant differences ($p > 0.05$) were detected for pH values among batches during refrigerated storage (Table 1). Samples with antioxidants displayed pH values lower than the

CON batch. This fact could be due to the acidity of the antioxidant extracts (pH of 4.29 and 5.74 for GRA and TEA extracts, respectively), conferring a lower pH on the samples containing antioxidants in their formulation. Regarding the dose used of each antioxidant, a reduction in pH was observed with increasing concentration (Table 1). Colour parameters were significantly ($p < 0.05$) influenced by the period of storage. Significant differences were found for L* values in the samples treated with BHT and with low concentrations of natural antioxidants ($p < 0.05$) within each group. However, no significant differences ($p > 0.05$) were observed for L* values among the antioxidant levels.

In our study, the lowest values for L* were found in the samples treated with 1000 mg/kg, obtaining values of 62.34 and 64.40 for samples treated with TEA and GRA extracts, respectively, compared to values of 70.56 and 70.03 observed in samples treated with 50 mg/kg of TEA and GRA extracts, respectively. These findings are in agreement with other authors who observed lower L* values in pâté samples with increasing amounts of natural antioxidant [18]. The effect was compared with that of BHT at the same dose (200 mg/kg), TEA and GRA extracts showed lower L* values than BHT (68.43 and 69.81 *vs.* 71.39 for TEA, GRA and BHT, respectively). Finally, similar values were found between CON and samples treated with the lowest concentration of natural antioxidants (50 mg/kg) (71.39 *vs.* 70.56 and 70.03 for TEA and GRA, respectively).

Regarding redness, its development was different depending on the batch. Many differences were observed among the antioxidant used, which could be due to the colour that the addition of the antioxidant may give to the final product. At the beginning of the period of storage, the values ranged between 1.59 and 4.63, with the highest values obtained for CON and BHT and the lowest obtained for TEA_{50} and GRA_{50} (4.63 and 4.61 *vs.* 3.66 and 4.12 for CON, BHT, TEA_{50} and GRA_{50} batches, respectively). At the end of the period of storage, samples treated with BHT showed the highest values, followed by CON, TEA and GRA at the lowest concentration (4.73 and 4.58 *vs.* 3.63 and 3.72 for BHT, CON, TEA_{50} and GRA_{50} batches, respectively). For the dose effect, as happened with the L* values, the higher the amounts of the natural antioxidants the lower the a* values of the pate samples. The trend observed for samples treated with TEA extract were similar to that observed for samples treated with GRA.

Concerning yellowness (b*), the obtained values decreased during the period of storage, although only significant ($p < 0.05$) differences were found in BHT, TEA_{50} and TEA_{1000} batches. Unlike other studies, the b* values were lower in pâtés to which the natural antioxidants were added than in CON and BHT samples [19]. As with L* and a* values, the greater the amount of natural antioxidant added to the pâté, the lower the b* values of the sample. TEA_{1000} and GRA_{1000} samples had the lowest values at the beginning and at the end of the storage period. At the end of the period of storage, TEA_{50} and GRA_{50} treatments showed similar b* values to those of the CON and BHT batches, with values of 17.77 and 18.11 *vs.* 18.35 and 18.71 for TEA_{50}, GRA_{50}, CON and BHT batches, respectively. Values obtained for the total color difference (ΔE) between pâtés at day 0 and each sampling point showed that samples treated with natural antioxidants had the lowest color differences during storage (Table 1). In fact, the changes observed in the color parameters were less in TEA_{200}, GRA_{200} and GRA_{1000} batches. These color differences can be considered noticeable when ΔE values are higher than 2 [20]. The color changes that occurred during the storage period could be associated with oxidation processes and are contrary to what other authors observed in porcine liver pâtés [19].

Table 1. Effect of antioxidants on physical properties of Celta pig liver pâtés (n = 3) during refrigerated storage.

		CON	BHT	TEA			GRA			*p*-Value	SEM
				50	200	1000	50	200	1000		
pH	0	6.34 ^{a,5}	6.27 ^{a,4}	6.29 ^{4}	6.21 ^{2,3}	6.18 ^{a,b,1,2}	6.23 ^{3}	6.20 ^{a,b,2,3}	6.16 ^{b,1}	0.000	0.02
	4	6.30 ^{b,c,2}	6.16 ^{b,1,2}	6.00 ^{1}	6.14 ^{1,2}	6.14 ^{b,1,2}	6.17 ^{1,2}	6.17 ^{b,1,2}	6.13 ^{c,1,2}	0.207	0.03
	8	6.27 ^{c}	6.18 ^{b}	6.13	6.06	6.07 ^{c}	6.22	6.19 ^{b}	6.12 ^{c}	0.069	0.02
	24	6.34 ^{a,b}	6.25 ^{a}	6.15	6.22	6.21 ^{a}	6.25	6.23 ^{a}	6.20 ^{a}	0.071	0.02
	p-value	0.025	0.009	0.306	0.296	0.003	0.217	0.031	0.001		
	SEM	0.01	0.02	0.05	0.03	0.02	0.01	0.01	0.01		
L*	0	67.64 ^{c,4}	74.50 ^{a,7}	76.06 ^{a,8}	66.88 ^{3}	58.01 ^{b,1}	74.38 ^{a,6}	69.71 ^{5}	65.21 ^{2}	0.000	1.45
	4	69.34 ^{b,c,4}	68.16 ^{b,3,4}	67.49 ^{b,3,4}	65.81 ^{2,3}	61.62 ^{a,1}	70.10 ^{b,4}	67.14 $^{2–4}$	63.87 ^{1,2}	0.004	0.73
	8	70.87 ^{a,b,2}	71.67 ^{a,2}	71.77 ^{b,2}	68.81 ^{2}	61.55 ^{a,1}	72.17 ^{a,b,2}	70.29 ^{2}	64.51 ^{1}	0.001	0.98
	24	71.39 ^{a,2}	71.39 ^{a,2}	70.56 ^{b,2}	68.43 ^{2}	62.34 ^{a,1}	70.03 ^{b,2}	69.81 ^{2}	64.40 ^{1}	0.000	0.85
	p-value	0.014	0.024	0.022	0.334	0.010	0.065	0.306	0.397		
	SEM	0.58	0.90	1.23	0.62	0.66	0.75	0.62	0.26		
a*	0	4.63 ^{a,7}	4.61 ^{a,b,7}	3.66 ^{a,5}	2.43 ^{3}	1.85 ^{a,2}	4.12 ^{6}	3.17 ^{4}	1.59 ^{1}	0.000	0.29
	4	4.43 ^{a,4}	4.36 ^{a,b,4}	3.37 ^{a,b,3}	2.47 ^{2}	1.47 ^{b,1}	3.38 ^{3}	3.25 ^{3}	1.21 ^{1}	0.000	0.29
	8	3.96 ^{b,4,5}	4.09 ^{b,5}	3.24 ^{b,3}	1.93 ^{2}	1.12 ^{c,1}	3.67 ^{4}	3.00 ^{3}	1.04 ^{1}	0.000	0.30
	24	4.58 ^{a,4,5}	4.73 ^{a,5}	3.63 ^{a,b,3,4}	2.46 ^{2}	1.13 ^{c,1}	3.72 ^{3,4}	3.38 ^{2,3}	1.20 ^{1}	0.000	0.34
	p-value	0.024	0.099	0.116	0.243	0.011	0.359	0.686	0.307		
	SEM	0.11	0.11	0.08	0.11	0.12	0.14	0.10	0.10		
b*	0	19.45 ^{a,7}	20.18 ^{a,8}	18.79 ^{a,6}	15.58 ^{3}	11.96 ^{a,1}	18.53 ^{5}	17.06 ^{4}	13.51 ^{2}	0.000	0.71
	4	18.78 ^{a,b,5}	19.22 ^{b,5}	17.44 ^{b,4}	15.86 ^{3}	11.35 ^{a,b,1}	17.57 ^{4}	17.49 ^{4}	12.99 ^{2}	0.000	0.68
	8	18.97 ^{a,b,5}	19.48 ^{a,b,5}	17.73 ^{b,4}	15.25 ^{3}	10.26 ^{c,1}	18.51 ^{4,5}	17.79 ^{4}	12.93 ^{2}	0.000	0.80
	24	18.35 ^{b,5}	18.71 ^{b,5}	17.77 ^{b,4,5}	15.63 ^{3}	10.45 ^{b,c,1}	18.11 ^{4,5}	17.21 ^{4}	12.86 ^{2}	0.000	0.73
	p-value	0.110	0.033	0.023	0.464	0.027	0.363	0.263	0.190		
	SEM	0.17	0.22	0.21	0.13	0.28	0.21	0.14	0.12		
Δ*E*	0-4	1.89 ^{1,2}	6.42 ^{3,4}	8.70 ^{4}	1.14 ^{1}	3.68 $^{1–3}$	4.52 ^{2,3}	2.62 ^{1,2}	1.50 ^{1,2}	0.005	0.68
	0-8	3.36	2.97	4.77	2.34	4.01	2.37	1.21	1.07	0.252	0.39
	0-24	3.91 ^{1,3}	3.47 ^{2,3}	5.59 ^{4}	1.87 ^{1}	4.65 ^{3,4}	4.42 ^{3,4}	2.06 ^{1,2}	1.24 ^{1}	0.001	0.40
	p-value	0.122	0.148	0.173	0.759	0.559	0.313	0.136	0.860		
	SEM	0.44	0.80	0.96	0.54	0.32	0.60	0.30	0.26		

$^{a–c}$ Mean values in the same column (same antioxidant in different weeks) with different letter indicating significant differences; $^{1–8}$ Mean values in the same row (different antioxidant in the same week) with different number indicating significant differences; SEM: standard error of mean; Batches: CON: control; BHT: tert-butyl-4-hydroxytoluene; TEA: tea and GRA: grape seed extracts.

On the contrary, TEA_{50} and GRA_{50} batches displayed higher ΔE_{0-24} than the CON batch. In this case, color changes could be associated with some compositional or physical changes that happened during period of storage which are not directly related to oxidative processes. In this regard, Estévez *et al.* [19] also observed this behavior in porcine liver pâtés prepared with rosemary and sage extracts. However, the color change (ΔE_{0-24}) of pâtés treated with TEA_{1000} and GRA_{1000} extracts were lower than those observed by the aforementioned authors in samples prepared with the same concentration (0.1%) of rosemary and sage extracts [19]. Finally, the values found for ΔE_{0-24} in BHT batches were lower than those found in other liver pâtés (3.47 *vs.* 5.45) [19], while the values observed for CON batches were higher (3.91 *vs.* 3.38). Among samples, no significant differences ($p > 0.05$) were found during the storage period; the lowest ΔE_{0-24} values were for samples with GRA extract followed by TEA and BHT batches.

Within the dose effect, as happened with L* and a* values in GRA batches, the greater the amount of natural antioxidant added to the pâté, the lower was ΔE value of the sample. This behavior was also observed in frankfurters treated with increasing levels of rosemary essential oil [21]. On the contrary, the results obtained for TEA batches showed an increase in ΔE_{0-24} for the highest concentration.

3.3. Oxidative Stability of Pig Liver Pâtés during Storage Time

The oxidative stability of pig liver pâtés was measured based on the TBARs index (Figure 1), which is frequently used as a marker of lipid oxidation [22]. The results obtained show that TBARs were unstable during processing, although the overall trend was an increase during the storage period. This rise was expected to occur during refrigerated storage of pâtés as a result of the onset of oxidative reactions following cooking [23]. The conditions used for the thermal treatment (80 °C for 30 min) (see the experimental section) are commonly used in the manufacture of liver pâtés [8,9,19,24].

Different values were observed among treatments during the period of storage. At day 0, TBARs values of all samples ranged from 0.4 to 3.3 mg malondialdehyde (MDA)/kg. At the end of the storage period, the lower TBARs values were obtained in samples with TEA: 2.00, 1.03 and 0.81 mg MDA/kg *vs.* 2.92 mg MDA/kg for TEA_{50}, TEA_{200}, TEA_{1000} and CON batches, respectively. This behavior was also observed by other authors in goat meat sausages treated with 0.025% and 0.05% of rosemary extract [25]. These results are in agreement with previously published studies reporting the greater effectiveness of the natural antioxidants compared to the synthetic antioxidant supports the possibility of using these extracts in place of commercially used synthetic antioxidants such as BHT [9,26].

On the other hand, results revealed that TBARs values were affected by the period of storage within each batch ($p < 0.05$), increasing at the beginning of storage period and decreasing after reaching halfway through the storage period (at week eight). This trend could be due to oxidative reactions beginning during processing and are directly correlated with the protective effects of the antioxidant and indicate that lipid oxidation begins during the processing of ingredients (scalding of fat and liver) and before addition of antioxidants. Similar TBARs graphs have been obtained for other meat products that have undergone thermal treatments, because the exposure to high temperatures (>70 °C) is a strong promoter of lipid oxidation [27]. Some authors suggested that the decrease in the TBARs index occurs when the reaction rate of the carbonyls in proteins becomes higher than the rate of TBARs formation [28].

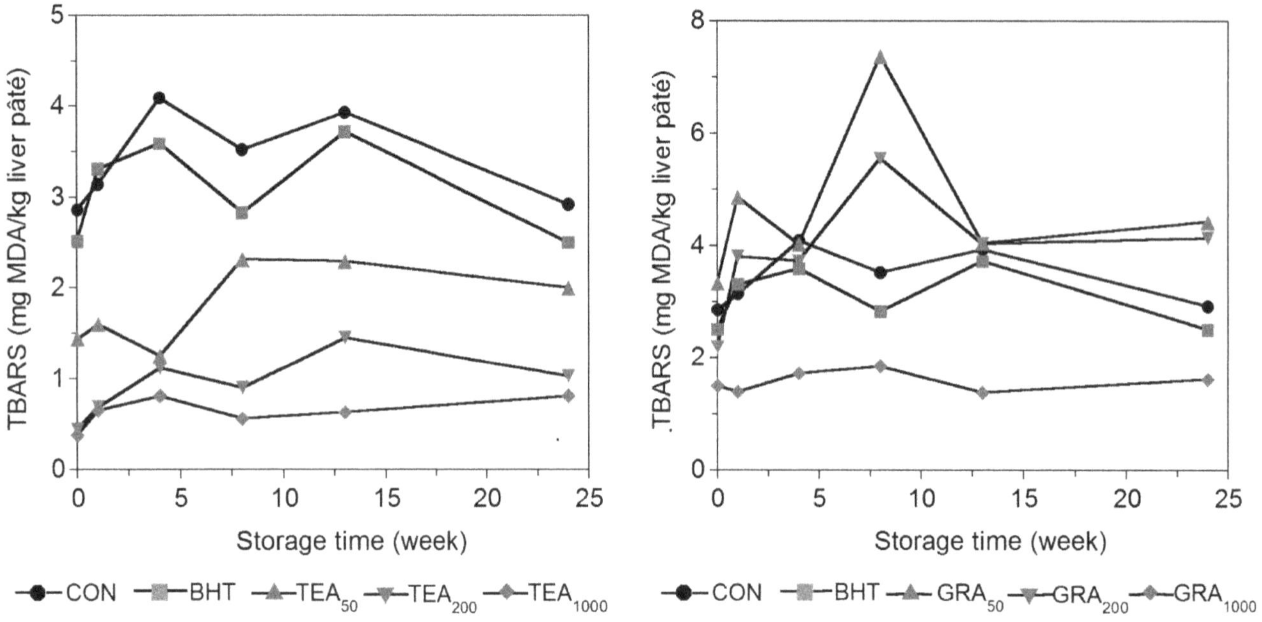

Figure 1. Evolution of TBARs during refrigerated storage of porcine liver pâté with added BHT and natural antioxidants.

3.4. Relationship between TBARs and METOX Production and the Concentration of the TEA and GRA Extracts

METOX values are related to oxidative stability, which is a marker of food deterioration [29]. METOX is also related to redness values [30]. In general, the higher the amount of natural antioxidant added to the pâté, the lower was TBARs and METOX values of the sample (Figure 2). The results obtained of TBARs evolution showed the ability of these natural antioxidants to reduce the oxidative deterioration of lipids. In some cases, TEA extracts improved the results obtained by BHT (even at a lower concentration). Although, with a low number of freedom degrees, the correlation of TBARs and METOX trends *vs.* extracts concentration was mainly linear (Figure 2). The degree of determination was high in almost all responses ($R^2 > 0.94$) less for the case of TBARs production influenced by TEA extracts. Several works showed similar capacity of natural extracts to prevent lipid oxidation on meat products [27,31]. As in the present study, in many reports the improving effects of extracts were concentration-dependent [6,32], descending the production of TBARs at increasing levels of antioxidants. Nevertheless, our results were not in agreement with those found by other authors who observed an increment of TBARs and METOX in minced beef patties with higher concentration of tea catechins [5]. The storage time also revealed to be a factor that enhanced the value of the slope from first order equations (Figure 2) and therefore the suitability to reduce the oxidation process in meat products. However, no significant ($p > 0.05$) effect of the storage time on the slope of the TBARs *vs.* grape extracts equation was observed. Finally, the non-linear relationship between data of TBARs (T) and METOX (M) were described by means of a logarithmic equation: $M = 55.71 + 2.19 \ln T$.

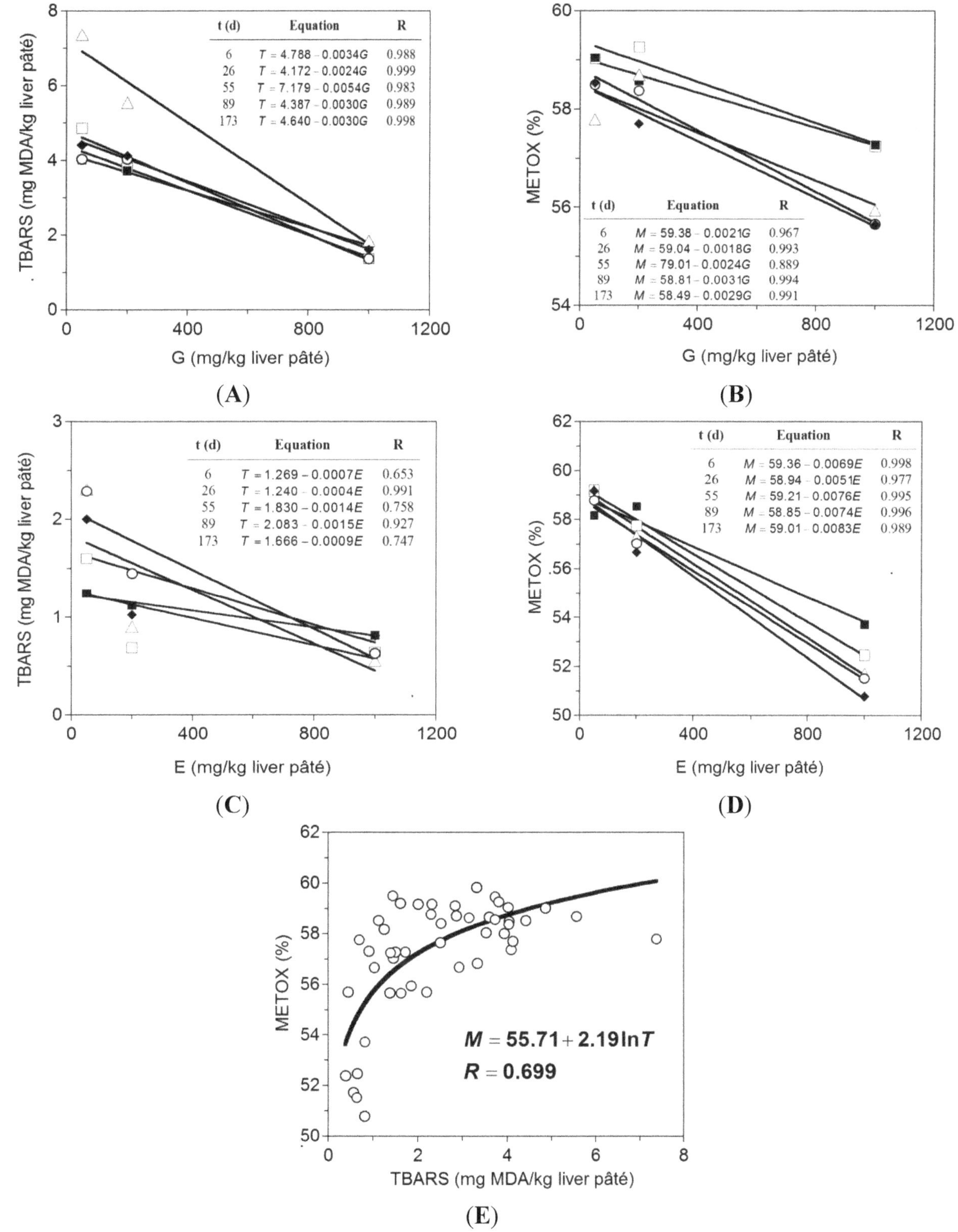

Figure 2. Linear relationships among TBARs and METOX production and extracts concentrations (tea: *E* and grape: *G*) at different storage times: 6 (□), 26 (■), 55 (Δ), 89 (○) and 173 days (♦), and non-linear relationship among TBARs and METOX production data. (**A**) TBARs for grape extracts application; (**B**) METOX for grape extracts application; (**C**) TBARs for tea extracts application; (**D**) METOX for tea extracts application and (**E**) TBARS *vs.* METOX production.

Although the coefficient of determination was poor ($R^2 = 0.488$), the consistency of the equation was good ($p < 0.05$) and the experimental data trend was clearly non-linear. This result is not in concordance with the affirmation, extensively reported in the literature, of linearity correlation among both variables [33,34]. However, similar low correlations and logarithmic relationships were also established for the variables antioxidant vitamins and TBARs in fresh muscle treated under different experimental conditions [35].

3.5. Fatty Acid Composition of Pig Liver Pâtés during Storage Time

The fatty acid composition of pig liver pâtés and the most important nutritional indexes are shown in Tables 2 and 3, respectively. The FA profile showed that MUFA were the predominant FAs, with values ranging from 50% to 60% of total methyl esters. This was followed by SFAs with values between 32.39% and 38.26% and then PUFAs with values in all cases less than 13%, following in importance.

Estévez *et al.* [4,9] also reported similar results for MUFA in porcine liver pâtés, and lower percentages (<48%) when they used white pigs for the manufacture of liver pâtés. Lower percentages were also displayed in other Spanish porcine liver pâtés [36], and when this product was manufactured using meat and liver from foals [24].

Regarding MUFA, oleic acid was the most abundant, followed by palmitoleic acid in agreement with other authors who found that oleic acid was the predominant FA in liver pâtés [4,24]. Little variations were observed for this FA based on the period of storage and among batches. Except for BHT, statistical analysis did not show significant ($p > 0.05$) differences for oleic acid concentrations during the period of storage. Among batches only significant differences ($p < 0.05$) were found at weeks 0 and 4 of the refrigerated storage. At the beginning of the period of storage, the highest values were observed in TEA_{1000} batch, with a mean value of 52.53%; while at week 24, the TEA_{200} batch had the highest values, with a mean value of 57.23%. During period of storage, CON batches had the lowest values for this FA. A similar pattern was observed for palmitoleic acid, although, in this case, BHT batch had the highest values for this FA.

Table 2. Effect of antioxidants on fatty acid profile of Celta pig liver pâtés (n = 3) during refrigerated storage.

		CON	BHT	TEA			GRA			*p*-Value	SEM
				50	200	1000	50	200	1000		
	0	20.46 ^{b,1}	20.43 ^{c,1}	20.70 ^{b,c,5}	20.45 ^{1}	20.77 ^{a,b,6}	20.50 ^{2}	20.58 ^{3}	20.62 ^{4}	0.000	0.03
Palmitic Acid	4	20.33 ^{b,2}	20.28 ^{c,2}	20.14 ^{c,2}	20.22 ^{2}	19.10 ^{b,1}	19.85 ^{2}	19.76 ^{2}	19.90 ^{2}	0.013	0.11
	8	23.23 ^{a}	23.59 ^{a}	22.89 ^{a}	22.33	22.61 ^{a}	22.26	22.34	22.67	0.176	0.14
C16:0	24	21.84 ^{a,b}	21.77 ^{b}	21.83 ^{a,b}	18.20	20.94 ^{a,b}	21.41	21.79	21.25	0.465	0.43
	p-value	0.039	0.002	0.026	0.225	0.042	0.103	0.228	0.068		
	SEM	0.48	0.51	0.43	0.70	0.51	0.40	0.48	0.43		
	0	2.45 ^{3}	2.58 ^{b,c,5}	2.51 ^{4}	2.38 ^{1}	2.54 ^{4}	2.41 ^{1,2}	2.39 ^{1}	2.44 ^{2,3}	0.000	0.02
Palmitoleic Acid	4	2.36 ^{1}	2.51 ^{c,2}	2.49 ^{2}	2.41 ^{1,2}	2.36 ^{1}	2.35 ^{1}	2.32 ^{1}	2.35 ^{1}	0.042	0.02
	8	2.40	2.72 ^{a,b}	2.52	2.49	2.53	2.44	2.46	2.57	0.396	0.03
C16:1	24	2.40	2.75 ^{a}	2.35	1.97	2.56	2.41	2.54	2.55	0.527	0.08
	p-value	0.842	0.025	0.692	0.657	0.373	0.164	0.338	0.121		
	SEM	0.03	0.04	0.05	0.14	0.04	0.01	0.04	0.04		
	0	12.15 ^{2}	12.12 ^{a,2}	12.24 ^{3}	12.38 ^{4}	11.42 ^{1}	12.45 ^{5}	12.36 ^{4}	12.25 ^{3}	0.000	0.08
Stearic Acid	4	12.73	12.15 ^{a}	12.14	12.23	12.45	12.18	12.07	12.04	0.496	0.08
	8	13.43	12.42 ^{a}	11.97	12.16	12.13	12.68	12.43	12.25	0.528	0.17
C18:0	24	12.73 ^{3}	11.42 ^{b,1,2}	12.05 ^{2,3}	10.73 ^{1}	11.40 ^{1,2}	12.46 ^{2,3}	12.28 ^{2,3}	11.81 $^{1-3}$	0.029	0.18
	p-value	0.557	0.020	0.695	0.083	0.077	0.352	0.916	0.067		
	SEM	0.28	0.15	0.07	0.29	0.19	0.09	0.16	0.08		
	0	50.30 ^{1}	50.60 ^{a,4}	50.62 ^{5}	50.66 ^{5}	52.53 ^{7}	50.80 ^{6}	50.48 ^{3}	50.45 ^{2}	0.000	0.18
Oleic Acid	4	49.75 ^{1}	50.86 ^{a,1,2}	50.91 ^{1,2}	50.78 ^{1,2}	50.51 ^{1,2}	51.80 ^{2,3}	52.35 ^{3}	51.55 ^{2,3}	0.020	0.22
	8	47.56	46.52 ^{b}	48.83	49.04	48.44	48.61	49.58	49.05	0.151	0.29
C18:1n9	24	49.49	50.70 ^{a}	50.05	57.23	50.86	50.19	49.59	50.72	0.412	0.89
	p-value	0.254	0.005	0.212	0.241	0.103	0.108	0.500	0.184		
	SEM	0.60	0.76	0.45	1.60	0.68	0.58	0.75	0.49		

Table 2. *Cont.*

		CON	BHT	TEA			GRA			*p*-Value	SEM
				50	200	1000	50	200	1000		
Linolei Acid	0	10.18 ^{a,5}	9.70 ^{b,3}	9.53 ^{2}	9.78 ^{4}	9.44 ^{1}	9.54 ^{2}	9.70 ^{3}	9.54 ^{2}	0.000	0.06
	4	10.40 ^{a,2}	9.70 ^{b,1}	9.62 ^{1}	9.75 ^{1}	9.85 ^{1}	9.65 ^{1}	9.63 ^{1}	9.84 ^{1}	0.014	0.07
C18:2n6	8	9.12 ^{c}	10.04 ^{a}	9.36	9.52	9.63	10.01	8.92	9.10	0.229	0.13
	24	9.72 ^{b}	9.61 ^{b}	9.87	9.76	9.59	9.46	9.66	9.56	0.639	0.05
	p-value	0.001	0.033	0.265	0.573	0.791	0.256	0.284	0.156		
	SEM	0.19	0.07	0.09	0.07	0.12	0.10	0.16	0.12		
Linolenic Acid	0	0.61 ^{a,3}	0.57 ^{c,2}	0.57 ^{2}	0.53 ^{1}	0.56 ^{1,2}	0.57 ^{2}	0.56 ^{1,2}	0.58 ^{b,2,3}	0.019	0.01
	4	0.53 ^{b}	0.55 ^{c}	0.35	0.56	0.54	0.59	0.57	0.59 ^{b}	0.493	0.03
C18:3n3	8	0.59 ^{a,b,1}	0.78 ^{a,3}	0.67 ^{1,2}	0.70 ^{2,3}	0.70 ^{2,3}	0.65 ^{1,2}	0.69 ^{2}	0.74 ^{a,2,3}	0.022	0.02
	24	0.56 ^{a,b}	0.63 ^{b}	0.61	0.31	0.67	0.62	0.69	0.58 ^{b}	0.540	0.04
	p-value	0.091	0.001	0.366	0.413	0.061	0.454	0.420	0.007		
	SEM	0.01	0.03	0.07	0.08	0.03	0.02	0.03	0.03		
Arachidonic Acid	0	1.33 ^{a,3}	1.55 ^{a,5}	1.33 ^{b,3}	1.43 ^{a,4}	0.42 ^{b,1}	1.21 ^{2}	1.43 ^{a,4}	1.33 ^{a,3}	0.000	0.09
	4	1.67 ^{a}	1.70 ^{a}	1.66 ^{a}	1.61 ^{a}	1.78 ^{a}	1.14	1.00 ^{a}	1.26 ^{a}	0.054	0.08
C20:4n6	8	0.25 ^{b}	0.39 ^{b}	0.31 ^{c}	0.27 ^{b}	0.32 ^{b}	0.63	0.34 ^{b}	0.29 ^{b}	0.066	0.03
	24	0.61 ^{b}	0.44 ^{b}	0.53 ^{c}	0.40 ^{b}	0.53 ^{b}	0.71	0.53 ^{b}	0.66 ^{a,b}	0.988	0.07
	p-value	0.003	0.004	0.001	0.014	0.018	0.173	0.009	0.052		
	SEM	0.22	0.24	0.21	0.24	0.24	0.12	0.17	0.18		

$^{a–c}$ Mean values in the same column (same antioxidant in different weeks) with different letter indicating significant differences; $^{1–6}$ Mean values in the same row (different antioxidant in the same week) with different number indicating significant differences; SEM: standard error of mean; Batches: CON: control; BHT: tert-butyl-4-hydroxytoluene; TEA: tea and GRA: grape seed extracts.

Table 3. Effect of antioxidants on nutritional properties of Celta pig liver pâtés (n = 3) during refrigerated storage.

		CON	BHT	TEA			GRA			*p*-Value	SEM
				50	200	1000	50	200	1000		
SFA	0	33.60 ^{b,3}	33.55 ^{b,c,2}	33.91 ^{b,5}	33.72 ^{4}	33.07 ^{b,1}	33.92 ^{a,b,5}	33.92 ^{5}	33.92 ^{b,5}	0.000	0.07
	4	33.92 ^{b,4}	33.32 ^{c,3,4}	33.25 $^{b,2–4}$	33.37 ^{3,4}	32.39 ^{b,1}	32.88 $^{b,1–3}$	32.63 ^{1,2}	32.88 $^{b,1–3}$	0.010	0.13
	8	38.26 ^{a}	37.74 ^{a}	36.50 ^{a}	36.00	36.38 ^{a}	36.26 ^{a}	36.27	36.49 ^{a}	0.375	0.27
	24	35.85 ^{a,b}	34.53 ^{b}	35.14 ^{a,b}	29.59	33.87 ^{b}	35.16 ,^{ab}	35.64	34.39 ^{b}	0.349	0.66
	p-value	0.082	0.001	0.058	0.225	0.033	0.058	0.312	0.032		
	SEM	0.79	0.68	0.52	1.10	0.61	0.53	0.73	0.53		
MUFA	0	53.85 ^{1}	54.22 ^{a,5}	54.25 ^{a,b,5}	54.12 ^{4}	56.16 ^{7}	54.32 ^{a,6}	53.96 ^{2}	54.07 ^{3}	0.000	0.18
	4	53.11 ^{1}	54.36 $^{a,1–3}$	54.51 $^{a,1–3}$	54.28 ^{1,2}	53.92 ^{1,2}	55.28 ^{a,2,3}	55.77 ^{3}	55.00 ^{2,3}	0.033	0.23
	8	51.16	50.40 ^{b}	52.56 ^{b}	52.83	52.20	52.08 ^{b}	53.21	52.83	0.265	0.30
	24	52.87	54.39 ^{a}	53.44 ^{a,b}	59.74	54.62	53.58 ^{a,b}	53.17	54.35	0.331	0.72
	p-value	0.362	0.001	0.129	0.234	0.094	0.054	0.550	0.137		
	SEM	0.52	0.65	0.34	1.28	0.62	0.49	0.65	0.35		
PUFA	0	12.55 ^{a,8}	11.87 ^{a,7}	11.84 ^{a,3}	12.16 ^{6}	10.77 ^{1}	11.76 ^{2}	12.11 ^{a,5}	11.92 ^{a,4}	0.000	0.13
	4	12.97 ^{a,3}	11.97 $^{a,1–3}$	12.05 ^{a,1,2}	12.34 $^{1–3}$	12.56 ^{2,3}	11.83 ^{1,2}	11.60 ^{a,b,1}	12.12 ^{a,1,2}	0.049	0.12
	8	10.51 ^{c}	11.39 ^{a}	10.94 ^{b}	11.15	11.27	11.66	10.52 ^{c}	10.68 ^{b}	0.117	0.15
	24	11.29 ^{b}	10.73 ^{b}	11.42 ^{a,b}	10.67	11.34	11.26	11.13^{b}	11.25 ^{b}	0.627	0.09
	p-value	0.000	0.015	0.036	0.068	0.100	0.537	0.008	0.011		
	SEM	0.37	0.19	0.17	0.29	0.29	0.13	0.23	0.22		
P/S	0	0.37 ^{a}	0.35 ^{a}	0.35 ^{a}	0.36	0.33 ^{b}	0.35	0.36 ^{a}	0.35 ^{a,b}	0.319	0.01
	4	0.39 ^{a}	0.36 ^{a}	0.36 ^{a}	0.37	0.39 ^{a}	0.36	0.36 ^{a}	0.37 ^{a}	0.060	0.01
	8	0.28 ^{c}	0.30 ^{b}	0.30 ^{b}	0.31	0.31 ^{b}	0.32	0.29 ^{b}	0.30 ^{c}	0.313	0.01
	24	0.32 ^{b}	0.31 ^{b}	0.33 ^{a,b}	0.36	0.34 ^{b}	0.32	0.31 ^{a,b}	0.33 ^{b,c}	0.488	0.01
	p-value	0.002	0.003	0.048	0.098	0.030	0.170	0.059	0.014		
	SEM	0.02	0.01	0.01	0.01	0.01	0.01	0.01	0.01		

Table 3. *Cont.*

		CON	BHT	TEA			GRA			*p*-Value	SEM
				50	200	1000	50	200	1000		
n6/n3	0	17.30 ^{a,b,2}	18.34 ^{a,b,4}	17.73 ^{3}	19.43 ^{a,6}	17.75 ^{a,b,3}	17.32 ^{2}	18.45 ^{5}	16.73 ^{1}	0.000	0.20
	4	22.12 ^{a}	20.26 ^{a}	18.59	18.82 ^{a,b}	19.99 ^{a}	16.91	17.99	17.46	0.504	2.60
	8	13.61 ^{b,2}	11.73 ^{c,1}	12.30 ^{1,2}	12.00 ^{c,1}	12.33 ^{b,1,2}	16.40 ^{3}	11.65 ^{1}	11.30 ^{1}	0.001	0.42
	24	17.27 ^{a,b}	14.93 ^{b,c}	16.81	17.18 ^{b}	14.04 ^{b}	15.61	14.66	16.63	0.976	0.78
	p-value	0.047	0.009	0.392	0.001	0.052	0.871	0.193	0.105		
	SEM	1.25	1.28	5.61	1.12	1.25	0.63	1.28	1.07		
NV	0	0.35 ^{b}	0.35 ^{b}	0.36 ^{b}	0.35	0.35 ^{b}	0.35	0.35	0.36 ^{a,b}	0.972	0.01
	4	0.35 ^{b,3}	0.35 ^{b,2,3}	0.35 ^{b,2,3}	0.35 ^{2,3}	0.33 ^{b,1}	0.33 $^{1-3}$	0.33 ^{1,2}	0.34 $^{b,1-3}$	0.033	0.01
	8	0.43 ^{a}	0.44 ^{a}	0.41 ^{a}	0.40	0.41 ^{a}	0.40	0.40	0.41 ^{a}	0.135	0.01
	24	0.38 ^{a,b}	0.38 ^{b}	0.38 ^{a,b}	0.28	0.37 ^{a,b}	0.37	0.39	0.37 ^{a,b}	0.532	0.01
	p-value	0.044	0.003	0.064	0.245	0.049	0.144	0.254	0.058		
	SEM	0.01	0.02	0.01	0.02	0.01	0.01	0.01	0.01		

Results expressed as fatty acid percentage composition (percent by weight of total fatty acids); $^{a-c}$ Mean values in the same column (same antioxidant in different weeks) with different letter indicating significant differences; $^{1-8}$ Mean values in the same row (different antioxidant in the same week) with different number indicating significant differences; SEM is the standard error of the mean; Batches: CON: control; BHT: tert-butyl-4-hydroxytoluene; TEA: tea and GRA: grape seed extracts; SFA = ∑(C14:0 + C16:0 + C17:0 + C18:0); MUFA = ∑(C16:1 + C17:1 + C18:1 + C20:1); PUFA = ∑(C18:2*n6* + C18:3*n3* + C20:2 + C20:3*n3* + C20:4*n6*); P/S = PUFA/SFA; NV: Nutritional value = ∑(C14:0 + C16:0)/∑(C18:1 + C18:2*n6*c).

The amount of SFA increased during the period of storage; palmitic and stearic acids were the predominant SFAs. Statistical analysis displayed significant ($p < 0.05$) differences for palmitic during period of storage in CON, BHT, TEA_{50} and TEA_{1000} batches. Regarding stearic acid, only significant differences ($p < 0.05$) were found in the BHT batch during the period of storage. These findings are in agreement with those reported by other authors who noticed similar percentages of SFAs in porcine liver pâtés [9]. The percentages of PUFAs were higher than those observed in other porcine liver pâtés also containing antioxidants [9] and lower than the percentages found in foal liver pâtés [24]. Linoleic acid was the predominant PUFA, with mean values around 10%. Only the BHT and CON batches showed significant ($p < 0.05$) differences during the period of storage. Among batches, significant ($p < 0.05$) differences were observed at the beginning of storage period (0 and 4 weeks). At the beginning of period of storage, the highest values were found in CON batch, with a mean value of 10.18%, while TEA_{1000} batch showed the lowest, with a mean value of 9.44%. Lower percentages of PUFA were found in arachidonic and linolenic acids, with values below 2% and 1%, respectively. Significant ($p < 0.05$) differences were observed for arachidonic acid, which decreased during the period of storage. Regarding linolenic acid, only significant ($p < 0.05$) differences were found among batches at the beginning and at week 8 of refrigerated storage.

Because of FAs contain double bonds they are targets for oxidative reactions therefore the amount of PUFAs measured is an indicator of the oxidative deterioration of meats [37]. As can be seen in Table 3, the oxidative degradation of PUFA mainly occurred after week 4. This behavior is in agreement with the results found by other authors [9] and could be attributed to the gradual degradation of endogenous antioxidants and the release of iron from the heme molecule [38]. In the present work, it can be observed that the addition of the antioxidants only protect the pâtés from oxidative degradation between week 4 and 8 of refrigerated storage, due to higher amount of PUFA found in treated samples compared to CON samples (10.51% *vs.*11.39%, 10.94%, 11.15%, 11.27%, 11.66%, 10.52% and 10.68% for CON, BHT, TEA_{50}, TEA_{200}, TEA_{1000}, GRA_{50}, GRA_{200} and GRA_{1000} batches, respectively).

To assess the nutritional properties, the ratios PUFA/SFA (P/S), *n*-6/*n*-3 and the nutritional value (NV) were determined (Table 3). PUFA/SFA ratio values were lower than the optimal values (0.5–0.7) recommended in the Mediterranean diet [39] and by the British Department of Health [40]. These ratio values decreased slightly during the period of storage, showing significant ($p < 0.05$) differences between CON and samples treated with BHT, TEA_{50}, TEA_{1000} and GRA_{1000}. At the beginning of the period of storage, pâtés that contained TEA_{1000} extract showed the lowest values, even though no significant ($p > 0.05$) differences were found among batches. Regarding *n*-6/*n*-3 ratio, significant ($p < 0.05$) differences were observed during refrigerated storage in CON and in the samples treated with BHT and TEA_{200}, showing a decrease that was more pronounced in the BHT batch. All the *n*-6/*n*-3 ratios were higher than the nutritional recommendations of the British Department of Health [40] and *Food and Agricultural Organization* [41] for human diet, which should not exceed 4. Our results showed in all cases values above 11, which are higher than those found in foal liver pâtés [24] and Iberian liver pâtés [4]. Nevertheless, ratios reported in the present study are similar to those observed in other liver pâtés manufactured using white pigs [4]. Finally, the NV, which gives an estimation of product healthiness could have in the diet regarding its lipid content, showed a slight increase during the period of storage, mainly in CON and in the samples treated with BHT and TEA_{1000} presented significant ($p < 0.05$) differences during the period of storage.

3.6. Volatile Profile of Pig Liver Pâtés during Storage Time

The analysis of volatile compounds gives an indication of the chemical and metabolic processes that occur during the manufacturing process [42] and period of storage [8,36]. Also, these compounds provide information about the oxidative stability and the aroma characteristics of this product [36]. Thirty eight volatile compounds were identified from Celta pig liver pâtés samples, nevertheless only the evolution of the most abundant lipid-derived volatiles of liver pâté are shown in Table 4, being hexanal; octen-1,3-ol; hexan-1-ol and heptanal the most abundant.

Aldehydes are probably the most interesting lipid-derived volatile compounds because they can produce a wide range of flavors and odors. In the present study, almost half of the identified volatiles belonged to this family (hexanal, heptanal, and octanal). Statistical analysis showed significant ($p < 0.001$) differences for this chemical family among batches at the beginning of the storage period. The predominant aldehyde detected was hexanal, which increased significantly ($p < 0.05$) during the storage period. Hexanal as with the TBARs index is frequently used as marker of lipid oxidation due to its high sensitivity [22,43]. In fact, a statistical correlation was found between them ($r = 0.65$, $p < 0.01$). Furthermore, it is mainly generated as a result of the oxidative decomposition of PUFAs and has been related to rancid flavors [44]. Batches that contained antioxidants showed a decrease in the amount of lipid-derived volatile compounds isolated from the liver pâté (Table 4). This finding is in agreement with those reported by other authors in porcine liver pâtés [9]. Furthermore, the greater the amount of natural antioxidant added to the pâté, the lower was the concentration of aldehydes in the sample, which could equate to greater product protection.

Except for octanal, significant ($p < 0.05$) changes were observed at the end of the storage period. As happened at day 0, the addition of antioxidants as well as their use at higher concentrations of them decreased the amount of volatiles compounds isolated. These results indicate that the addition of BHT and natural antioxidants had a significant ($p < 0.05$) effect on the generation of the most volatiles compounds (Table 4). Furthermore, TEA and GRA extracts even improved the results obtained by BHT. These results are in agreement with previously published studies [9], which reported higher effectiveness of natural products compared to synthetic antioxidants and suggest the possibility of using these extracts as replacements for commercially used synthetic compounds. Compared to the CON batch, pâtés with TEA provided the most favorable results, with smaller amounts of heptanal, hexanal, hexan-1-ol and octen-1,3-ol. Furthermore, a concentration of 50 mg/kg of the natural antioxidants was enough to significantly improve the results obtained for CON and BHT batches: heptanal (6.13 and 6.02 *vs.* 7.60 and 6.70 $\times 10^6$ AU for TEA_{50} and GRA_{50} *vs.* CON and BHT, respectively) and hexanal (231.12 and 227.21 *vs.* 374.33 and 330.40 $\times 10^6$ AU for TEA_{50} and GRA_{50} *vs.* CON and BHT, respectively).

Table 4. Evolution of lipid-derived volatiles of Celta pig liver pâtés ($n = 3$) at day 0 and week 24 of refrigerated storage.

		CON	BHT	TEA			GRA			p-Value	SEM
				50	200	1000	50	200	1000		
Week 0	Hexanal	171.58 ^{b}	159.38 ^{c}	116.70 ^{e}	49.22 ^{g}	31.79 ^{h}	233.34 ^{a}	154.90 ^{d}	62.33 ^{f}	0.000	16.92
	2-heptanona	0.00 ^{d}	0.94 ^{c}	0.00 ^{d}	0.00 ^{d}	0.00 ^{d}	2.80 ^{a}	2.38 ^{b}	0.00 ^{d}	0.000	0.28
	Heptanal	9.34 ^{a}	2.69 ^{d}	2.56 ^{e}	1.41 ^{g}	1.22 ^{h}	4.57 ^{b}	4.18 ^{c}	1.77 ^{f}	0.000	0.64
	Octanal	0.00 ^{d}	0.00 ^{d}	0.00 ^{d}	0.00 ^{d}	0.00 ^{d}	7.50 ^{a}	6.94 ^{b}	4.04 ^{c}	0.000	0.81
	Hexan-1-ol	8.77 ^{c}	5.06 ^{d}	1.94 ^{f}	0.00 ^{g}	0.00 ^{g}	9.03 ^{b}	10.46 ^{a}	2.89 ^{e}	0.000	1.02
	Octen-1,3-ol	20.81 ^{b}	17.21 ^{c}	13.43 ^{d}	5.94 ^{f}	0.00 ^{g}	0.00 ^{g}	22.59 ^{a}	7.44 ^{e}	0.000	2.15
	Furan-2-penthyl	3.50 ^{a}	0.74 ^{c}	0.43 ^{d}	0.00 ^{e}	0.00 ^{e}	0.75 ^{c}	1.02 ^{b}	0.75 ^{c}	0.000	0.27
Week 24	Hexanal	374.33 ^{a}	330.40 ^{a,b}	231.12 ^{b,c}	86.10 ^{d,e}	24.91 ^{e}	227.21 ^{b,c}	182.72 ^{c,d}	82.08 ^{d,e}	0.001	31.22
	2-heptanona	2.90 ^{a}	2.70 ^{a}	1.04 ^{a}	0.00 ^{b}	0.00 ^{b}	2.01 ^{a}	2.02 ^{a}	0.00 ^{b}	0.002	0.32
	Heptanal	7.60 ^{a}	6.70 ^{a,b}	6.13 ^{a,b}	2.86 ^{b,c}	0.94 ^{c}	6.02 ^{a,b}	6.74 ^{a,b}	3.36 ^{b,c}	0.035	0.65
	Octanal	5.90	0.00	0.00	0.00	0.00	5.77	4.75	0.00	0.211	0.88
	Hexan-1-ol	9.47 ^{a}	6.87 ^{b}	8.97 ^{c}	0.98 ^{c}	0.95 ^{c}	5.82 ^{b}	6.83 ^{b}	2.26 ^{c}	0.000	0.80
	Octen-1,3-ol	47.41 ^{a}	29.58 ^{a,b}	14.10 ^{b,c}	4.30 ^{b,c}	2.31 ^{b,c}	0.00 ^{c}	10.47 ^{b,c}	7.17 ^{b,c}	0.025	4.53
	Furan-2-penthyl	0.87 ^{a}	1.05 ^{a}	0.96 ^{a}	0.00 ^{c}	0.00 ^{c}	0.93 ^{a}	1.03 ^{a}	0.67 ^{b}	0.000	0.11

Results expressed as AU $\times 10^6$; $^{a–h}$ Mean values in the same row (different batches on the same storage week) with different letter indicating significant differences; Batches: CON: control; BHT: tert-butyl-4-hydroxytoluene; TEA: tea and GRA: grape seed extracts.

4. Conclusions

From the obtained results, it can be concluded that the addition of the natural extracts improved the color stability during the period of storage, with grape extract giving the smallest color differences at the end of the period of storage. The addition of both extracts resulted in minor increases in the TBARs index and metmyoglobin percentage. Furthermore, the oxidative stability (measured in terms of amount of hexanal and other lipid-derived aldehydes such as heptanal, octanal) of the liver pates treated with natural antioxidants was significantly higher than in control sample.

Acknowledgements

Authors are grateful to Consellería de Innovación e Industria of Xunta de Galicia (The Regional Government) (Project 09TAL006CT) for the financial support.

Author Contributions

Mirian Pateiro and José Antonio Vazquez have performed the measurements and statistical analysis of the data, respectively. José Manuel Lorenzo and Daniel Franco have designed and supervised the research. All authors have contributed to the writing of the manuscript and have approved the final paper.

Conflicts of Interest

The authors declare no conflict of interest.

References

1. Lorenzo, J.M.; Sineiro, J.; Amado, I.; Franco, D. Influence of natural extracts on the shelf life of modified atmosphere-packaged pork patties. *Meat Sci.* **2014**, *96*, 526–534.
2. Doolaege, E.H.A.; Vossen, E.; Raes, K.; de Meulenaer, B.; Verhé, R.; Paelinck, H.; de Smet, S. Effect of rosemary extract dose on lipid oxidation, colour stability and antioxidant concentrations, in reduced nitrite liver pâtés. *Meat Sci.* **2012**, *90*, 925–931.
3. Moure, A.; Cruz, J.M.; Franco, D.; Domínguez, J.M.; Sineiro, J.; Domínguez, H.; Núñez, M.J.; Parajó, J.C. Natural antioxidants from residual sources. *Food Chem.* **2001**, *72*, 145–171.
4. Estévez, M.; Ventanas, S.; Ramírez, R.; Cava, R. Analysis of volatiles in porcine liver pâtés with added sage and rosemary essential oils by using SPME-GC-MS. *J. Agric. Food Chem.* **2004**, *52*, 5168–5174.
5. Tang, S.Z.; Ou, S.Y.; Huang, X.S.; Li, W.; Kerry, J.P.; Buckley, D.J. Effects of added tea catechins on colour stability and lipid oxidation in minced beef patties held under aerobic and modified atmospheric packaging conditions. *J. Food Eng.* **2006**, *77*, 248–253.
6. Carpenter, R.; O'Grady, M.N.; O'Callaghan, Y.C.; O'Brien, M.N.; Kerry, J.P. Evaluation of the antioxidant potential of grape seed and bearberry extracts in raw and cooked pork. *Meat Sci.* **2007**, *76*, 604–610.

7. Siripatrawan, U.; Noipha, S. Active film from chitosan incorporating green tea extract for shelf life extension of pork sausages. *Food Hydrocoll.* **2012**, *27*, 102–108.
8. Pateiro, M.; Lorenzo, J.M.; Amado, I.; Franco, D. Effect of addition of green tea, chestnut and grape extract on the shelf-life of pig liver pate. *Food Chem.* **2014**, *147*, 386–394.
9. Estévez, M.; Ramírez, R.; Ventanas, S.; Cava, R. Sage and rosemary essential oils *versus* BHT for the inhibition of lipid oxidative reactions in liver pâté. *LWT Food Sci. Technol.* **2007**, *40*, 58–65.
10. Lorenzo, J.M.; González-Rodríguez, R.M.; Sánchez, M.; Amado, I.R.; Franco, D. Effects of natural (grape seed and chestnut extract) and synthetic antioxidants (buthylatedhydroxytoluene, BHT) on the physical, chemical, microbiological and sensory characteristics of dry cured sausage "chorizo". *Food Res. Int.* **2013**, *54*, 611–620.
11. Yudd, D.B.; Wyszecki, G. *Color in Business, Science and Industry*, 3rd ed.; Wiley: New York, NY, USA, 1975.
12. Krzywicki, K. Assessment of relative content of myoglobin, oxymyoglobin and metmyoglobin at the surface of beef. *Meat Sci*. **1979**, *3*, 1–10.
13. Vyncke, W. Evaluation of the direct thiobarbituric acid extraction method for determining oxidative rancidity in mackerel (*Scomber scombrus* L.). *Fette Seifen Anstrichm.* **1975**, *77*, 239–240.
14. Jayaprakasha, G.K.; Selvi, T.; Sakariah, K.K. Antibacterial and antioxidant activities of grape (*Vitis vinifera*) seed extracts. *Food Res. Int.* **2003**, *36*, 117–122.
15. Rubilar, M.; Pinelo, M.; Shene, C.; Sineiro, J.; Nunez, M.J. Separation and HPLC−MS identification of phenolic antioxidants from agricultural residues: Almond hulls and grape pomace. *J. Agric. Food Chem.* **2007**, *55*, 10101–10109.
16. Van der Hooft, J.J.; Akermi, M.; Unlu, F.M.; Mihaleva, V.; Gomez-Roldan, V.; Bino, R.J.; de Vos, R.C.H.; Vervoort, J. Structural annotation and elucidation of conjugated phenolic compounds in black, green, and white tea extracts. *J. Agric. Food Chem.* **2012**, *60*, 8841–8850.
17. Fontecave, M.; Lepoivre, M.; Elleingand, E.; Gerez, C.; Guittet, O. Resveratrol, a remarkable inhibitor of ribonucleotide reductase. *FEBS Lett.* **1998**, *421*, 277–279.
18. Zhang, L.; Lin, Y.H.; Leng, X.J.; Huang, M.; Zhou, G.H. Effect of sage (*Salvia officinalis*) on the oxidative stability of Chinese-style sausage during refrigerated storage. *Meat Sci.* **2013**, *95*, 145–150.
19. Estévez, M.; Ventanas, S.; Cava, R. Effect of natural and synthetic antioxidants on protein oxidation and colour and texture changes in refrigerated stored porcine liver pâté. *Meat Sci*. **2006**, *74*, 396–403.
20. Francis, F.J.; Clydesdale, F.M. *Food Colorimetry: Theory and Applications*; Avi Publishing Company Inc.: Westport, CT, USA, 1975.
21. Estévez. M.; Ventanas, S.; Cava, R. Protein oxidation in frankfurters with increasing levels of added rosemary essential oil: Effect on color and texture deterioration. *J. Food Sci*. **2005**, *70*, 427–432.
22. Visessanguan, W.; Benjakul, S.; Riebroy, S.; Yarchai, M.; Tapingkae, W. Changes in lipid composition and fatty acid profile of Nham, a Thai fermented pork sausage, during fermentation. *Food Chem.* **2006**, *94*, 580–588.

23. Ganhão, R.; Estévez, M.; Morcuende, D. Suitability of the TBA method for assessing lipid oxidation in a meat system with added phenolic-rich materials. *Food Chem.* **2011**, *126*, 772–778.
24. Lorenzo, J.M.; Pateiro, M. Influence of fat content on physico-chemical and oxidative stability of foal liver pâté. *Meat Sci.* **2013**, *95*, 330–335.
25. Nassu, R.T.; Gonçalves, L.A.G.; da Silva, M.A.A.P.; Beserra, F.J. Oxidative stability of fermented goat meat sausage with different levels of natural antioxidant. *Meat Sci.* **2003**, *63*, 43–49.
26. Formanek, Z.; Kerry, J.P.; Higgins, F.M.; Buckley, D.J.; Morrissey, P.A.; Farkas, J. Addition of synthetic and natural antioxidants to α-tocopheryl acetate supplemented beef patties: Effects of antioxidants and packaging on lipid oxidation. *Meat Sci.* **2001**, *58*, 337–341.
27. Jayawardana, B.C.; Hirano, T.; Han, K.-H.; Ishii, H.; Okada, T.; Shibayama, S.; Fukushima, M.; Sekikawa, M.; Shimada, K.-I. Utilization of adzuki bean extract as a natural antioxidant in cured and uncured cooked pork sausages. *Meat Sci.* **2011**, *89*, 150–153.
28. Racanicci, A.M.C.; Danielsen, B.; Menten, J.F.M.; Reginato-d'Arce, M.A.B.; Skibsted, L.H. Antioxidant effect of dittany (*Origanum dictamnus*) in pre-cooked chicken meat balls during chill-storage in comparison rosemary (*Rosmarinus officinalis*). *Eur. Food Res. Technol.* **2004**, *218*, 521–524.
29. Vossen, E.; Utrera, M.; de Smet, S.; Morcuende, D.; Estévez, M. Dog rose (*Rosa canina* L.) as a functional ingredient in porcine frankfurters without added sodium ascorbate and sodium nitrite. *Meat Sci.* **2012**, *92*, 451–457.
30. Juárez, M.; Polvillo, O.; Gómez, M.D.; Alcalde, M.J.; Romero, F.; Valera, M. Breed effect on carcass and meat quality of foals slaughtered at 24 months of age. *Meat Sci.* **2009**, *83*, 224–228.
31. Sebranek, J.G.; Sewalt, V.J.H.; Robbins, K.L.; Houser, T.A. Comparison of a natural rosemary extract and BHA/BHT for relative antioxidant effectiveness in pork sausage. *Meat Sci.* **2005**, *69*, 289–296.
32. O'Grady, M.N.; Carpenter, R.; Lynch, P.B.; O'Brien, N.M.; Kerry, J.P. Addition of grape seed extract and bearberry to porcine diets: Influence on quality attributes of raw and cooked pork. *Meat Sci.* **2008**, *78*, 438–446.
33. O'Grady, M.N.; Monahan, F.J.; Bailey, J.; Allen, P.; Buckley, D.J.; Keane, M.G. Colour-stabilising effect of muscle vitamin E in minced beef stored in high oxygen packs. *Meat Sci.* **1998**, *1*, 73–80.
34. Renerre, M. Review: Factors involved in the discoloration of beef meat. *Int. J. Food Sci. Technol.* **1990**, *25*, 613–630.
35. Descalzo, A.M.; Insani, E.M.; Biolatto, A.; Sancho, A.M.; García, P.T.; Pensel, N.A.; Josifovich, J.A. Influence of pasture or grain-based diets supplemented with vitamin E on antioxidant/oxidative balance of Argentine beef. *Meat. Sci.* **2005**, *70*, 35–44.
36. Estévez, M.; Ventanas, J.; Cava, R.; Puolanne, E. Characterisation of a traditional Finnish liver sausage and different types of Spanish liver pâtés: A comparative study. *Meat Sci.* **2005**, *71*, 657–669.
37. Gray, J.I.; Gomaa, E.A.; Buckley, D.J. Oxidative quality and shelf life of meats. *Meat Sci.* **1996**, *43*, 111–123.

38. Estévez, M.; Cava, R. Lipid and protein oxidation, release of iron from heme molecule and colour deterioration during refrigerated storage of liver pâté. *Meat. Sci.* **2004**, *68*, 551–558.
39. Ulbricht, T.L.V.; Southgate, D.A.T. Coronary heart disease: Seven dietary factors. *Lancet* **1991**, *338*, 985–992.
40. British Department of Health. Nutritional aspects of cardiovascular diseases: Report of the cardiovascular review group Committee on Medical Aspects of Food Policy. In *Report on Health and Social Subjects n°46*; H.M. Stationery Office: London, UK, 1994.
41. FAO (Food and Agriculture Organization of the United Nations). Fat and fatty acid requirements for adults. In *Fats and Fatty Acids in Human Nutrition*; FAO: Rome, Italy, 2010, pp. 55–62.
42. Lorenzo, J.M.; Montes, R.; Purriños, L.; Franco, D. Effect of pork fat addition on the volatile compounds of foal dry-cured sausage. *Meat Sci.* **2012**, *91*, 506–512.
43. Ahn, D.U.; Sell, J.L.; Jo, C.; Chen, X.; Wu, C.; Lee, J.I. Effects of dietary vitamin E supplementation on lipid oxidation and volatiles content of irradiated, cooked turkey meat patties with different packaging. *Poult. Sci.* **1998**, *77*, 912–920.
44. Shahidi, F.; Pegg, R.B. Hexanal as an indicator of meat flavor deterioration. *J. Food Lipids* **1994**, *1*, 177–186.

9

Oil Content, Fatty Acid Composition and Distributions of Vitamin-E-Active Compounds of Some Fruit Seed Oils

Bertrand Matthäus [1] and Mehmet Musa Özcan [2,*]

[1] Max Rubner-Institut, Federal Research Institute for Nutrition and Food, Schützenberg 12, 32756 Detmold, Germany; E-Mail: bertrand.matthaus@mri.bund.de

[2] Department of Food Engineering, Faculty of Agriculture, University of Selcuk, 42079 Konya, Turkey

* Author to whom correspondence should be addressed; E-Mail: mozcan@selcuk.edu.tr

Academic Editor: Adrianne Bendich

Abstract: Oil content, fatty acid composition and the distribution of vitamin-E-active compounds of selected Turkish seeds that are typically by-products of the food processing industries (linseed, apricot, pear, fennel, peanut, apple, cotton, quince and chufa), were determined. The oil content of the samples ranged from 16.9 to 53.4 g/100 g. The dominating fatty acids were oleic acid (apricot seed oil, peanut oil, and chufa seed oil) in the range of 52.5 to 68.4 g/100 g and linoleic acid (pear seed oil, apple seed oil, cottonseed oil and quince seed oil) with 48.1 to 56.3 g/100 g, while in linseed oil mainly α-linolenic acid (53.2 g/100 g) and in fennel seed oil mainly 18:1 fatty acids (80.5 g/100 g) with petroselinic acid predominating. The total content of vitamin-E-active compounds ranged from 20.1 (fennel seed oil) to 96 mg/100 g (apple seed oil). The predominant isomers were established as α- and γ-tocopherol.

Keywords: fruit seed; oil content; fatty acids; vitamin-E-active compounds; by-products; oil characterisation

1. Introduction

Large amounts of different seeds are discarded yearly as by-products of food processing [1,2]. These seeds are often rich sources of oil and other interesting minor compounds. Since some time an increasing interest in oils from unconventional seeds has been noted [3–5], and the by-products generated by fruit and vegetable industry may contribute as important sources for such oils. The added value of the by-products and the potential use of these resulting seed oils depends on the fatty acid composition and the content of minor components in the oil. It is desirable that the quantities of seeds which accumulate during fruit processing as residues from various fruits and vegetables are utilized as sources for the production of oils. The continued expansion in the food industry indeed calls attention to different fruit seeds as a potential source for oil. To ensure the utilization of these by-products more information about the oil content and the composition of different fruits is required showing the economical and efficient usability of these seeds [1,6]. The aim of the present study was to compare the oil contents as well as the composition of fatty acids and vitamin-E-active compounds of oils extracted from selected seeds.

2. Material and Methods

2.1. Material

The samples were collected by hand in May and August in 2007 from plants growing at several locations of Turkey. Seeds and kernels were obtained from fruits by hand-processing and then transferred to laboratory in polypropylene bags under cool conditions. Then they were stored in glass jars at 6°C until analysis. Detailed information related to the samples is given in Table 1.

Table 1. Seeds used in experiment.

Sample No.	General Name	Botanical Name	Family	Locations
1	Linseed	*Linum usitatissimum*	FLinaceae	Akören-Konya
2	Apricot (sweet)	*Prunus armeniaca*	Rosacea	Beybes-Konya
3	Apricot (bitter)	*Prunus armeniaca*	Rosaceae	Beybes-Konya
4	Pear	*Pyrus communis*	Rosaceae	Ankara
5	Fennel (dulce)	*Foeniculum vulgare*	Apicaceae	Konya
6	Peanut	*Arachis hypogaea*	Leguminaceae	Silifke-Mersin
7	Apple (Golden)	*Malus communis*	Rosaceae	Eğridir-Isparta
8	Apple (Starking)	*Malus communis*	Rosaceae	Hadim-Konya
9	Cotton	*Gossypium hirsutum*	Malvaceae	Adana
10	Quince	*Cydonia vulgaris*	Malvaceae	Beybes-Konya
11	Chufa	*Cyperus esculentus*	Cyperaceae	Çumra-Konya
12	Apple (Starking)	*Malus communis*	Rosaceae	Karaman

2.2. Reagents

Petroleum ether (40–60) was of analytical grade (>98%; Merck, Darmstadt, Germany). Heptane and tert-butyl methyl ether were of HPLC grade (Merck, Darmstadt, Germany). Tocopherol and tocotrienol standard compounds were purchased from CalBiochem (Darmstadt, Germany).

2.3. Oil Content

The oil content was determined according to the method ISO 659:1998 [7]. About 2 g of the seeds were ground in a ball mill and extracted with petroleum ether in a Twisselmann apparatus for 6 h. The solvent was removed by a rotary evaporator at 40 °C and 25 Torr. The oil was dried by a stream of nitrogen and stored at −20 °C until use.

2.4. Fatty Acid Composition

The fatty acid composition was determined following the ISO standard ISO 5509:2000 [8]. In brief, one drop of the oil was dissolved in 1 mL of *n*-heptane, 50 µg of sodium methylate (Merck, Darmstadt, Germany) was added, and the closed tube was agitated vigorously for 1 min at room temperature. After addition of 100 µL of water, the tube was centrifuged at 4500 g for 10 min and the lower aqueous phase was removed. Then 50 µL of HCl (1 mol with methyl orange (Merck, Darmstadt, Germany)) was added, the solution was shortly mixed, and the lower aqueous phase was rejected. About 20 mg of sodium hydrogen sulphate (monohydrate, extra pure; Merck, Darmstadt, Germany) was added, and after centrifugation at 4500 g for 10 min, the top *n*-heptane phase was transferred to a vial and injected in a HP5890 gas chromotograph (Agilent Technologies Sales & Services GmbH & Co. KG, Waldbronn, Germany), with a capillary column, CP-Sil 88 (100 m long, 0.25 mm ID, film thickness 0.2 µm). The temperature program was as follows: From 155 °C; heated to 220 °C (1.5 °C/min), 10 min isotherm; injector 250 °C, detector 250 °C; carrier gas 36 cm/s hydrogen; split ratio 1:50; detector gas 30 mL/min hydrogen; 300 mL/min air and 30 mL/min nitrogen; manual injection volume less than 1 µL. The peak areas were computed by the integration software, and percentages of fatty acid methyl esters (FAME) were obtained as weight percent by direct internal normalization.

2.5. Vitamin-E-Active Compounds

For determination of vitamin-E-active compounds, a solution of 250 mg of oil in 25 mL of *n*-heptane was directly used for the HPLC. The HPLC analysis was conducted using a Merck-Hitachi low-pressure gradient system, fitted with a L-6000 pump (Merck-Hitachi, Darmstadt, Germany), a Merck-Hitachi F-1000 fluorescence spectrophotometer (Darmstadt, Germany; detector wavelengths for excitation 295 nm, for emission 330 nm), and a ChemStation integration system (Agilent Technologies Deutschland GmbH, Böblingen, Germany). The samples in the amount of 20 µL were injected by a Merck 655-A40 autosampler (Merck-Hitachi, Darmstadt, Germany) onto a Diol phase HPLC column 25 cm × 4.6 mm ID (Merck, Darmstadt, Germany) used with a flow rate of 1.3 mL/min. The mobile phase used was 99 mL *n*-heptane + 1 mL tert-butyl methyl ether [9]. The mean values were given in the tables, without the standard deviation, because this value would represent only the deviation of the method and not the variation of the appropriate sample.

3. Results and Discussion

The results for oil content and fatty acid composition of the seeds are shown in Table 2. The oil contents of the seeds were found between 16.9 g/100 g (dw) (quince) and 53.6 g/100 g (dw) (apricot seeds, sweet). While most of the samples showed oil contents below or near 30 g/100 g, the oil content of apricot seeds (sweet and bitter) was remarkably higher with showing values of 53.4 g/100 g (dw) and 45.2 g/100 g (dw), respectively. From an economical point of view, this high oil content would justify oil extraction of the seeds, whereas for seeds with oil contents below 20 g/100 g an economical extraction of oil is only meaningful by solvent when the meal contains valuable protein, e.g., soybean or if the oil can be marketed as high-quality cold-pressed edible oil. The oil content of linseed was found relatively low in comparison to the literature with 34.2 to 44.4 g/100 g [10], probably due to season, environment and locations.

Results for the fatty acid composition showed that the oils can be divided into two groups with one group predominant in 18:1 fatty acids and the other group high in linoleic acid (18:2). Only in linseed oil was α-linolenic acid predominating with amounts of 53.2 g/100 g, while its content in the other samples was between 0.1 and 1.0 g/100 g, only. The highest contents of oleic acid were found in apricot (sweet) (68.3 g/100 g), chufa (62.4 g/100 g) and apricot (bitter) (57.8 g/100 g) seed oils. Fennel (dulce) seed oil is characterized by a high content of 18:1 fatty acids (80.5 g/100 g) with petroselinic acid as predominant fatty acid. The used GLC-method was not able to separate petroselinic acid from oleic acid, but from literature it is known that petroselinic acid is predominant in members of the family Apiaceae. From a nutritional point of view this high content of oleic acid favors these seed oils for human nutrition. The content of linoleic acid of the oils ranged from 12.0 g/100 g (fennel seed oil) to 56.3 g/100 g (cotton seed oil), with half (six) of the selected seed oils comprising of at least 50 g/100 g linoleic acid. The relatively low content of saturated fatty acids such as palmitic acid or stearic acid with amounts below 10 g/100 g is interesting from a nutritional point of view. Only chufa seed oil (17.2 g/100 g) and cottonseed oil (24.2 g/100 g) had higher amounts of nutritionally unfavourable saturated fatty acids. The fatty acid composition found in the present investigation for apricot seed oil was similar to the fatty acid composition published by Dubois *et al.* [11] with 4.4 g/100 g palmitic acid, 0.5 g/100 g stearic acid, 66.3 g/100 g oleic acid, and 28.6 g/100 g linoleic acid.

The results from Dubois *et al.* [11] also confirm the presented results for chufa seed oil with 13.8 g/100 g palmitic acid, 3.2 g/100 g stearic acid, 72.6 g/100 g oleic acid, and 8.9 g/100 g linoleic acid. Similarly, Kim *et al.* [12] published the high amount of oleic acid (65.5 g/100 g) and the moderate amounts of palmitic acid (15.2 g/100 g) and linoleic acid (16.2 g/100 g) for chufa seed oil. A little different was the fatty acid composition for chufa seed oil reported by Eteshola and Oraedu [13] with 28.1 g/100 g myristic acid, 14.5 g/100 g palmitic acid, 3.4 g/100 g stearic acid, 44.8 g/100 g oleic acid and 8.8 g/100 g linoleic acids. Especially the finding of myristic acid and the low amount of linoleic acid do not agree with the results from other authors or the presented work. Different varieties or locations of cultivation can be reasons for these differences.

Reiter *et al.* [14] determined 4.4 g/100 g palmitic acid, 73.9 g/100 g petroselinic, 4.8 g/100 g oleic acid, and 16.3 g/100 g linoleic acid in fennel seed oil. The amount of 18:1 fatty acids found in the present investigation was very similar to the results from Reiter *et al.* [14].

Table 2. Oil content and fatty acid composition of seed oils (g/100 g).

	Oil	16:0	16:1-δ-7	16:1-δ-9	18:0	Sum 18:1	18:1-δ-9	18:1-δ-11	18:2-δ-9,12	20:1-δ-11	18:3-δ-9,12,15	18:4-δ-6,9,12,15
Linseed	33.6	4.9	n.d.	0.1	4.8	n.d.	20.2	0.6	15.0	0.4	53.2	n.d.
Apricot (sweet)	53.4	4.9	n.d.	0.6	1.1	n.d.	68.3	1.4	23.1	0.1	0.1	0.1
Apricot (bitter)	45.2	6.4	1.0	0.9	1.0	n.d.	57.8	1.8	31.4	0.1	0.2	0.1
Pear (Ankara)	31.7	9.0	n.d.	0.2	2.1	n.d.	31.8	0.5	53.6	1.2	0.4	0.4
Fennel (Dulce)	18.2	4.3	0.2	0.2	1.6	80.5			12.0	0.3	0.4	n.d.
Peanut	38.3	9.5	n.d.	0.1	3.2	n.d.	52.5	n. d.	28.3	1.5	0.1	1.0
Apple (Golden)	21.9	7.0	0.1	0.1	1.9	n.d.	35.7	0.5	51.7	1.1	0.6	0.5
Apple (Starking)	25.6	6.8	0.1	0.10	2.0	n.d.	40.4	n. d.	48.1	1.2	0.3	0.5
Cotton	27.0	21.9	n.d.	0.5	2.3	n.d.	14.6	0.7	56.3	0.3	0.2	0.1
Quince	16.9	6.8	0.1	0.2	1.5	n.d.	33.8	0.6	55.4	0.6	0.2	0.3
Chufa	17.3	14.5	n.d.	0.1	2.7	n.d.	62.4	0.9	17.0	0.6	0.6	0.3
Apple (Starking)	23.5	6.3	0.1	0.1	2.1	n.d.	38.8	0.3	49.6	1.3	0.3	0.5

16:0, palmitic acid; 16:1-δ-7, 7c-hexadecenoic acid; 16:1-δ-9, 9c-hexadecenoic acid 18:0, stearic acid; Sum 18:1, 18:1-δ-6 (petroselinic acid) + 18:1-δ-9 (oleic acid) + 18:1-δ-11 (*cis*-vaccenic acid); 18:1-δ-9, oleic acid; 18:1-δ-11, *cis*-vaccenic acid; 18:2-δ-9,12, linoleic acid; 20:1-δ-11, gondoic acid; 18:3-δ-9,12,15, α-linolenic aicd; 18:4-δ-6,9,12,15, stearidonic acid; n.d., not detectable.

In linseed oil, the major fatty acids were reported as 5.5 ±1.5 g/100 g palmitic acid, 3.5 ±1.2 g/100 g stearic acid, 22.1 ±5 g/100 g oleic acid, 20.5 ±1.5 g/100 g linoleic acid and 47.5 ±5.6 g/100 g α-linolenic acid [15]. Also, Seher and Gundlach [16] established 6.7 g/100 g palmitic acid, 3.42 g/100 g stearic acid, 16.8 g/100 g oleic acid, 16.5 g/100 g linoleic acid and 54.7 g/100 g α-linolenic acid in linseed oil. Other authors found 4.0–7.0 g/100 g palmitic acid, 2.0–4.0 g/100 g stearic acid, 14.0–38.0 g/100 g oleic acid, 7.0–19.0 g/100 g linoleic acid and 35.0–66.0 g/100 g α-linolenic acid [17], while Overeem *et al.* [18] reported 5.3 g/100 g palmitic acid, 3.1 g/100 g stearic acid, 18.1 g/100 g oleic acid, 15.2 g/100 g linoleic acid and 54.10 g/100 g α-linolenic acid in linseed oil. Ryan *et al.* [19] determined 6.6 g/100 g palmitic acid, 4.1 g/100 g stearic acid, 24.0 g/100 g oleic acid, 19.90 g/100 g linoleic acid and 43.3 g/100 g α-linolenic acids. Within a certain variation the published results agree with the results of the present paper considering linseed oil as an important source of polyunsaturated α-linolenic acid with comparable low amounts of saturated fatty acids.

Palmitic acid and linoleic acid were the dominant fatty acids, constituting 70–80 g/100 g of the total fatty acids in the apple fruit [20]. As seen in Table 2, the fatty acid composition of apple seed oils in this investigation show similar results.

For peanut oil Dubois *et al.* [11] found 11.2–16.1 g/100 g palmitic acid, 3.2–4.1 g/100 g stearic acid, 35.9–58.7 g/100 g oleic acid, 20.7–37.3 g/100 g linoleic acid and 1.2–1.7 g/100 g eicosanoic acid which is in the range observed in the presented results. Other authors reported 6.0–16.0 g/100 g palmitic acid, 1.3–6.5 g/100 g stearic acid, 35.0–72.0 g/100 g oleic acid and 13.0–45.0 g/100 g linoleic acids for peanut oil [21].

In cottonseed oil, 21.0 and 27.2 g/100 g palmitic acid, 1.8 and 2.2 g/100 g stearic acid, 16.9 and 16.9 g/100 g oleic acid, 50.5 and 56.7 g/100 g linoleic acid and 0.9 and 1.2 g/100 g α-linolenic acids were determined [16], statistically within the range observed in the current study. According to Yazıcıoğlu and Karaali [21] cottonseed oil contained 17.0–31.0 g/100 g palmitic acid, 13.0–44.0 g/100 g oleic acid and 33.0–59.0 g/100 g linoleic acids. Also, Dubois *et al.* [11] determined 24.2 g/100 g palmitic acid, 2.3 g/100 g stearic acid, 17.4 g/100 g oleic acid and 53.2 g/100 g linoleic acids in cottonseed oil.

In vegetable oils four different derivatives of tocopherols and tocotrienols (α-, β-, γ-, δ-), respectively, can be found as vitamin-E-active compounds differing in the methylation of the chroman ring. Vitamin-E-active compounds have two functions: On one side they stabilize the oil against oxidative deterioration during storage or heat-treatment and on the other side they have some biological effects to humans. The main biochemical function of the tocopherols is believed to be the protection of polyunsaturated fatty acids against peroxidation [22,23]. The biological effect is derived from the inhibitory effect of vitamin-E-active compounds on the oxidation of low-density lipoprotein which may be responsible for the formation of atherosclerotic plaque, one reason for myocardial infarction and cardiovascular death [24]. Whereas the antioxidant and the anti-inflammatory properties of α-tocopherol have been well established, it is becoming increasingly evident that other isomers, especially γ-tocopherol are equally potent or possess additional biological properties [25,26]. Recent research has also shown that γ-tocopherol is a better negative risk factor for certain types of cancer and myocardial infarctions than α-tocopherol, whereas a high supplementation of α-tocopherol can deplete the body of γ-tocopherol [27].

In general the antioxidant activity increases for tocopherols and tocotrienols in the order α to δ, whereas the biological activity is opposite to the antioxidant activity [28,29]. Another member of this class of compounds is plastochromanol-8 (P-8). Its structure is similar to that of γ-tocopherol, but its basic molecular structure is a chromanol-6 ring in which methyl groups are located at positions 2, 7 and 8. The number of isoprene units on the side chain of position 2 is variable. Plastochromanol-8 is described as being more effective against oxidative deteriorations than α-tocopherol [30], while tocotrienols are less helpful in inhibiting autoxidation than tocopherols [31].

Due to the antioxidant properties the content and the composition of vitamin-E-active compounds is an important feature for the assessment of seed oils. The total content of vitamin-E-active compounds in the different seeds varied between 20.1 (fennel) and 94.0 mg/100g (quince) (Table 3). In comparison, the total content of vitamin-E-active compounds in rapeseed oil ranges between 43 and 268.0 mg/100 g and in soybean oil between 60.0 and 337.0 mg/100 g [32].According to this, the contents of vitamin-E-active compounds in the different seed oils from fruits and vegetables could be a good source for antioxidants.

Table 3. Distribution of vitamin-E-active compounds of seed oils (mg/100 g).

	α-T	α-T3	β-T	γ-T	P-8	γ-T3	δ-T
Linseed	0.8	n.d.	n.d.	39.0	13.5	n.d.	0.5
Apricot (Sweet)	2.8	n.d.	n.d.	67.3	n.d.	n.d.	2.2
Apricot (Bitter)	3.1	n.d.	n.d.	81.0	n.d.	n.d.	2.5
Pear (Ankara)	5.4	n.d.	0.2	55.6	0.6	n.d.	1.3
Fennel (Dulce)	n.d.	1.4	n.d.	0.5	n.d.	18.2	n.d.
Peanut	14.9	n.d.	0.5	16.9	n.d.	0.5	0.7
Apple (Golden)	51.4	n.d.	28.3	6.8	n.d.	n.d.	3.5
Apple (Starking)	54.4	n.d.	30.9	0.5	n.d.	n.d.	1.7
Cotton	36.2	0.3	0.2	48.7	1.1	n.d.	0.3
Quince	49.6	3.2	27.8	8.2	n.d.	n.d.	5.2
Chufa	68.5	n.d.	1.4	0.5	0.3	n.d.	n.d.
Apple (Starking)	60.5	0.3	34.3	n.d.	n.d.	n.d.	n.d.

n.d., not detectable; T, tocopherol; T3, tocotrienol; P-8, Plastochromanol-8.

All the oils investigated exhibited differences in their tocopherol contents. The predominant tocopherol compounds in the different seeds oils were α- and γ-tocopherol. Additionally in some oils lower amounts of α-tocotrienol, β-tocopherol, plastochromanol-8 and δ-tocopherol were found. The main vitamin-E-active compound of fennel seed oil was γ-tocotrienol with 18.2 mg/100 g. Also seed oils of other members of the family *Apiaceae* are characterized by a higher content of tocotrienols, such as dill (*Anethum graveolens)* (10.2 mg/100 g α-tocotrienol) or coriander (*Coriandrum sativum* L.) (23.1 mg/100 g γ-tocotrienol) [33].

The α-tocopherol group comprised of apple (golden) (51.4 mg/100 g), apple (starking) (54.4 mg/100 g, 60.5 mg/100 g), quince (49.6 mg/100 g) and chufa (68.5 mg/100 g) seed oil. Except for chufa seed oil, these oils also contained higher amounts of β-tocopherol with an average amount of about 30 mg/100 g ranging between 27.8 (quince seed oil) and 34.3 mg/100 g (apple seed oil (starking)). γ-tocopherol was predominant in linseed oil (39.0 mg/100 g), apricot seed oil (sweet, 67.3 mg/100 g), apricot seed oil (bitter, 81.1 mg/100 g), pear seed oil (55.6 mg/100 g), and cottonseed oil (48.7 mg/100 g).Among theseed oils in this group only cottonseed oil contained considerably high amount of α-tocopherol (36.2 mg/100 g) while for the other seed oils γ-tocopherol accounted for between 72% and 93% of the total vitamin-E-active compounds. In linseed oil remarkable amounts of plastochromanol-8 were found (13.5 mg/100 g).

Peanut oil neither belongs to the α-tocopherol nor to the γ-tocopherol group. The oil showed the lowest total amount of vitamin-E-active compounds of all investigated oils with almost equal amounts of α-tocopherol (14.9 mg/100 g) and γ-tocopherol (16.9 mg/100 g).

4. Conclusions

Linseeds, apricot seeds, pear seeds and peanut seeds contained more than 30 g/100 g oil making them interesting as valuable sources for oil production. Due to the lower oil content the other seeds can be interesting for oil processing if the raw material also contains protein with valuable amino acid composition, such as soybeans. The value of the oil depends on the fatty acid composition and the composition of vitamin-E-active compounds. Here the oils can be divided into two groups, one predominant in oleic acid with contents higher than 50 g/100 g (apricot seed oil, peanut oil, chufa seed oil) and one group mainly comprised of linoleic acid (pear seed oil, apple seed oil, cottonseed oil and quince seed oil). Linseed oil is characterized by a high amount of α-linolenic acid and fennel seed oil contained higher amounts of 18:1 fatty acids with petroselinic acid as predominant fatty acid which is usable in technical applications. Especially the fatty acid composition of apricot seed oil with low amount of saturated fatty acids, high amounts of oleic acid and moderate amounts of polyunsaturated fatty acids seems to be an interesting sources as edible oils for human nutrition. The composition of vitamin-E-active compounds of the different oils is characterized by either high amounts of α-tocopherol together with moderate amounts of β-tocopherol or by high amounts of γ-tocopherol. The predominant vitamin-E-amount in fennel seed oil is γ-tocotrienol and peanut oil contains more or less equal amounts of α- and γ-tocopherol. High sources of vitamin-E-active compounds with more than 70 mg/100 g are apricot seed oil, apple seed oil, cottonseed oil and quince seed oil.

In summary it can be concluded that raw materials such as seeds from fruit and vegetable processing are interesting and valuable sources for the production of vegetable oils usable as suppliers of fatty acids and vitamin-E-active compounds in human nutrition or technical applications.

Acknowledgments

This work was supported by The Scientific and Technical Research of Turkey (TÜBITAK) and Deutsche Forschungsgemeinschaft (DFG, Germany). The authors also thank to C.E., U.E. and B.B. for skilful technical assistance with gas liquid chromatography and high performance liquid chromatography.

Author Contributions

The experimental design was conceived and designed by M. M. Özcan and B. Matthäus. M. M. Özcan was responsible for the choice, collection, drying and purification of the samples. B. Matthäus carried out the analysis of the samples regarding oil content as well as composition of fatty acids and vitamin-E-active compounds. M. M. Özcan contributed the sections on introduction, methods and parts of the results, while B. Matthäus wrote the results and conclusions.

Conflicts of Interest

The authors declare no conflict of interest.

References

1. Femenia, A.; Rossello, C.; Mulet, A.; Canellas, J. Chemical composition of bitter and sweet apricot kernels. *J. Agric. Food Chem.* **1995**, *43*, 356–361.
2. Kamel, B.S.; Kakuda, Y. Characterization of the seed oil and meal from apricot, cherry, nectarine, peach and plum. *J. Am. Oil Chem. Soc.* **1992**, *69*, 492–494.
3. Nehdi, I.A. Characteristics and composition of *Washingtonia filifera* (Linden ex André) H Wendl. seed and seed oil. *Food Chem.* **2011**, *126*, 197–202.
4. Nehdi, I.A. Characteristics, chemical composition and utilisation of *Albizia julibrissin* seed oil. *Ind. Crops Prod.* **2011**, *33*, 30–34.
5. Nogala-Kalucka, M.; Rudzinska, M.; Zadernowski, R.; Siger, A.; Krzyzostaniak, I. Phytochemical content and antioxidant properties of seeds of unconventional oil plants. *J. Am. Oil Chem. Soc.* **2010**, *87*, 1481–1487.
6. Abdel-Rahaman, A.-H.Y. A study on some Egyptian citrus seed oils. *Grasas Aceit.* **1980**, *31*, 331–333.
7. *Oilseeds—Determination of Hexane Extract (or Light Petroleum Extract), Called "Oil Content"*; International Standard ISO 659:1998; ISO: Geneva, Switzerland, 1998.
8. *Animal and Vegetable Fats and Oils—Preparation of Methyl Esters of Fatty Acids*; International Standard ISO 5509:2000; ISO: Geneva, Switzerland, 2000.
9. Balz, M.; Schulte, E.; Thier, H.-P. Trennung von Tocopherolen und Tocotrienolen durch HPLC. *Eur. J. Lipid Sci. Technol.* **1992**, *94*, 209–213.
10. Teneva, O.T.; Zlatanov, M.D.; Antova, G.A. Lipid composition of flaxseeds. *Bulg. Chem. Comm.* **2014**, *36*, 157–163.
11. Dubois, V.; Breton, S.; Linder, M.; Fani, J.L.; Parmentier, M. Fatty acid profiles of 80 vegetable oils with regard to their nutritional potential. *Eur. J. Lipid Sci. Technol.* **2007**, *109*, 710–732.
12. Kim, M.; No, S.; Yoon, S.H. Stereospecific analysis of fatty acid composition of chufa (*Cyperus esculentus* L.) tuber oil. *J. Am. Oil Chem. Soc.* **2007**, *84*, 1079–1080.
13. Eteshola, E.; Oraedu, A.C.I. Fatty acid compositions of tiger nut tubers (*Cyperus esculentus* L.) baobab (*Adansonia digitata* L.) and their mixture. *J. Am. Oil Chem. Soc.* **1996**, *73*, 255–257.
14. Reiter, B.; Lechner, M.; Lorbeer, E. The fatty acid profiles—including petroselinic and *cis*-vaccenic acid—of different Umbelliferae seed oils. *Lipid Fett* **1998**, *100*, 498–502.
15. Kostik, V.; Memeti, S.; Bauer, B. Fatty acid composition of edible oils and fats. *J. Hyg. Eng. Des.* **2013**, *4*, 112–116.
16. Seher, V.A.; Gundlach, U. Isomere monoensauren in Pflanzenölen. *Fette Seifen Anstrichm.* **1982**, *84*, 342–349.
17. Dambroth, V.M.; Kluding, H.; Seehuber, R. Vegetable oils as raw materials of industry—A contribution of agriculture to secure the raw materials. *Fette Seifen Anstrichm.* **1982**, *84*, 173–178.

18. Overeem, A.; Buisman, G.J.H.; Derksen, J.T.P.; Cuperus, F.P.; Molhoek, L.; Grisnich, W.; Goemans, C. Seed oils rich in linolenic acid as renewable feedstock for environment-friendly in powder coatings. *Ind. Crops Prod.* **1999**, *10*, 157–165.
19. Ryan, E.; Galvin, K.; O'Connor, T.P.; Maguire, A.R. Phytosterol, squalene, tocopherol content and fatty acid profile of selected seeds, rains, and legumes. *Plant Foods Hum. Nutr.* **2007**, *62*, 85–91.
20. Wu, J.; Gao, H.; Zhao, L.; Liao, X.; Chen, F.; Wang, Z.; Hu, X. Chemical compositional characterization of some apple cultivars. *Food Chem.* **2007**, *103*, 88–93.
21. Yazıcıoğlu, T.; Karaali, A. On the fatty acid composition of Turkish vegetable oils. *Fette Seifen Anstrichm.* **1983**, *85*, 23–29.
22. Beringer, H.; Dompert, W.U. Fatty acid and tocopherol pattern in oil seeds. *Fette Seifen Anstrichm.* **1976**, *78*, 228–231.
23. Kamal-Eldin, A.; Andersson, R.A. A multivariate study of the correlation between tocopherol content and fatty acid compostion in vegetable oils. *J. Am. Oil Chem. Soc.* **1997**, *74*, 375–380.
24. Stephens, N.G.; Parsons, A.; Schofield, P.M.; Kelly, F.; Cheeseman, K.; Mitchinson, M.J. Randomised controlled trial of vitamin E in patients with coronary disease: Cambridge Heart Antioxidant Study (CHAOS). *Lancet* **1996**, *347*, 781–786.
25. Jiang, Q.; Elson-Schwab, I.; Courtemanche, C.; Ames, B.N. γ-tocopherol and its major metabolite, in contrast to α-tocopherol, inhibit cyclooxygenase activity in macrophages and epithelial cells. *Proc. Natl. Acad. Sci.* **2000**, *97*, 11494–11499.
26. Jiang, Q.; Lykkesfeldt, J.; Shigenaga, M.K.; Shigeno, E.T.; Christen, S.; Ames, B.N. γ-tocopherol supplementation inhibits protein nitration and ascorbate oxidation in rats with inflammation. *Free Radic. Biol. Med.* **2002**, *33*, 1534–1542.
27. Hensley, K.; Benaksas, E.J.; Bolli, R.; Comp, P.; Grammas, P.; Hamdheydari, L.; Mou, S.; Pye, Q.N.; Stoddard, M.F.; Wallis, G.; *et al.* New perspectives on vitamin E: γ-tocopherol and carboxyethylhydroxylchroman metabolites in biology and medicine. *Free Rad. Bio. Med.* **2004**, *36*, 1–15.
28. Olejnik, D.; Gogolewski, M.; Nogala-Kalucka, M. Isolation and some properties of plastochromanol-8. *Nahrung* **1997**, *41*, 101–104.
29. Pongracz, G.; Weiser, H.; Matzinger, D. Tocopherole—Antioxidantien der natur. *Fat Sci. Technol.* **1995**, *97*, 90–104.
30. Papas, A.M. Oil-soluble antioxidants in foods. *Toxicol. Ind. Health* **1993**, *9*, 123–149.
31. Elmadfa, I.; Wagner, K.-H. Vitamin E und Haltbarkeit von Pflanzenölen. *Fett Lipid* **1997**, *99*, 234–238.
32. AOCS. *Official Methods and Recommended Practices of the American Oil Chemists' Society: Physical and Chemical Characteristics of Oils, Fats and Waxes*; AOCS Press: Champaign, IL, USA, 1996.
33. Matthäus, B.; Vosmann, K.; Long Quoc, P.; Aitzetmüller, K. FA and tocopherol composition of Vietnamese oilseeds. *J. Am. Oil Chem. Soc.* **2003**, *80*, 1013–1020.

The Effect of *Convolvulus arvensis* Dried Extract as a Potential Antioxidant in Food Models

Nurul Aini Mohd Azman [1,2], Maria Gabriela Gallego [1], Luis Juliá [3], Lluis Fajari [3] and MaríaPilar Almajano [1,*]

[1] Chemical Engineering Department, Technical University of Catalonia, Av. Diagonal 647, 08028 Barcelona, Spain; E-Mails: ainiazman@gmail.com (N.A.M.A.); maria.gabriela.gallego (M.G.G.)

[2] Chemical and Natural Resources Engineering Faculty, University Malaysia Pahang, Lebuhraya Tun Razak, 26300 Pahang, Malaysia

[3] Química Biològica i Modelització Molecular, Institut de Química Avançada de Catalunya (CSIC), Jordi Girona 18-26, 08034 Barcelona, Spain; E-Mails: ljbmoh@cid.csic.es (L.J.); lluis.fajari@iqac.csic.es (L.F.)

* Author to whom correspondence should be addressed; E-Mail: m.pilar.almajano@upc.edu

Academic Editors: Maria G. Miguel, João Rocha and Ehab A. Abourashed

---

Abstract: In this study, the antioxidant activity of the *Convolvulus arvensis Linn* (CA) ethanol extract has been evaluated by different ways. The antioxidant activity of the extract assessed by 2,2′-azino-bis-3-ethylbenzothiazoline-6-sulphonic acid (ABTS) radical cation, the oxygen radical absorbance capacity (ORAC) and the ferric reducing antioxidant power (FRAP) was 1.62 mmol Trolox equivalents (TE)/g DW, 1.71 mmol TE/g DW and 2.11 mmol TE/g DW, respectively. CA ethanol extract exhibited scavenging activity against the methoxy radical initiated by the Fenton reaction and measured by Electron Paramagnetic Resonance (EPR). The antioxidant effects of lyophilised CA measured in beef patties containing 0.1% and 0.3% (w/w) CA stored in modified atmosphere packaging (MAP) (80% O_2 and 20% CO_2) was determined. A preliminary study of gelatine based film containing CA showed a strong antioxidant effect in preventing the degradation of lipid in muscle food. Thus, the present results indicate that CA extract can be used as a natural food antioxidant.

Keywords: *Convolvulus arvensis*; lipid oxidation; active packaging film; antioxidant activity

1. Introduction

Free radicals produced in the human body result from natural biochemical reactions and, together with external attacks due to stress, smoke and unbalanced diets, among other factors, could cause an imbalance between oxidants and antioxidants. For this reason, it is necessary to supplement the diet with antioxidant based food. This excess of radicals is associated with aging and many diseases such as heart problems, diabetes, neurodegenerative disorder and cancers. Previous studies indicate that the consumption of plant foods rich in antioxidants is beneficial for health and helps to prevent degenerative processes which contribute to many diseases [1–3]. Due to the increasing awareness of the benefits of consuming healthy food, many food companies are using antioxidants as an alternative approach, instead of using synthetic preservatives which at high doses may have toxic effects on the consumer.

Natural antioxidants are compounds, generally from plants, that are used as food additives with the aim of inhibiting oxidation of the product [4]. Thus, the use of natural antioxidants as preservatives to maintain quality and nutritional traits is increasingly widespread, mainly in food that contains high levels of lipids, such as meat products. Therefore, the incorporation of natural antioxidants such as herbs could be an economical strategy to develop healthier meat products. Moreover, they can improve technological properties, as well as increase the eco-efficiency [5] in the food industry. Besides formulation of food with a natural antioxidant strategy, active packaging is also gaining interest for its potential to provide food quality and safety benefits. The combination of natural preservatives and biodegradable plastic into one food packaging formulation is a promising approach to extending product shelf life [6].

Plants rich in polyphenol constituents possess antioxidant activity by free radical scavenging. For instance, green tea can inhibit lipid peroxidation and chelate transition metals, consequently helping to prevent degenerative diseases. If incorporated into an edible film, it could help to maintain the quality of food products [7].

Convolvulus arvensis Linn (CA) is an annual (or sometimes perennial climber), commonly found as a weed throughout Europe and Asia. This plant is being used for many purposes. The root and the resin are cholagogue, diuretic, laxative and purgative [8]. The flower is laxative, used as a tea infusion and also in treatment of wounds and fever, whereas the leaf can be helpful during the menstrual period [9]. Meanwhile, Meng *et al.* showed that the ubiquitous CA extract could be considered as a promising anti-cancer agent, with over 50% inhibition of tumor growth activity at non-toxic doses [10]. CA also provided an immunostimulant effect when tested on rabbits and turned out to have cytotoxic effects on human cancerous cells [11,12]. In a preliminary study, Thrakal *et al.* reported the antioxidant activity of CA extract using the DPPH method, nitric oxide scavenging activity and the reducing power assay [13]. Furthermore, the CA extract showed abundant traces of phenolic compounds including *p*-hydrobenzoic acid, syringic acid, vanillin, benzoic acid and ferulic acid [14]. This high content of phenolic compounds may allow it to serve as an antioxidant source for the food industry. However, the antioxidant activity of the CA extract towards lipid oxidation has not been fully determined yet. Thus, our goals were (1) to evaluate the antioxidant activity of CA using *in vitro* assays including FRAP,

TEAC, ORAC and EPR scavenging activity and (2) to demonstrate the ability of CA extract to inhibit lipid deterioration in beef meat, by adding the dry extract directly in the patty composition or in the formulation with active packaging. One of the components in CA is an alkaloid, which is a compound that exhibits anti-cancer activity but may display toxic effects in the host at high doses. Therefore, the extraction of CA has been carried out according to the method described by Meng *et al.* [10] to reduce the presence of alkaloid in the extract before adding the lyophilized extract directly into the beef.

2. Experimental Section

2.1. Plant Material

Commercial dried CA was kindly supplied by Pàmies Hortícoles (Balaguer, Spain), a registered herbal company. All reagents and solvents used were of analytical grade and obtained from Panreac (Barcelona, Spain) and Sigma Aldrich (Gillingham, England).

2.2. Extraction of CA Extract

Dried roots of CA were finely ground using a standard kitchen food processor. Ground CA was extracted in three different ways: (1) with 50:50 (v/v) ethanol:water; (2) with 75:25 (v/v) ethanol:water and (3) with 90:10 (v/v) ethanol:water, always in the ratio 1:30 (w/v). The extractions were performed at 4 °C ± 1 °C for 24 h, in the dark with constant stirring. The extract solutions of CA were recovered by filtration using Whatman Filter paper, 0.45 μm (Whatman, GE healthcare, Wauwatosa, WI, USA). Part of the supernatant was taken for subsequent use to determine the antiradical capacity. The volume of the remaining supernatant was measured and the excess of ethanol was removed under vacuum using a rotary evaporator (Buchi Re111, Switzerland) and kept frozen at −80 °C for 24 h. All extracts were dried in a freeze dryer (Unicryo MC2L −60 °C, Germany) under vacuum at −60 °C for three days to remove moisture. Finally, lyophilised CA was weighed to determine the soluble solids concentration (g/L) as described by Zhang *et al.* [15]

2.3. Determination of the Total Phenolic Content (TPC)

The Folin-Ciocalteu method was used to determine the total phenolic content (TPC) as reported by Santas *et al.* [16].

2.4. Determination of Free Radical Scavenging Activity Assays

2.4.1. *In-Vitro* Antioxidant Capacity Determination

Three different methods were used for the evaluation of the antioxidant activity of the extracts: 2,2′-azino-bis-(3-ethylbenzthiazoline)-6-sulphonic acid TEAC assay [17], Oxygen Radical Absorbance Capacity (ORAC) assay [18] and Ferric Reducing Antioxidant Power (FRAP) method [19]. Results were expressed as μM of Trolox equivalent (TE) per gram of dry weight of plant (DW).

2.4.2. Electron Paramagnetic Resonance (EPR) Spectroscopy Radical Scavenging Assay

EPR radical scavenging activity was measured following the method described by Azman *et al.* [20]. The extraction was executed in MeOH in 1:10 (w/v) ratio and the soluble concentration of CA was determined according to the procedure above. The spin-trapping reaction mixture consisted of 100 μL of DMPO (35 mM); 50 μL of H_2O_2 (10 mM); 50 μL CA extract at different concentrations or 50 μL of ferulic acid used as reference (0–20 g/L) or 50 μL of pure MeOH used as a control; and, finally, 50 μL of $FeSO_4$ (2 mM), added in this order. The final solutions (125 μL) were passed through a narrow (inside diameter = 2 mm) quartz tube and introduced into the cavity of the EPR spectrometer. The spectrum was recorded 10 min after the addition of the $FeSO_4$ solution, when the radical adduct signal is greatest.

X-band EPR spectra were recorded with a Bruker EMX-Plus 10/12 spectrometer under the following conditions: microwave frequency, 9.8762 GHz; microwave power, 30.27 mW; center field, 3522.7 G; sweep width, 100 G; receiver gain, 5.02×10^4; modulation frequency, 100 kHz; modulation amplitude, 1.86 G; time constant, 40.96 ms; conversion time, 203.0 ms.

2.5. Determination of Antioxidant Activity in Food Model

2.5.1. Preparation of Beef Patties

The meat consisted of flank of beef provided by "Embutidos La Masia", Barcelona. It was collected seven days after slaughter to allow it to mature and was kept at approximately −20 °C for further treatment. The extraction of CA was carried out according to the method used by Meng *et al.* to remove alkaloid compounds [10]. Fat and joint tissues were trimmed off lean meat (2000 g) and the meat was minced through 8 mm industrial plates. Then, the minced meat was divided into four batches and mixed with 1.5% of NaCl and either (i) control (no addition), (ii) 0.1% BHT, (iii) 0.1% lyophilised CA, (iv) 0.3% lyophilised CA. All batches were mixed vigorously for 2 min to attain an even distribution of additives throughout the meat. Each sample was moulded into smaller portions (about 20 g each), stuffed and packed with polystyrene B5-37 (Aerpack) trays and placed in BB4L bags (Cryovac) of low gas permeability (8–12 $cm^3 \cdot m^{-2}$ per 24 h). The air in the packaged trays was flushed with 80:20 (v/v) O_2:CO_2 by EAP20 mixture (Carburos Metalicos, Barcelona). Samples were stored in the dark at 4 °C ± 2 °C for 10 days and the samples were analysed for oxidation by thiobarbituric acid reactive substances (TBARS) method, % metmyoglobin, colour, pH and microbial quality. Every measurement was carried out in triplicate each day for 10 days (except for microbiological analysis which was done every three days).

2.5.2. TBARS Assay

The TBARS method was used to measure the extent of lipid oxidation over the storage period as described by Grau *et al.* [21]. Samples (1 g) were weighed in a tube and mixed with 3 g/L aqueous EDTA. Then, the sample was immediately mixed with 5 mL of thiobarbituric acid reagent using an Ultra-Turrax (IKA, Germany); at 32,000 rpm speed, for 2 min. All procedures were carried out in the dark and all samples were kept in ice. The mixture was incubated at 97 ± 1 °C in hot water for 10 min and shaken for 1 min during the process to form a homogeneous mixture. The liquid sample was recovered by filtration (Whatman Filter paper, 0.45 μm), and then it was cooled for 10 min. The absorbance value of each sample was measured at 531 nm using a spectrophotometer. The TBARS value was calculated from a malonaldehyde (MDA) standard curve prepared with 1,1,3,3-tetraethoxypropane and analysed by linear regression. All results were reported in mg malonaldehyde per kg of sample (mg MDA/kg sample).

2.5.3. Colour Measurement

Objective measurements of colour were performed using a CR 400 colorimeter (Minolta, Osaka, Japan). Each patty was cut and the colour of the slices was measured three times at each point. A portable colorimeter with the settings: pulsed xenon arc lamp, 0° viewing angle geometry and aperture size 8 mm, was used to measure meat colour in the CIELAB space (Lightness, L*; redness, a*; yellowness, b* (CIE, 1978). Before each series of measurements, the instrument was calibrated using a white ceramic tile.

2.5.4. Percentage of Metmyoglobin

The metmyoglobin method was based on that developed by Xu *et al.* [22]. Five grams of beef patties were homogenized with 25 mL of ice-cold 0.04 M phosphate buffer (pH 6.8) for 15 s using a homogenizer (Ultra-Turrax, IKA, Germany), which was set at speed setting 2 (18,000 rpm). The homogenised patty was allowed to stand at 4 °C for 1 h and centrifuged at 4500 g for 20 min at 4 °C using a high-speed freezing centrifuge (GI-20G, Anke, Shanghai, China). The absorbance of the filtered supernatant was read at 572, 565, 545, and 525 nm with a spectrometer (Fluostar Omega, BMG Labtech, Germany). The percentage of metmyoglobin was determined using the formula: MetMb (%) = [−2.514 (A572/A525) + 0.777 (A565/A525) + 0.8 (A545/A525) + 1.098] × 100

2.5.5. Development of Gelatin-Film with Antioxidant Coating

The fabrication of gelatin based film with antioxidant coating was adapted and characterized from Bodini *et al.* [23]. While the filmogenic solution was cooled after the solubilization of sorbitol, 0.75% (w/w) of CA extract / gelatin and 0.1% (w/w) BHT/gelatin were added.

2.6. Statistical Analysis

A one-way analysis of variance (ANOVA) was performed using Minitab 16 software program (Minitab Pty Ltd., Sydney, Australia) ($\alpha = 0.05$). The results were presented as mean values ($n \geq 3$).

3. Results and Discussion

3.1. Analysis of Total Polyphenols and Free Radical Activity Assays

On average, a higher weight of soluble solids was extracted from CA with 50% ethanol than with 75% and 90% of ethanol. The use of ethanol as extraction solvent is due to the fact that the solvent is recognized as a GRAS (Generally Recognized as Safe) component which can be safely used for applications in the food industry [24]. Ethanol also turned out to be effective in the extraction of flavonoids and their glycosides, catechols and tannins from raw plant materials. Generally, CA extracted with 50% ethanol showed higher phenolic content and antioxidant activity values in ORAC, FRAP and TEAC. Our results showed that the total phenolic content correlated with the antioxidant activity determined by the assays. Nevertheless, the values obtained in the ORAC assay were higher than the ones in the FRAP and TEAC assays, which also showed the extract scavenging activity against peroxy radicals (OOH•) generated in the assay. Total phenolic content reported for the plant extract with ethyl acetate turned out to be higher than our present results with 244 mg GAE/g DW [24]. The presence of compounds with antioxidant potential in the ethanol extract (Table 1) was revealed in the measurement of total antioxidant capacity in this study. In previous studies, the antioxidant activity of CA has been analyzed using the DPPH method, nitric oxide scavenging activity and reducing power assay applied to both methanol and ethyl acetate solvent extracts [13,25]. To the best of our knowledge, this is the first report of the antioxidant activity of CA extracts assessed using the TEAC, ORAC and FRAP methods.

Table 1. Soluble solids concentration, total phenolic content (TPC) and antioxidant activity of *Convolvulus arvensis Linn* (CA) extract.

Activity *Convolvulus arvensis*	**Extraction Solvent**		
	50:50 EtOH:H$_2$O	**75:25 EtOH:H$_2$O**	**90:10 EtOH:H$_2$O**
Soluble concentration (g/L)	13.76 ± 0.05	13.61 ± 0.02	11.43 ± 0.05
Total phenolic content (g GAE/g DW)	13.0 ± 0.05	12.1 ± 0.03	9.9 ± 0.02
FRAP (mmol of TE/g DW)	1.62 ± 0.02	1.51 ± 0.06	0.98 ± 0.01
TEAC (mmol of TE/g DW)	1.71 ± 0.01	1.68 ± 0.01	1.41 ± 0.04
ORAC (mmol of TE/g DW)	2.11 ± 0.05	2.05 ± 0.05	1.71 ± 0.03

* Mean value $n = 3$. The standard deviation for each assay is less than 5%. Gallic Acid Equivalent (GAE), Trolox Equivalent (TE), Dry Weight (DW).

3.2. EPR Scavenging Radical Assay

The EPR radical scavenging method has been developed by Azman *et al.* to evaluate the concentration of free methoxy radicals ($CH_3O^•$) generated in the Fenton reaction with the CA extract [20]. Figure 1 shows the decreasing signal of EPR with the increase of CA extract concentration. The free radical scavenging activity of CA extracts was investigated against methoxy ($CH_3O^•$) radical by a competitive method in the presence of DMPO as spin trap, using EPR spectroscopy. $CH_3O^•$ was generated according to the Fenton procedure with a relatively short half-life that was identified by EPR because of its ability to form a stable nitroxide adduct with DMPO, DMPO-OCH_3 (hyperfine splitting constants, $a_N = 13.9$ G and $a_H = 8.3$ G). This stable DMPO-OCH_3 compound can be detected by the double integration value of the signal from EPR. The presence of CA extract at different concentrations may compete with the spin trap in the scavenging of methoxy radicals. Thus, the effect reduces the amount of radical adducts and, accordingly, reduces the intensity of the EPR signal. The best fitting with intensity of EPR signal was shown as an exponential function (Figure 1) that, if concentration values are in g/L, corresponds to Equation (1):

$$y = 48.856\, e^{-0.001\, x};\ R^2 = 0.953 \quad (1)$$

The graph indicates that the exponential value of the signal of the spectrum decreased as the amount of CA increased. This study confirmed that the scavenging activity of the *Convolvulus arvensis* extracts containing polyphenol constituents could be measured by the decrease of the intensity of the spectral bands of the adduct DMPO-OCH_3 in the EPR spectrum with the amount of antioxidant.

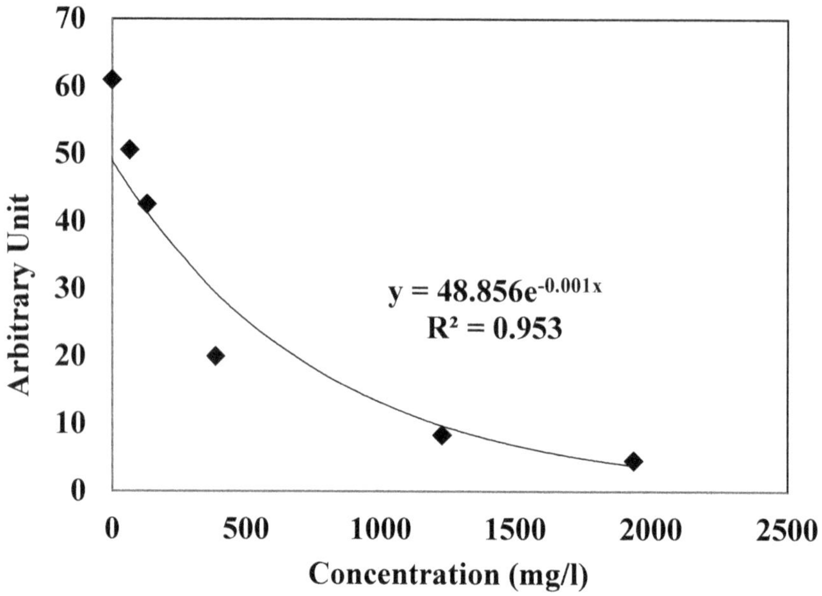

Figure 1. Antioxidant activity determined by the Electron paramagnetic resonance (EPR) spectrum of the radical adduct DMPO-OCH_3 generated from a solution of H_2O_2 (2 mM) and $FeSO_4$ (0.04 mM) with DMPO (14 mM) as spin trap in MeOH as solvent. The EPR signal decreases with the higher antioxidant activity.

3.3. Antioxidant Activity in Model Food

3.3.1. Colour and % Metmyoglobin

Meat colour is one of the most important traits that reflect the meat freshness and quality for consumers. The colour parameters representing lightness (L*), redness (a*), and yellowness (b*) are shown in Table 2. Generally, the value of colour (L*, a* and b*) decreased as the storage time increased. Initial mean lightness (CIE L*) was 38.68 ± 0.87, and control sample showed the lowest value of L* at the end of 10 days storage. There are marginally differences in L* with all samples throughout storage times. The slight change of L* value in meat storage was addressed by few authors [26,27]. The decrease of L* value indicates that a darkening developed, which may be due to the Maillard reaction or the effect of moisture content, which influences lightness values [28,29].

Table 2. Effect of CA extract and BHT on instrumental colour value (L*, a*, b*) of beef patties during 10 days of refrigerated storage at 4 °C. (Mean ± SE).

Assay	Sample	Days of Storages					
		0	2	4	6	8	10
L*	Control	$38.68 \pm 0.87^{a,1}$	$38.68 \pm 1.50^{a,1}$	$37.89 \pm 0.32^{b,3}$	$37.10 \pm 1.23^{b,2}$	$36.23 \pm 0.45^{c,2}$	$35.61 \pm 2.22^{d,1}$
	0.1% BHT	$38.68 \pm 0.87^{a,1}$	$39.06 \pm 1.08^{b,2}$	$38.25 \pm 0.97^{a,2}$	$38.43 \pm 1.06^{a,1}$	$37.09 \pm 1.19^{c,1}$	$36.18 \pm 0.46^{c,2}$
	0.1% CA	$38.68 \pm 0.87^{a,1}$	$38.60 \pm 1.05^{a,1}$	$39.26 \pm 1.46^{b,1}$	$38.63 \pm 0.55^{a,1}$	$37.11 \pm 1.02^{c,1}$	$37.06 \pm 1.22^{c,3}$
	0.3 % CA	$38.68 \pm 0.87^{a,1}$	$39.94 \pm 0.71^{b,2}$	$39.79 \pm 1.23^{b,1}$	$38.25 \pm 1.40^{a,1}$	$38.91 \pm 1.47^{a,3}$	$38.84 \pm 1.13^{a,4}$
a*	Control	$7.49 \pm 0.27^{a,1}$	$7.77 \pm 0.29^{a,1}$	$6.54 \pm 0.33^{b,1}$	$6.27 \pm 0.16^{b,2}$	$4.71 \pm 0.02^{c,1}$	$2.09 \pm 0.01^{d,1}$
	0.1% BHT	$7.49 \pm 0.27^{a,1}$	$8.18 \pm 0.42^{b,2}$	$9.28 \pm 0.28^{c,2}$	$7.05 \pm 0.31^{a,1}$	$6.36 \pm 0.37^{d,2}$	$2.87 \pm 0.01^{e,1}$
	0.1% CA	$7.49 \pm 0.27^{a,1}$	$7.61 \pm 0.33^{a,1}$	$5.57 \pm 0.26^{b,3}$	$6.25 \pm 0.19^{c,2}$	$6.60 \pm 0.33^{c,2}$	$3.31 \pm 0.02^{d,2}$
	0.3 % CA	$7.49 \pm 0.27^{a,1}$	$7.64 \pm 0.21^{a,1}$	$7.20 \pm 0.47^{a,4}$	$7.50 \pm 0.20^{a,1}$	$7.61 \pm 0.37^{a,3}$	$4.08 \pm 0.01^{b,3}$
b*	Control	$7.42 \pm 0.32^{a,1}$	$4.86 \pm 0.01^{b,1}$	$7.68 \pm 0.36^{a,1}$	$8.55 \pm 0.19^{c,1}$	$9.95 \pm 0.21^{d,1}$	$6.77 \pm 0.02^{e,1}$
	0.1% BHT	$7.42 \pm 0.32^{a,1}$	$6.68 \pm 0.16^{b,2}$	$8.40 \pm 0.27^{c,1}$	$8.39 \pm 0.37^{c,1}$	$8.38 \pm 0.24^{c,2}$	$6.10 \pm 0.01^{d,1}$
	0.1% CA	$7.42 \pm 0.32^{a,1}$	$8.00 \pm 0.37^{b,3}$	$8.19 \pm 0.33^{b,1}$	$5.17 \pm 0.13^{c,2}$	$7.49 \pm 0.07^{a,3}$	$4.35 \pm 0.09^{d,2}$
	0.3 % CA	$7.42 \pm 0.32^{a,1}$	$7.14 \pm 0.49^{a,4}$	$7.59 \pm 0.29^{a,2}$	$7.01 \pm 0.21^{a,3}$	$7.99 \pm 0.27^{a,3}$	$3.25 \pm 0.01^{b,3}$

Control: 1.5% salt (w/w); 0.1% BHT: 1.5% salt with 0.1% BHT (w/w); 0.1% CA: 1.5% salt with 0.1% CA (w/w) 0.3% CA: 1.5% salt with 0.3% CA (w/w). $^{a–d}$: Means within a row with different letters are significantly different ($p < 0.05$). $^{1–4}$: For each attribute, means within a column with different number are significantly different ($p < 0.05$). Mean value $n = 6$ and the standard deviation for each assay is less than 5%.

A reduction of the a* value was experienced by all samples in 10 days' storage ($p < 0.05$), indicating that a decrease in redness occurred in the meat. The 0.1% BHT displayed the highest value of a* during three days' storage and declined gradually afterwards ($p < 0.05$). This finding was expected due to the role of BHT as a synthetic antioxidant which is used to retain colour and delay lipid oxidation in the meat [30]. The redness of 0.3% CA was maintained around a value of 7 during the eight days before the colour faded rapidly in 10 days' storage ($p > 0.05$). At the end of storage, 0.3% CA showed the highest a* value followed by 0.1% CA ($p < 0.05$) and 0.1% BHT and control exhibited a low value with no significant difference between both samples ($p > 0.05$). Many features contributed to the red colour in the meat such as the influence of salt and oxygen composition that enhanced the red colour of beef patties [31,32]. The samples had an initial yellowness (b*) value of 7.42 ion that enhanced the red in both samples (eight days before $p > 0.05$). In general, no significant difference ($p > 0.05$) was observed in b* values in all samples throughout storage. The present findings seem to be consistent with other research which found that yellowness in meat patties is not influenced by storage time and packaging conditions [26,33].

The effect of CA extracts and BHT on relative MetMb percentage in beef patties are presented in Table 3. A significant correlation between MetMb (%) and the instrumental colour features was reported previously [22]. The MetMb percentage increased as the storage time increased throughout the 10 days' refrigeration, whereas the control showed the highest MetMb compared to all samples. The treated groups of CA extract and BHT had lower ($p < 0.05$) proportions of MetMb compared to the control at the end of storage. The acceleration of colour deterioration and lipid oxidation depended on many causes, including storage time, type of packaging and test system. Free radicals produced by lipid oxidation in meat are susceptible to initiating the reaction of oxidizing oxymyoglobin (red colour) to metmyoglobin (brown colour) which results in the discolouration of meat during storage. Previous research has indicated a relationship between lipid oxidation and myoglobin oxidation or discolouration in meat products [22,34]. A sufficient amount of antioxidant in the sample can delay the formation of metmyoglobin. The scavenging ability of samples treated with antioxidant can reduce the oxidation of metmyoglobin acting as scavengers of hydroxyl radicals produced from oxidation of oxymyoglobin. The 0.3% of CA extract displayed the lowest metmyoglobin percentage compared to all samples, and the change of % metmyoglobin was inversely proportional to the value of redness (a*).

3.3.2. TBARS Analysis in Beef Patties

In general, the levels of lipid oxidation in beef patties increased over time and the values followed the order: 0.3% CA < 0.1% BHT < 0.1% CA < Control (Figure 2). The presence of a controlled atmosphere with high oxygen packaging (MAP) resulted in higher TBARS values and increased the oxidation rate in muscle food [32,35]. No statistical difference was observed between 0.1% BHT and 0.1% CA on any of the storage days However, the TBARS values of both samples showed significant differences compared to those of the control samples ($p < 0.05$). From seven days onwards, the control reached the highest TBARS values of all samples, with values greater than 1.2 mg malonaldehyde/kg sample. The levels of lipid oxidation were the lowest in 0.3% CA in beef patties throughout storage and significantly lower than for all other samples. The oxidation rate of meat patties was more reduced for a higher concentration of CA extract, as shown by comparison of the rates for 0.1% and 0.3% addition.

The 0.1% BHT was added for comparison with the natural antioxidant bearing in mind the FDA guidelines for using BHT is ≤200 ppm in meat products. The effect of CA extract on lipid oxidation in meat has never been reported. The active properties of CA reported by Hegab and Ghareib [14] have been attributed to various phenolic acids such as ferulic acid, cinnamic acid and *p*-coumaric acid. The antioxidant activity of phenolic compounds is closely related to the hydroxyl group linked to the aromatic ring which is capable of donating hydrogen atoms with electrons and neutralizing free radicals. This mechanism blocks further degradation by oxidation to form MDA, which can be measured by the TBARS method [36]. This study confirmed the potential of CA extract to inhibit lipid degradation in beef patties.

3.3.3. TBARS Analysis in Meat under Active Packaging

The TBARS index (Figure 3) revealed that the coating of beef patties with edible films enriched with antioxidants lowered the oxidation rate during 17 days' storage. By comparison, the gelatin film without any added antioxidants did not display any protective effect. Lipid oxidation with respect to TBARS values of control, meat patties sample and those wrapped with CA and BHT incorporated film showed a significantly different TBARS value ($p < 0.05$) than the control sample. This result suggested that lipid oxidation in meat samples could be minimized by the use of a gelatin film containing CA probably due to the antioxidant activity of the CA extract. However, BHT and CA coated in gelatin film did not show any significant difference between the values for the different periods of storage.

Duthie *et al.* demonstrated the presence of phenolic acids measured using LC-MS in chicken patties mixed with vegetable powders including ferulic acid, *p*-hydrobenzoic acid, *p*-coumaric acid, caffeic acid and cinnamic acid [37]. In reviewing the literature, CA contained a great amount of phenolic compounds that may be responsible for its strong antioxidant activity in many assays. The constituents included *p*-hydroxybenzoic acid, syringic acid, vanillin, benzoic acid, ferulic acid found by Elzaawely and Tawata [25]. HPLC analysis done by Hegab and Ghareib showed traces of eight phenolic constituents including pyrogallic acid, protocatechuic acid, resorcinol, chologenic acid, caffeic acid, salicylic acid, *p*-coumaric acid and cinnamic acid [14]. These compounds lead to many pharmacological benefits to human health. Benzoic acid and its derivatives showed antimicrobial potential [38] while gallic acid and caffeic acid showed 50% inhibitory effects on cancer cell proliferation [39]. *p*-coumaric, ferulic acid and cinnamic acid and their derivatives bring many pharmacological benefits to humans including, anticancer and antioxidant effects [40,41]. Moreover, many constituents detected in the CA extract correlated significantly with antioxidant activity measured by ORAC and TEAC assays and have played an important role in the detoxification of endogenous compounds in humans [42].

Table 3. Effects of CA extract and BHT on metmyoglobin changes in beef patties during 10 days of refrigerated storage at 4 °C. (Mean ± SE).

Assay	Sample	Day of Storages 0	2	4	6	8	10
% Metmyoglobin	Control	23.38 ± 0.46 [1]	29.19 ± 0.71 [2]	37.56 ± 1.31 [2]	47.84 ± 1.21 [1]	53.9 ± 1.16 [1]	60.03 ± 2.82 [2]
	0.1% BHT	23.38 ± 0.46 [1]	25.69 ± 1.04 [4]	29.98 ± 0.81 [1]	37.5 ± 1.85 [2]	45.7 ± 1.53 [2]	57.6 ± 1.24 [1]
	0.1% CA	23.38 ± 0.46 [1]	27.96 ± 0.33 [1]	33.48 ± 0.79 [3]	44.81 ± 1.29 [3]	48.7 ± 1.67 [3]	57.1 ± 1.18 [1]
	0.3 % CA	23.38 ± 0.46 [1]	28.91 ± 0.81 [1,2]	30.71 ± 0.29 [1]	33.37 ± 0.94 [4]	40.1 ± 1.53 [4]	50.78 ± 1.56 [3]

Control: 1.5% salt (w/w); 0.1% BHT: 1.5% salt with 0.1% BHT (w/w); 0.1% CA: 1.5% salt with 0.1% CA (w/w) 0.3% CA: 1.5% salt with 0.3% CA (w/w). All samples values are significantly different throughout the storage time ($p < 0.05$) [1–4]: Means within a column with different numbers are significantly different ($p < 0.05$). Mean value $n = 6$ and the standard deviation for each assay is less than 5%.

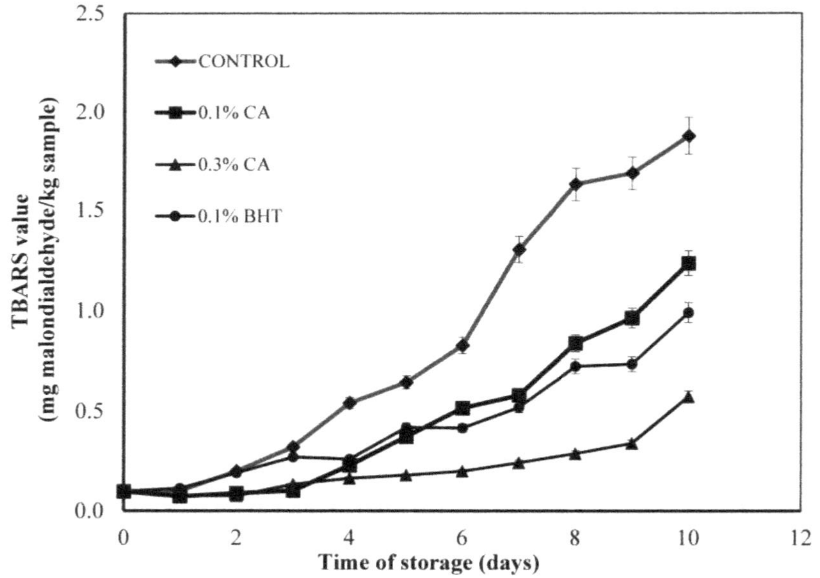

Figure 2. Changes in TBARS values (mg malondialdehyde/kg sample) of control and sample containing different concentrations (0.1% and 0.3% w/w) of CA extract in MAP atmosphere during 10 days storage at 4 ± 1 °C without light. Each sample was measured in triplicate and the average standard deviation for each sample was less than 5%.

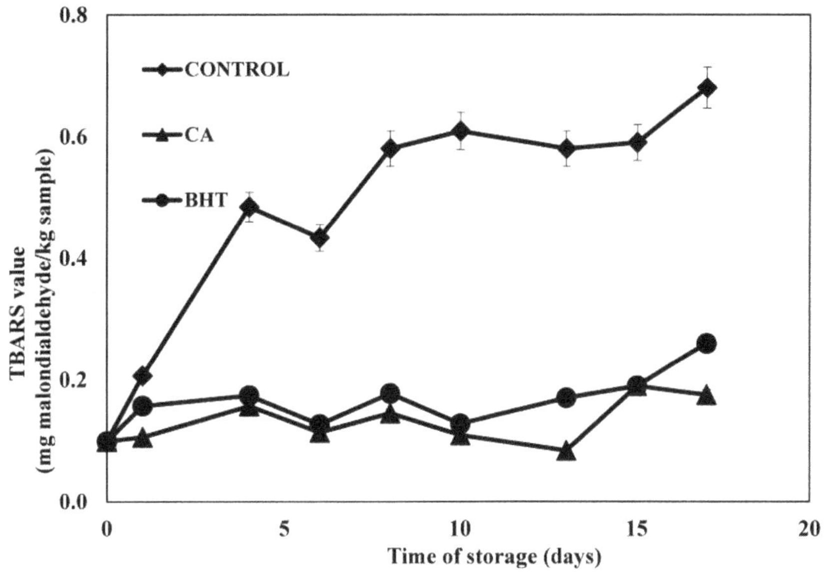

Figure 3. Changes in TBARS values (mg malondialdehyde/kg sample) of control and sample containing BHT and CA extract in MAP atmosphere during 17 days' storage at 4 ± 1 °C without light. Each sample was measured in triplicate and the average standard deviation for each sample was less than 5%.

4. Conclusions

The CA extract showed an excellent antioxidant activity in 50% aqueous ethanol measured by FRAP, TEAC and ORAC assays. This is also the first time that the radical scavenging activity has been evaluated in a CA extract against methoxy radical generated in the Fenton Reaction assessed by EPR.

The CA extract also showed a protective effect against lipid degradation in the muscle food model. Lyophilised CA (0.1% and 0.3% w/w) can be applied as an antioxidant in meat patties. It showed inhibition of lipid oxidation in MAP. 0.3% of CA retained meat redness and browning colour measured by the metmyoglobin assay which was much better than the control ($p < 0.05$) during 10 days' storage. A preliminary study of gelatin based film coated with CA showed there was a significant delay in the lipid degradation in beef ($p < 0.05$). Therefore, this study confirmed that CA could be used by the food industry as a source of antioxidants.

Acknowledgments

The authors acknowledge the special fund from The Technical University of Catalonia, University Malaysia Pahang, Malaysia Government and the Government of Andorra in the framework of the research grants of Working Community of the Pyrenees (ACTP016-AND/2012) for the financial support of this research. The authors acknowledge Monica Blanco, María Rosal and Michael H. Gordon for their help with the English.

Author Contributions

María Pilar Almajano and Nurul Aini Mohd Azman conceived and designed the study. Maria Gabriela Gallego performed the LCMS analysis whereas Luis Juliá and Lluis Fajari supervise the experimental work on the EPR study. Nurul Aini Mohd Azman wrote the paper with assistance of Maria Pilar Almajano who reviewed all the manuscript and the final version to be submitted.

Conflicts of Interest

The authors declare no conflict of interest.

References

1. Alvarez-Suarez, J.M.; Giampieri, F.; González-Paramás, A.M.; Damiani, E.; Astolfi, P.; Martinez-Sanchez, G.; Bompadre, S.; Quiles, J.L.; Santos-Buelga, C.; Battino, M. Phenolics from monofloral honeys protect human erythrocyte membranes against oxidative damage. *Food Chem. Toxicol.* **2012**, *50*, 1508–1516.
2. Zhao, G.; Yin, Z.; Dong, J. Antiviral efficacy against hepatitis B virus replication of oleuropein isolated from *Jasminum officinale* L. var. grandiflorum. *J. Ethnopharmacol.* **2009**, *125*, 265–268.
3. Chang-Liao, W.-L.; Chien, C.-F.; Lin, L.-C.; Tsai, T.-H. Isolation of gentiopicroside from Gentianae Radix and its pharmacokinetics on liver ischemia/reperfusion rats. *J. Ethnopharmacol.* **2012**, *141*, 668–673.
4. Halliwell, B. Antioxidants in human health and disease. *Annu. Rev. Nutr.* **1996**, *16*, 33–50.
5. Martín-Sánchez, A.M.; Navarro, C.; Pérez-Álvarez, J.A.; Kuri, V. Alternatives for Efficient and Sustainable Production of Surimi: A Review. *Compr. Rev. Food Sci. Food Saf.* **2009**, *8*, 359–374.
6. Suppakul, P.; Miltz, J.; Sonneveld, K.; Bigger, S.W. Active Packaging Technologies with an Emphasis on Antimicrobial Packaging and Its Applications. *J. Food Sci.* **2003**, *68*, 408–420.
7. Siripatrawan, U.; Noipha, S. Active film from chitosan incorporating green tea extract for shelf life extension of pork sausages. *Food Hydrocoll.* **2012**, *27*, 102–108.
8. Chopra, R.N.; Nayar, S.L.; Chopra, I.C. *Glossary of Indian Medicinal Plants (Including Supplement)*; CSIR: New Delhi, India, 1986.
9. Foster, S.; Duke, J.A. *A Field Guide of Medicinal Plants*; Houghton Mifflin Co.: New York, NY, USA, 1990.
10. Meng, X.L.; Riordan, N.H.; Casciari, J.J.; Zhu, Y.; Gonzalez, J.M.; Miranda-Massari, J.R.; Riordan, H.D. Effect of High Molecular Mass *Convolvulus arvensis* Extract on Tumor Growth and Angiogenesis. *Pharmacognosis* **2002**, *21*, 323–328.
11. Ghasemi, N.; Kohi, M. Cytotoxic effect of *Convolvulus arvensis* extracts on human cancerous cell line. *Iran. J. Pharm. Res.* **2008**, *3*, 31–34.
12. Al-Bowait, M.E.; Albokhadaim, I.F.; Homeida, A.M. Immunostimulant Effect of Binweed (*Convolvulus arvensis*) Extract in Rabbit. *Res. J. Pharmacol.* **2010**, *4*, 51–54.
13. Thakral, J.; Borar, S.; Kalia, A.N. Antioxidant Potential Fractionation from Methanol Extract of Aerial Parts of *Convolvulus arvensis* Linn (Convolvulaceae). *Int. J. Pharm. Sci. Drug Res.* **2010**, *2*, 219–223.

14. Hegab, M.M.; Ghareib, H.R. Methanol Extract Potential of Field Bindweed (*Convolvulus arvensis* L) for Wheat Growth Enhancement. *Int. J. Bot.* **2010**, *6*, 334–342.
15. Zhang, S.; Bi, H.; Liu, C. Extraction of bio-active components from *Rhodiola sachalinensis* under ultrahigh hydrostatic pressure. *Sep. Purif. Technol.* **2007**, *57*, 277–282.
16. Santas, J.; Carbo, R.; Gordon, M.; Almajano, M. Comparison of the antioxidant activity of two Spanish onion varieties. *Food Chem.* **2008**, *107*, 1210–1216.
17. Almajano, M.P.; Carbó, R.; Jiménez, J.A.L.; Gordon, M.H. Antioxidant and antimicrobial activities of tea infusions. *Food Chem.* **2008**, *108*, 55–63.
18. Skowyra, M.; Falguera, V.; Gallego, G.; Peiró, S.; Almajano, M.P. Antioxidant properties of aqueous and ethanolic extracts of tara (*Caesalpinia spinosa*) pods *in vitro* and in model food emulsions. *J. Sci. Food Agric.* **2014**, *94*, 911–918.
19. Gallego, M.G.; Gordon, M.H.; Segovia, F.J.; Skowyra, M.; Almajano, M.P. Antioxidant Properties of Three Aromatic Herbs (Rosemary, Thyme and Lavender) in Oil-in-Water Emulsions. *J. Am. Oil Chem. Soc.* **2013**, *90*, 1559–1568.
20. Azman, N.A.M.; Peiró, S.; Fajarí, L.; Julià, L.; Almajano, M.P. Radical scavenging of white tea and its flavonoid constituents by electron paramagnetic resonance (EPR) spectroscopy. *J. Agric. Food Chem.* **2014**, *62*, 5743–5748.
21. Grau, A.; Guardiola, F.; Boatella, J.; Barroeta, A.; Codony, R. Measurement of 2-thiobarbituric acid values in dark chicken meat through derivative spectrophotometry: Influence of various parameters. *J. Agric. Food Chem.* **2000**, *48*, 1155–1159.
22. Xu, Z.; Tang, M.; Li, Y.; Liu, F.; Li, X.; Dai, R. Antioxidant properties of Du-zhong (*Eucommia ulmoides* Oliv.) extracts and their effects on color stability and lipid oxidation of raw pork patties. *J. Agric. Food Chem.* **2010**, *58*, 7289–7296.
23. Bodini, R.B.; Sobral, P.J.A.; Favaro-Trindade, C.S.; Carvalho, R.A. Properties of gelatin-based films with added ethanol–propolis extract. *LWT Food Sci. Technol.* **2013**, *51*, 104–110.
24. Fernández-Agulló, A.; Pereira, E.; Freire, M.S.; Valentão, P.; Andrade, P.B.; González-Álvarez, J.; Pereira, J.A. Influence of solvent on the antioxidant and antimicrobial properties of walnut (*Juglans regia* L.) green husk extracts. *Ind. Crops Prod.* **2013**, *42*, 126–132.
25. Elzaawely, A.A.; Tawata, S. Antioxidant Activity of Phenolic Rich Fraction Obtained from *Convolvulus arvensis* L. leaves Grown in Egypt. *Asian J. Crop Sci.* **2012**, *4*, 32–40.
26. Cachaldora, A.; García, G.; Lorenzo, J.M.; García-Fontán, M.C. Effect of modified atmosphere and vacuum packaging on some quality characteristics and the shelf-life of "morcilla", a typical cooked blood sausage. *Meat Sci.* **2013**, *93*, 220–225.
27. Martín-Sánchez, A.M.; Ciro-Gómez, G.; Sayas, E.; Vilella-Esplá, J.; Ben-Abda, J.; Pérez-Álvarez, J.Á. Date palm by-products as a new ingredient for the meat industry: Application to pork liver pâté. *Meat Sci.* **2013**, *93*, 880–887.
28. Pérez-Álvarez, J.A.; Fernández-López, J. Color Characteristics of Meat and Poultry Analysis, Handbook of Processed Meats and Poultry Analysis; CRC Press: Boca Raton, FL, USA, 2009; pp. 355–373.
29. Sidhu, J.S. Date Fruits Production and Processing, Handbook of Fruits and Fruit Processing; Blackwell Publishing: Ames, IA, USA, 2006, pp. 391–420.

30. Sánchez-Escalante, A.; Djenane, D.; Torrescano, G.; Beltrán, J.; Roncalés, P. The effects of ascorbic acid, taurine, carnosine and rosemary powder on colour and lipid stability of beef patties packaged in modified atmosphere. *Meat Sci.* **2001**, *58*, 421–429.
31. Azman, N.A.M.; Gordon, M.H.; Skowyra, M.; Segovia, F.; Almajano, M.P. Use of lyophilised and powdered *Gentiana lutea* root in fresh beef patties stored under different atmospheres. *J. Sci. Food Agric.* **2014**, doi:10.1002/jsfa.6878.
32. Martínez, L.; Djenane, D.; Cilla, I.; Beltrán, J.A.; Roncalés, P. Effect of varying oxygen concentrations on the shelf-life of fresh pork sausages packaged in modified atmosphere. *Food Chem.* **2006**, *94*, 219–225.
33. Triki, M.; Herrero, A.M.; Rodríguez-Salas, L.; Jiménez-Colmenero, F.; Ruiz-Capillas, C. Chilled storage characteristics of low-fat, *n*-3 PUFA-enriched dry fermented sausage reformulated with a healthy oil combination stabilized in a konjac matrix. *Food Control* **2013**, *31*, 158–165.
34. O'Grady, M.N.; Monahan, F.; Brunton, N.P. Oxymyoglobin Oxidation and Lipid Oxidation in Bovine Muscle—Mechanistic Studies. *J. Food Sci.* **2001**, *66*, 386–392.
35. Azman, N.; Segovia, F.; Martínez-Farré, X.; Gil, E.; Almajano, M. Screening of Antioxidant Activity of Gentian Lutea Root and Its Application in Oil-in-Water Emulsions. *Antioxidants* **2014**, *3*, 455–471.
36. Kim, S.J.; Cho, A.R.; Han, J. Antioxidant and antimicrobial activities of leafy green vegetable extracts and their applications to meat product preservation. *Food Control* **2013**, *29*, 112–120.
37. Duthie, G.; Campbell, F.; Bestwick, C.; Stephen, S.; Russell, W. Antioxidant effectiveness of vegetable powders on the lipid and protein oxidative stability of cooked Turkey meat patties: Implications for health. *Nutrients* **2013**, *5*, 1241–1252.
38. Oloyede, G.K.; Willie, I.E.; Adeeko, O.O. Synthesis of Mannich bases: 2-(3-Phenylaminopropionyloxy)-benzoic acid and 3-Phenylamino-1-(2,4,6-trimethoxy-phenyl)-propan-1-one, their toxicity, ionization constant, antimicrobial and antioxidant activities. *Food Chem.* **2014**, *165*, 515–521.
39. Tao, L.; Wang, S.; Zhao, Y.; Sheng, X.; Wang, A.; Zheng, S.; Lu, Y. Phenolcarboxylic acids from medicinal herbs exert anticancer effects through disruption of COX-2 activity. *Phytomedicine* **2014**, *21*, 1473–1482.
40. Kiliç, I.; Yeşiloğlu, Y. Spectroscopic studies on the antioxidant activity of *p*-coumaric acid. *Spectrochim. Acta A Mol. Biomol. Spectrosc.* **2013**, *115*, 719–724.
41. Nithiyanantham, S.; Siddhuraju, P.; Francis, G. A promising approach to enhance the total phenolic content and antioxidant activity of raw and processed *Jatropha curcas* L. kernel meal extracts. *Ind. Crops Prod.* **2013**, *43*, 261–269.
42. Yeh, C.-T.; Yen, G.-C. Effects of phenolic acids on human phenolsulfotransferases in relation to their antioxidant activity. *J. Agric. Food Chem.* **2003**, *51*, 1474–1479.

11

Betalains, Phenols and Antioxidant Capacity in Cactus Pear [*Opuntia ficus-indica* (L.) Mill.] Fruits from Apulia (South Italy) Genotypes

Clara Albano [1,†], Carmine Negro [2,†], Noemi Tommasi [1], Carmela Gerardi [1], Giovanni Mita [1], Antonio Miceli [2], Luigi De Bellis [2] and Federica Blando [1,†,*]

[1] Institute of Sciences of Food Production (ISPA), CNR, Lecce Unit, 73100 Lecce, Italy; E-Mails: clara.albano@ispa.cnr.it (C.A.); noetom@hotmail.it (N.T.); carmela.gerardi@ispa.cnr.it (C.G.); giovanni.mita@ispa.cnr.it (G.M.)

[2] Department of Biological and Environmental Sciences and Technologies (DISTeBA), Salento University, 73100 Lecce, Italy; E-Mails: carmine.negro@unisalento.it (C.N.); antonio.miceli@unisalento.it (A.M.); luigi.debellis@unisalento.it (L.B.)

† These authors contributed equally to this work.

* Author to whom correspondence should be addressed; E-Mail: federica.blando@ispa.cnr.it

Academic Editors: Antonio Segura-Carretero and David Arráez-Román

---

Abstract: Betacyanin (betanin), total phenolics, vitamin C and antioxidant capacity (by Trolox-equivalent antioxidant capacity (TEAC) and oxygen radical absorbance capacity (ORAC) assays) were investigated in two differently colored cactus pear (*Opuntia ficus-indica* (L.) Mill.) genotypes, one with purple fruit and the other with orange fruit, from the Salento area, in Apulia (South Italy). In order to quantitate betanin in cactus pear fruit extracts (which is difficult by HPLC because of the presence of two isomers, betanin and isobetanin, and the lack of commercial standard with high purity), betanin was purified from *Amaranthus retroflexus* inflorescence, characterized by the presence of a single isomer. The purple cactus pear variety showed very high betanin content, with higher levels of phenolics, vitamin C, and antioxidant capacity (TEAC) than the orange variety. These findings confirm the potential for exploiting the autochthonous biodiversity of cactus pear fruits. In particular, the purple variety could be an interesting source of colored bioactive compounds which not only have coloring potential, but are also an excellent source of dietary antioxidant components which may have beneficial effects on consumers' health.

Keywords: *Opuntia ficus-indica*; cactus pear fruit; betanin; phenolic content; antioxidant activity; ORAC and TEAC assays

---

1. Introduction

Betalains are water-soluble nitrogen-containing pigments that are responsible for the bright red or yellow color of fruits, flowers, roots and leaves of plants belonging to the order of *Caryophyllales*. One of these plants, *Opuntia ficus-indica* (L.) Mill. (cactus or prickly pear) contains betalains in the fruits, particularly betacyanins in the purple variety and betaxanthins in the orange variety [1]. Cactus pear is native to Mexico and was subsequently brought to Europe, Africa and Middle East, showing remarkable adaptation to arid and semi-arid climates in tropical and sub-tropical regions of the globe. In Italy, it is spontaneous and also cultivated in the South, namely in Sicily, Sardinia, Calabria and Apulia [2].

Biodiversity of the *Opuntia* sp. have been since years the subject of investigations, both at taxonomic and molecular level. Taxonomical genotype assignments, based only on morphological features, demonstrated some inconsistencies, since the continuous morphological variability within the genus [3]. To overcome these problems, molecular markers (SSR) were used, in order to approach the assessment of genetic diversity in *Opuntia* germoplasm, classified solely on the morphological basis [4].

Cactus pear fruits are commercially quite important as they are flavorsome and well appreciated by consumers. The fruit is usually consumed fresh, during the ripening period, July–October, but the increasing market demand for health-promoting food has prompted food technologists to develop procedures to increase cactus pear fruit shelf life [5,6]. Cactus pear fruit has attracted attention due to its nutritional and health-promoting benefits, being rich in bioactive antioxidant compounds (betalains, ascorbic acid and polyphenols). The nutritional and chemical composition of the prickly pear fruit has already been reviewed [7,8]. Moreover, cactus pear fruit extract has been shown to have antiulcerogenic, antioxidant, anticancer, neuroprotective, hepatoprotective, and antiproliferative activities [9–13]. Cactus pears have also been considered as a good source for red and yellow food colorings. Since betalains are particularly stable in the range of pH 4 to 7, they are preferably indicated for coloring non-acid foods; moreover, the presence of betacyanins and betaxanthins together provides a wide color interval [1,14].

Over the last year there has been an abundance of scientific papers on cactus pear as a source of bioactive compounds for nutrition, health and disease [14–18], underlining the interest in the numerous properties (both its bioactivity and coloring potential) of this plant species, well adapted to extreme growing conditions in arid and semi-arid zones.

Mexico is the main cactus pear fruit producer at world level, and accounts for 45% of worldwide production, followed by Italy with 7400 Ha and 78,000 tons [19]. Sicily is the main Italian producer (90%), while Apulia recorded 2013 production of 2650 tons from 320 Ha, mainly grown in the province of Foggia (North Apulia) with selected (spineless) cultivars [19]. At Italian level, intensive orchards mainly grow the yellow variety (which is spineless). In Apulia, particularly in the Salento peninsula

(South Apulia) there is an equal distribution of the two colored fruits, from spiny genotypes, often growing wild or in private gardens, with a different balance of purple-red betacyanin (in the purple variety) and yellow-orange betaxanthin (Figure 1) (in the orange variety).

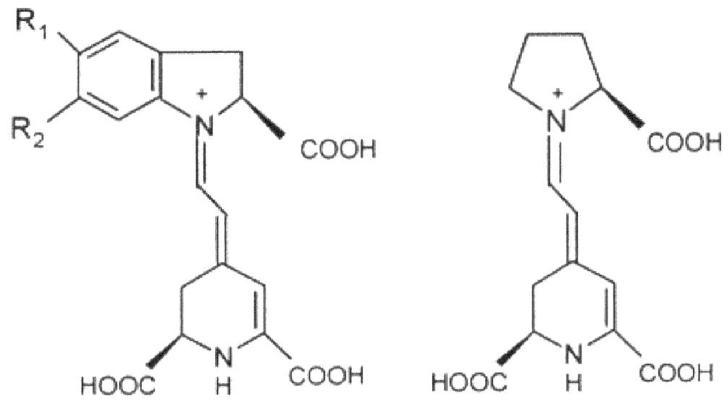

Betacyanins **Indicaxanthin**

Betanidin: R_1 and R_2=OH
Betanin (5-*O*-glucose betanindin): R_1=glucose; R_2=OH

Figure 1. Structures of the principal betacyanins (betanidin and betanin) (**left**) and the betaxanthin indicaxanthin (**right**) in *O. ficus-indica* (L.) Mill. fruits, predominant in the purple and orange variety, respectively.

The purpose of the present study was to characterize the two coloured cactus pear fruits of local origin for their phytochemical and antioxidant properties, in order to describe the potential nutraceutical characteristics of cactus pear found in Apulia, particularly the genotypes largely found in Salento (South Apulia), often growing wild or in private gardens. The results of this study could be useful to maximise the potential of the fruits' autochthonous biodiversity for their nutraceutical added value.

2. Experimental Section

2.1. Sample Material, Standards and Chemical Reagents

Prickly pear fruits [*Opuntia ficus-indica* (L.) Mill.] were collected in September 2013 and September 2014 at full maturity, without being overripe, from non-irrigated plants grown in private gardens in the Salento countryside (N 40° 21′18″; E 17° 59′47″, Apulia, Italy). Two local genotypes (with purple and orange pulp, respectively) were considered, and 15 fruits (five fruits from three plants of each variety) were collected. The fruits from each plant were considered as an experimental replicate, for a total of three replicates.

After manual separation of the peel from the pulp, fruits of each color were briefly homogenized in a kitchen-type blender, and used to measure physico-chemical parameters. Then the pulp was separated from the seeds, portioned and stored at −20° until analysis (performed within two weeks). Standard betanin, Folin-Ciocalteu's phenol reagent, gallic acid, trolox [(*S*)-(−)-6-Hydroxy-2,5,7,8-tetramethylchroman-2-carboxylic acid], fluorescein disodium (FL), ABTS [2, 2′-azino-bis (3-ethylbenzothiazoline-6-sulfonic acid)], AAPH [2,2′-Azobis (2-methyl-propionamidine) dihydrochloride], meta-phosphoric acid, ethanol, methanol and formic acid were purchased from Sigma-Aldrich (Milan, Italy).

2.2. Physico-Chemical Parameters

Soluble solids content (SSC), pH and sucrose/glucose/fructose contents were measured in triplicate in cactus pear juice, obtained from fresh fruit pulp by centrifugation for 10 min at 2000 *g*. SSC was measured using a temperature-adjusted digital refractometer (DBR95, G. Bormac, Carpi, Italy). For the determination of sucrose/D-glucose/ D-fructose, the R-Biopharm (Melegnano, Italy) kit was used.

2.3. Fruit Extract Preparation

Fresh fruit pulp (6 g) without seeds of each variety (purple and orange) was extracted in triplicate with 10 mL ethanol: formic acid: water (50:5:45 v/v/v). The mixtures were allowed to stand, without stirring, for 60 min at 4 °C before centrifugation (10 min at 2000 *g*). After centrifugation, the supernatant was recovered and the extraction repeated with the same volume of solvent. The combined hydrophilic extracts were subjected to rotary evaporation (BÜCHI Labortechnik AG, Flawil, Switzerland) and re-suspended in acidified water (0.5% formic acid) to a final concentration of 1g fruit pulp *per* mL of extract. All the samples were portioned and stored at −20 °C before the HPLC/DAD/MS and total phenols/antioxidant activity assays, which were performed within two weeks.

2.4. Phenols, Vitamin C and Antioxidant Capacity: Assessment of O. Ficus-Indica (L.) *Fruit Extracts*

Total phenolics were measured in both colored hydrophilic extracts according to Magalhães *et al.* [20]. Briefly, 50 µL of gallic acid standard or sample and 50 µL of Folin-Ciocalteu Reagent (FCR) (1:5, v/v) were placed in each well, then 100 µL of sodium hydroxide (0.35 M) was added. The absorbance of blue complexes at 760 nm was monitored after 5 min. The intrinsic absorption of the sample was determined by replacing FCR with 50 µL of 0.4 M acetic acid and the reagent blank was evaluated by replacing standard or sample with 50 µL of double-distilled water. The absorbance of samples was compared with that of a gallic acid standard curve ($R^2 \geq 0.997$; concentration range, 2.5–40.0 mg/L) and F-C reducing capacity was expressed as gallic acid equivalents (GAE) mg/100 g FW.

In order to quantify the vitamin C (ascorbate, AA, plus dehydroascorbate, DHAA) content, cactus fruit pulp (200 mg) was homogenized with 900 µL 5% meta-phosphoric acid at 4 °C, following the spectrophotometric methodology reported by Paradiso *et al.* [21].

Antioxidant capacities using the ABTS assay [22] and the ORAC assay, as described in Blando *et al.* [23] were evaluated in both colored hydrophilic extracts. A rapid microplate methodology, using a microplate reader (VictorX5, Perkin Elmer, Waltham, MA, USA) and 96-well plates (Costar, 96-well clear round bottom plate, Corning, USA) were used, both for total phenolic content and antioxidant capacity evaluation.

All experiments were performed in triplicate, and two independent assays were performed for each sample.

2.5. LC/MS Analysis

The separation and identification of betalain compounds in the sample extracts were performed using a HPLC/DAD/MS system (Agilent 1100 series, Agilent Technologies, Santa Clara, CA, USA), equipped with a reversed phase column SB-C18 column (250 × 4.6 mm, i.d. 4.6 μm, Zorbax, Agilent Technologies, Santa Clara, CA, USA). The chromatographic conditions were defined as follows: mobile phase: 1% formic acid in water (solvent A) and 1% formic acid in methanol (solvent B); flow rate: 0.8 mL/min; program: 15 min isocratic elution step with 15% B, 10 min linear gradient from 15% to 25% B, 10 min linear gradient to 75% B, 15 min linear gradient to 85% B, 5 min linear decreasing gradient from 85% to 15%, then 10 min of equilibration time before the next injection (20 μL). This method provided optimal separation of betacyanin (betanin) (DAD at 535 nm) and betaxanthin peaks (DAD at 484 nm).

HPLC/DAD was coupled to a mass spectrometer (Agilent G6120B, Agilent Technologies, Santa Clara, CA, USA) equipped with an electrospray ionization (ESI) source operating in positive ionization mode. Nitrogen was used as carrier gas at a flow rate of 13 L/min with nebulizing (35 psi). The spectra were taken in the presence of formic acid to promote $[M + H]^+$ ion production (electrospray voltage 3.5 kV), and nebulizer temperature was fixed at 350 °C.

In order to quantify betacyanins, pure natural betanin was isolated from *Amaranthus retroflexus* L. inflorescence, a rich source of the betanin isomer and with a negligible presence of iso-betain [24]. To this purpose, 50 g inflorescences were ground in a mortar in the presence of liquid nitrogen. The resulting powder was extracted with acidified (0.1% HCl) aqueous (80%) methanol, at a ratio of 1:10 (w/v). After two-hour extraction (with stirring), the extract was vacuum paper filtrated, then extracted two more times in the same conditions. The total extract was concentrated *in vacuo* at 32 °C to 1/10 of the volume, then filtered (0.45 μm) before injection (200 μL). Betanin was purified by means of semipreparative HPLC, using the same analytical conditions described before, except for the column (Zorbax ODS 9.4 × 250 mm i.d. 5 μm, Agilent Technologies, Santa Clara, CA, USA) and flux (2 mL/min). Fractions were collected and those containing betanin in a pure form were lyophilized. The purity of the betanin fractions was verified by HPLC-DAD (UV-Vis) and LC/MS according to Cai *et al.* [24]. After solubilization in water, betanin was used as a standard to build a calibration curve ($R^2 \geq 0.997$; concentration range, 0.05–0.2 mg/L), used to quantify the betanin isomers in the cactus pear extracts.

2.6. Statistical Analysis

All experiments were the result of two runs averaged together. The value of each sample was expressed as the mean (of triplicate measurements) ± standard deviation. Mean comparisons were performed by Student's test, using the GraphPad Prism version 5.0 software (GraphPad Software, Inc., La Jolla, CA, USA).

3. Results and Discussion

3.1. Physico-Chemical Parameters

Cactus pear fruits are low acid fruits (pH > 4.5). The pH of the two varieties from Apulia were similar, slightly less acid than reported in Stintzing *et al.* [1] for other Italian *O. ficus-indica* varieties. Dry weight (%) was not significantly different between the two varieties, while TSS was significantly higher ($p < 0.05$) in the purple variety than in the orange (Table 1). For the purple variety, TSS was lower than that reported in Stintzing *et al.* [1], who considered the Sicilian cultivated variety cv "*Rossa*". Also for the orange variety, TSS was lower than reported by Piga *et al.* [5] in a Sardinian orange variety. However, TSS is a variable parameter depending on maturity stage and fruit metabolism. During ripening, the pH of cactus pear fruit usually rises from 5 to a range around 5.5–6.5. The fruits considered here showed a pH of around 6, so we can assume they were fully mature. Carbohydrate composition was different in both samples, glucose being significantly higher ($p < 0.01$) than fructose, as in Stintzing *et al.* [1] and Abdel-Hameed *et al.* [16]. According to Stintzing *et al.* [25] but contrary to Abdel-Hameed *et al.* [16], the orange variety was the sweeter (Table 1).

Table 1. Physico-chemical parameters measured in the two prickly pear fruit varieties. Results are means ± S.D. ($n = 3$).

Variety	pH	DW (%)	TSS (°Brix)	Sucrose/Glucose/Fructose (g/100 g FW)
Purple	5.89 ± 0.24	16.70 ± 1.20	12.67 ± 0.15	ND/1.88 ± 0.04/0.78 ± 0.01
Orange	6.02 ± 0.01	16.10 ± 0.48	12.10 ± 0.14	ND/2.14 ± 0.10/1.04 ± 0.12
Significance	n.s.	n.s.	*	n.s.

* significant at $p < 0.05$; n.s. not significant.

Consistent with previous findings [1,16,26], both varieties were devoid of sucrose (Table 1). High invertase activity, hydrolyzing sucrose to glucose and fructose, has been reported in *O. ficus-indica* fruits [27].

3.2. Antioxidant Capacity of Hydrophilic Cactus Pear Extract

The nutritional and health properties of cactus pear fruit are associated with the antioxidant compounds it contains, namely ascorbic acid, phenolics and a mixture of purple-red betacyanin and yellow-orange betaxanthin pigments. The aim of our investigation was to assess antioxidant features in different coloured cactus pear fruits found in the countryside of the Salento peninsula (Apulia, Italy).

Total phenol content was significantly higher ($p < 0.01$) in the purple fruits than in the orange ones (Table 2), as also already reported in prickly pear cultivars from Saudi Arabia, California, and Tunisia [16,25,28], while Yahia *et al.* [29] reported opposite findings in cultivars and lines belonging to *Opuntia* sp. from Mexico (orange cultivar > purple cultivar). Even if Folin-Ciocalteu reagent is not specific for phenols, and can be reduced by many non-phenolic compounds, this assay has been used extensively to produce a large body of data, becoming a routine assay in phenolic antioxidants measurements [30].

The phenolic content found in both fruit extracts was similar to that reported by Stintzing *et al.* [25], and lower than that reported by Yeddes *et al.* [28]. Fernandez-Lopez *et al.* [31] reported a much higher phenolic content but it referred to a whole (skin and pulp) red-skinned fruit.

Concerning ascorbic acid, as expected, a high concentration was detected [32,33]. Both cactus pear extracts showed a good amount of vitamin C (AA + DHAA) (Table 2) that was significantly higher ($p < 0.05$) in the purple variety than in the orange one. For both genotypes, values are higher than those reported for the Californian varieties [25], but similar to varieties from Sicily [32] and Sardinia [5]. However, the difference in the analytical procedure used in our study for ascorbic acid determination must be taken into account. The assay [21] is a spectrophotometric method relying on the determination of AA and DHAA, thus defining the "Redox State", that is the AA/DHAA ratio, as 0.88 for the purple variety, and 0.86 for the orange one. These values reveal a good antioxidant environment at cellular level ('Redox State') for the prickly pear fruits considered in this study. Nevertheless, it has already been reported that ascorbic acid is responsible for nearly one third of the total antioxidant capacity [25,32].

Table 2. Betacyanin, Vitamin C (AA + DHAA), Total Phenols, TEAC and ORAC values in a hydrophilic extract of cactus pear fruit pulp of different colors. Results are means ± S.D. ($n = 3$).

	Betacyanin [1]	Vitamin C	Total Phenols	TEAC	ORAC
Variety	mg/100 g FW	mg/100 g FW	mg GAE/100 g FW	mmol/100 g FW	mmol/100 g FW
Purple	39.3 ± 5.2	36.6 ± 1.5	89.2 ± 3.6	0.61 ± 0.02	1.28 ± 0.02
Orange	3.6 ± 0.9	30.2 ± 0.3	69.8 ± 1.7	0.37 ± 0.02	0.98 ± 0.01
Significance	***	*	**	***	n.s.

[1] as betanin equivalent; ***, ** and * significant at $p < 0.001$, $p < 0.01$ and $p < 0.05$, respectively; n.s., not significant.

To assess the antioxidant potential of bioactive compounds, it has been recommended to apply at least two different assays varying in their mechanisms of antioxidant action [30]. We assessed the antioxidant capacity of the hydrophilic extract of cactus pear fruits of purple and orange varieties by both TEAC assay, which is a single electron transfer (ET) reaction-based assay and ORAC assay, which is a hydrogen atom transfer (HAT) reaction-based assay. Folin-Ciocalteu assay can even be considered an electron transfer assay, as it actually measures the reducing capacity of the sample [34].

It has been reported that the antioxidant activity of cactus fruits is twice as high as pears, apples, tomatoes, bananas, white grapes and is comparable to red grapes, pink grapefruit and red orange [32].

The TEAC value of the extracts from the purple fruit was higher ($p < 0.001$) than the orange one (0.61 *vs.* 0.37 mmol/100 g FW) (Table 2). Interestingly, these TEAC values were higher than those found in the literature [25,32], and similar to those reported by Fernandez-Lopez *et al.* [31], who investigated the whole (skin plus pulp) red-skinned fruit. Purple fruit extract confirmed a significantly higher ($p < 0.01$) reducing capacity, measured by Folin Ciocalteu assay.

Conversely, the purple variety had similar antioxidant activity to the orange one when assessed by ORAC assay (1.28 *vs.* 0.98 mmol/100 g FW) (Table 2), and both are slightly higher than the already reported values [25]. ORAC is a very sensitive assay, relying on fluorescein decay, and different compound classes may account for the antioxidant capacity observed. The different results from TEAC and ORAC assays could rely in the different reactions involved. TEAC assay is a single electron transfer reaction-based assay and uses a cation radical ($ABTS^{\bullet+}$) as a reference [30]. ORAC assay, a competition method, is widely used because it takes into account both reactivity and stoichiometry of the antioxidant under analysis [35], and measures the hydrophilic antioxidant capacity towards peroxyl radicals by a hydrogen atom transfer reaction mechanism [30]. In this way the multifaceted nature of antioxidant is revealed: results from ORAC assay on prickly pear extract indicated that purple betanin has a hydrogen atom donating capacity similar to orange indicaxanthin. Theoretical ORAC values have been published [25,36] for betaxanthin and betacyanin, the former being more active than betacyanin (1.73 ORAC *vs.* 1.54 ORAC/μmol of reference compound). The good antioxidant performance of the orange variety could be related to its high betaxanthin content, in spite of its lower phenol content. Contradictory results with ABTS radical cation decolorisation assay on pure betanin and indicaxanthin showed higher activity with betanin than with indicaxanthin [32].

Free radical-scavenging antioxidants play important roles in the physiological defense network against oxidative stress. However, when dealing with antioxidant capacity, we must take into account, the debate in recent years on the reliability of the methods for assessing antioxidant capacity by radical scavenging *in vitro* [35]. This feature is important because of the inconsistent results obtained in many studies on the antioxidant capacity assessed using the different methods. We applied two of the most commonly used methods to evaluate the capacity of scavenging radicals. However, the free radical scavenging capacity measured by *in vitro* methods could be different from the capacity of antioxidants *in vivo* against oxidative stress, which is the primary concern of measuring antioxidant capacity. Therefore, the effects of antioxidants supplementation on appropriate biomarkers on biological fluids and tissues are considered the next frontiers of 'antioxidant capacity' in food science [35].

3.3. Quantification of Betanin in Cactus Pear Fruit Extracts

Since the purple variety is widespread in the Salento area, (South Apulia), our interest focused on the quantification of betacyanin pigments present in the purple fruits. Analytical quantification of betanin in *O. ficus-indica* (L.) Mill. fruit extract is complicated by the presence of two isomers (betanin and isobetanin) with identical MS spectra but different RT. We used the commercial standard betanin for quantification, but this made HPLC analysis rather difficult, due to the presence of impurities. For this reason we isolated betanin from *A. retroflexus* L. inflorescence, characterized by the presence of a single betanin isomer [24]. The approach of purifying reference substance from plant matrix where the metabolite is abundantly bio-synthesized is not new, even for betacyanin. Stintzing and collaborators [25] extracted and purified betacyanin from *Gomphrena globosa*, another *Amaranthaceae* species. Those authors used the isolated reference substances for co-injection experiments. Photometric quantification [25,29,37] has generally been used for betalain quantification, based on methodology first reported by Cai *et al.* [38]. We used *A. retroflexus* purified betanin as the standard to build a calibration curve. This made the quantification of the compound more precise than spectrophotometric quantification based on molar absorbance, where co-absorption may occur [1,32]. Betacyanin content in purple fruit from Salento area resulted much higher (39.3 mg/100 g FW) than the value reported for the Sicilian prickly pear fruit [32]. Nevertheless, as reported above, the difference in analytical procedure must be taken into account. As expected, the level of betacyanin in orange fruit was limited (3.6 mg/100 g FW) (Table 2).

HPLC/DAD separation coupled with mass spectra of cactus pear fruit extract (purple variety) is reported in Figure 2.

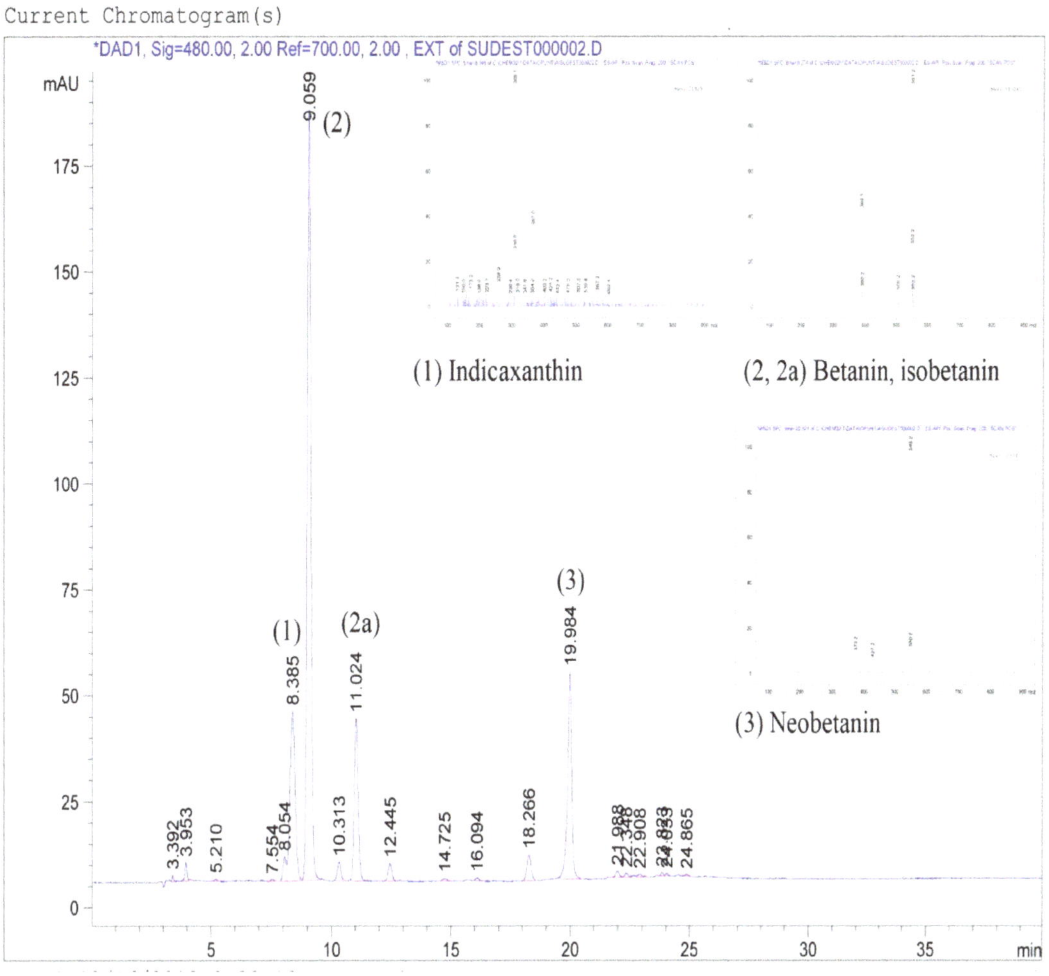

Figure 2. HPLC chromatogram of the cactus pear fruit extract (purple variety); MS fragmentation of the main peaks is also shown.

4. Conclusions

The present study confirms the potential of cactus pear [*Opuntia ficus-indica* (L.) Mill.] fruits, particularly the purple variety, as an important source of colored bioactive compounds which not only has coloring potential, but is also an excellent source of dietary antioxidant which may have beneficial effects on consumers' health.

Cactus pear and related species are characterized by their minimal water requirements, hardiness and adaptability to high temperatures. For this reason, it is expected to have enormous potential as a niche crop for desert zones in Mexico, Arizona and developing countries in Africa and Asia, where conventional crops are difficult to grow. Moreover, the cultivation of these species could be taken into account in order to deal with the global climate changes in several parts of the globe. Since the nutritional value of cactus pear fruits and other *Opuntia* sp. compare favorably with other fruit crops, they could nutritionally improve the diet of both rural and urban consumers and could be proposed as low-cost functional foods [39].

The various properties of this fruit outlined in this paper would therefore seem to indicate that the cactus pear is a neglected species worthy of greater consideration at world level. The characterization of autochthonous varieties could give more information on the presence of bioactive compounds and on the possible exploitation of existing biodiversity.

Acknowledgments

The authors wish to thank the local growers "Albano" and "To Giuseppe" for kindly providing the prickly pear fruits, Annalisa Paradiso for assistance with ascorbate/DHA assays, Paolo Inglese for his useful suggestions and Anthony Green for kindly revising the English of the manuscript.

Author Contributions

F.B., G.M., C.G., A.M. and L.B. conceived and designed the experiments; C.A., N.T., C.N. and F.B. performed the experiments; C.A., C.N. and F.B. analyzed the data; F.B. wrote the paper.

Conflicts of Interest

The authors declare no conflict of interest.

References

1. Stintzing, F.C.; Schieber, A.; Carle, R. Evaluation of colour properties and chemical quality parameters of cactus juices. *Eur. Food Res. Technol.* **2003**, *216*, 303–311.
2. Barbera, G. History, economics and agro-ecological importance. In *Agro-Ecology, Cultivation and Uses of Cactus Pear*; Barbera, G., Inglese, P., Pimienta, E., Eds.; FAO Plant Production and Protection: Rome, Italy, 1995; Volume 132, pp. 1–11.
3. Mondragón-Jacobo, C. Cactus pear breeding and domestication. *Plant Breed. Rev.* **2001**, *20*, 135–166.
4. Caruso, M.; Currò, S.; Las Casas, G.; La Malfa, S.; Gentile, A. Microsatellite markers help to assess genetic diversity among *Opuntia fiicus indica* cultivated genotypes and their relation with related species. *Plant Syst. Evol.* **2010**, *290*, 85–97.
5. Piga, A.; del Caro, A.; Pinna, I.; Agabbio, M. Changes in ascorbic acid, polyphenol content and antioxidant activity in minimally processed cactus pear fruits. *LWT—Food Sci. Technol.* **2003**, *36*, 257–262.
6. Cefola, M.; Renna, M.; Pace, B. Marketability of ready-to-eat cactus pear as affected by temperature and modified atmosphere. *J. Food Sci. Technol.* **2014**, *51*, 25–33.
7. Stintzing, F.C.; Schieber, A.; Carle, R. Phytochemical and nutritional significance of cactus pear. *Eur. Food Res. Technol.* **2001**, *212*, 396–407.
8. Magloire, J.F.; Konarski, P.; Zou, D.; Stintzing, F.C.; Zou, C. Nutritional and medicinal uses of cactus pear (*Opuntia* sp.) cladodes and fruits. *Front. Biosci.* **2006**, *11*, 2574–2589.
9. Galati, E.M.; Mondello, M.R.; Giuffrida, D.; Dugo, G.; Miceli, N.; Pergolizzi, S.; Taviano, M.F. Chemical characterization and biological effects of Sicilian *Opuntia ficus-indica* (L.) Mill. fruit juice: Antioxidant and antiulcerogenic activity. *J. Agric. Food Chem.* **2003**, *51*, 4903–4908.

10. Tesoriere, L.; Butera, D.; Pintaudi, A.M.; Allegra, M.; Livrea, M.A. Supplementation with cactus pear (*Opuntia ficus-indica)* fruit decreases oxidative stress in healthy humans: A comparative study with vitamic C. *Am. J. Clin. Nutr.* **2004**, *80*, 391–395.
11. Zou, D.-M.; Brewer, M.; Garcia, F.; Feugang, J.M.; Wang, J.; Zang, R.; Liu, H.; Zou, C. Cactus pear: A natural product in cancer chemoprevention. *Nutr. J.* **2005**, *4*, 25.
12. Kaur, M.; Kaur, A.; Sharma, R. Pharmacological actions of *Opuntia ficus indica*: A review. *J. Appl. Pharm. Sci.* **2012**, *2*, 15–18.
13. Madrigal-Santillán, E.; García-Melo, F.; Morales-González, J.A.; Vázquez-Alvarado, P.; Muñoz-Juárez, S.; Zuñiga-Pérez, C.; Sumaya-Martínez, M.T.; Madrigal-Buiaidar, E.; Hernández-Ceruelos, A. Antioxidant and anticlastogenic capacity of prickly pear juice. *Nutrients* **2013**, *5*, 4145–4158.
14. Esatbeyoglu, T.; Wagner, A.E.; Schini-Kerth, V.B.; Rimbach, G. Betanin—A food colorant with biological activity. *Mol. Nutr. Food Res.* **2015**, *59*, 36–47.
15. El-Mostafa, K.; El Karrassi, Y.; Badreddine, A.; Andreoletti, P.; Vamecq, J.; El Kebbaj, M.S.; Latruffe, N.; Lizard, G.; Nasser, B.; Cherkaoui-Malki, M. Nopal cactus (*Opuntia ficus-indica*) as a source of bioactive compounds for nutrition, health and disease. *Molecules* **2014**, *19*, 14879–14901.
16. Abdel-Hameed, E.-S.S.; Nagaty, M.A.; Salman, M.S.; Bazaid, S.A. Phytochemicals, nutritionals and antioxidant properties of two prickly pear cactus cultivars (*Opuntia ficus-indica* Mill.) growing in Taif, KSA. *Food Chem.* **2014**, *160*, 31–38.
17. Cejudo-Bastante, M.J.; Chaalal, M.; Louaileche, H.; Parrado, J.; Heredia, F.J. Betalain profile, phenolic content and color characterization of different parts and varieties of *Opuntia ficus-indica. J. Agric. Food Chem.* **2014**, *62*, 8491–4499.
18. Matias, A.; Nunes, S.L.; Poejo, J.; Mecha, E.; Serra, A.T.; Amorim Madeira, P.J.; Bronze, M.R.; Duarte, C.M.M. Antioxidant and anti-inflammatory activity of a flavonoid-rich concentrate recovered from *Opuntia ficus-indica* juice. *Food Funct.* **2014**, *5*, 3269–3280.
19. ISTAT Data Bank, 2013. Available online: http://agri.istat.it/sag_is_pdwout/excel/Dw302013.xls (accessed on 5 February 2015).
20. Magalhães, L.M.; Santos, F.; Segundo, M.A.; Reis, S.; Lima, J.L.F.C. Rapid microplate high-throughput methodology for assessment of Folin-Ciocalteu reducing capacity. *Talanta* **2010**, *83*, 441–447.
21. Paradiso, A.; Cecchini, C.; de Gara, L.; D'Egidio, M.G. Functional and rheological properties of dough from immature durum wheat. *J. Cereal Sci.* **2006**, *43*, 216–222.
22. Re, R.; Pellegrini, N.; Proteggente, A.; Pannala, A.; Yang, M.; Rice-Evans, C. Antioxidant activity applying an improved ABTS radical cation decolorization assay. *Free Radic. Biol. Med.* **1999**, *26*, 1231–1237.
23. Blando, F.; Spirito, R.; Gerardi, C.; Durante, M.; Nicoletti, I. Nutraceutical properties in organic strawberry from South Italy. *Acta Hort.* **2012**, *926*, 683–690.
24. Cai, Y.; Sun, M.; Corke, H. Identification and distribution of simple and acylated betacyanins in the *Amaranthaceae*. *J. Agric. Food Chem.* **2001**, *49*, 1971–1978.
25. Stintzing, F.C.; Herbach, K.M.; Mosshammer, M.R.; Carle, R.; Yi, W.; Sellappan, S.; Akoh, C.; Bunch, R.; Felker, P. Color, betalain pattern, and antioxidant properties of cactus pear (*Opuntia* spp.) clones. *J. Agric. Food Chem.* **2005**, *53*, 442–451.

26. El Kossori, R.L.; Villaume, C.; El Boustani, E.; Sauvaire, Y.; Mejean, L. Composition of pulp, skin and seeds of prickly pears fruit (*Opuntia ficus-indica* sp.). *Plant Foods Hum. Nutr.* **1998**, *52*, 263–270.
27. Ouelhazi, N.K.; Ghrir, R.; Diêp Le, K.H.; Lederer, F. Invertase from *Opuntia ficus-indica* fruits. *Phytochemistry* **1992**, *31*, 59–61.
28. Yeddes, N.; Cherif, J.K.; Guyot, S.; Sotin, H.; Ayadi, M.T. Comparative study of antioxidant power, polyphenols, flavonoids and betacyanins of the peel and pulp of three Tunisian *Opuntia* forms. *Antioxidants* **2013**, *2*, 37–51.
29. Yahia, E.M.; Mondragón-Jacobo, C. Nutritional components and anti-oxidant capacity of ten cultivars and lines of cactus pear fruit (*Opuntia* spp.). *Food Res. Int.* **2011**, *44*, 2311–2318.
30. Huang, D.; Ou, B.; Prior, R. The chemistry behind antioxidant capacity assays. *J. Agric. Food Chem.* **2005**, *53*, 1841–1856.
31. Fernandez-Lopez, J.A.; Almela, L.; Obón, J.M.; Castellar, M.R. Determination of antioxidant constituents in cactus pear fruits. *Plant Foods Hum. Nutr.* **2010**, *65*, 253–259.
32. Butera, D.; Tesoriere, L.; di Gaudio, F.; Bongiorno, A.; Allegra, M.; Pintaudi, A.M.; Kohen, R.; Livrea, M.A. Antioxidant activities of Sicilian prickly pear (*Opuntia ficus-indica*) fruit extracts and reducing properties of its betalains: Betanin and indicaxanthin. *J. Agric. Food Chem.* **2002**, *50*, 6895–6901
33. Castellar, M.R.; Solano, F.; Obón, J.M. Betacyanin and other antioxidants production during growth of *Opuntia stricta* (Haw.) fruits. *Plant Foods Hum. Nutr.* **2012**, *67*, 337–343.
34. MacDonald-Wicks, L.K.; Wood, L.G.; Garg, M.L. Methodology for the determination of biological antioxidant capacity *in vitro*: A review. *J. Sci. Food Agric.* **2006**, *86*, 2046–2056.
35. Niki, E. Antioxidant capacity: Which capacity and how to assess it? *J. Berry Res.* **2011**, *1*, 169–176.
36. Cai, Y.; Sun, M.; Corke, H. Antioxidant activity of betalains from plants of the Amaranthaceae. *J. Agric. Food Chem.* **2003**, *51*, 2288–2294.
37. Castellanos-Santiago, E.; Yahia, E.M. Identification and quantification of betalains from the fruits of 10 Mexican prickly pear cultivars by high-performance liquid chromatography and electrospray ionization mass spectrometry. *J. Agric. Food Chem.* **2008**, *56*, 5758–5764.
38. Cai, Y.; Sun, M.; Wu, H.; Huang, R.; Corke, H. Characterization and quantification of betacyanin pigments from diverse Amaranthus species. *J. Agric. Food Chem.* **1998**, *46*, 2063–2070.
39. Patel, S. Reviewing the prospect of *Opuntia* pears as low cost functional foods. *Rev. Environ. Sci. Biotechnol.* **2013**, *12*, 223–234.

12

In Vitro Cultivars of *Vaccinium corymbosum* L. (Ericaceae) are a Source of Antioxidant Phenolics

Rodrigo A. Contreras *, Hans Köhler †, Marisol Pizarro † and Gustavo E. Zúñiga *

Laboratorio de Fisiología y Biotecnología Vegetal, Departamento de Biología, Facultad de Química y Biología, Universidad de Santiago de Chile. L. B. O'Higgins Ave. #3363, Estación Central, Santiago of Chile, 9170022, Chile; E-Mails: hans.kohler@usach.cl (H.K.); marisol.pizarror@usach.cl (M.P.)

† Ph.D. Program in Biotechnology.

* Authors to whom correspondence should be addressed;
E-Mails: rodrigo.contrerasar@usach.cl (R.A.C.); gustavo.zuniga@usach.cl (G.E.Z.)

Academic Editors: Antonio Segura-Carretero and David Arráez-Román

---

Abstract: The antioxidant activity and phenolic composition of six *in vitro* cultured blueberry seedlings were determined. Extracts were prepared in 85% ethanol from 30 days old *in vitro* cultured plants and used to evaluate the antioxidant capacities that included Ferric reducing antioxidant power (FRAP) and 1,1-diphenyl-2-picrylhydrazin (DPPH•) scavenging ability, total polyphenols (TP) and the partial phenolic composition performed by high performance liquid chromatography with diode array detector (HPLC-DAD), liquid chromatography coupled to tandem mass spectrometry (LC-MS/MS (ESI-QqQ)). All ethanolic extracts from *in vitro* blueberry cultivars displayed antioxidant activity, with Legacy, Elliott and Bluegold cultivars being the most active. In addition, we observed a positive correlation between phenolic content and antioxidant activity. Our results suggest that the antioxidant activity of the extracts is related to the content of chlorogenic acid myricetin, syringic acid and rutin, and tissue culture of blueberry seedlings is a good tool to obtain antioxidant extracts with reproducible profile of compounds.

Keywords: high-bush blueberry *in vitro* cultivars; HPLC-DAD and LC-MS/MS; DPPH• and FRAP; phenolic composition

---

1. Introduction

Epidemiological studies have demonstrated that a diet rich in fruits and vegetables reduces the risk of certain types of cancer, cardiovascular, and other chronic diseases. Blueberries have been traditionally used for either fresh consumption or processed into jams, jellies and juices. Phenolic acids (benzoic acids and cinnamic acids), also known as non-flavonoids and their derivatives, are found in blueberries. The benzoic acids reported in blueberries are vanillic acid, syringic acid, gallic acid, protocatechuic acid, *m*-hydroxybenzoic acid, hydroxybenzoic acid and ellagic acid [1,2]. While the cinnamic acids found were chlorogenic acid, caffeic acid, ferulic acid, *p*-coumaric acid, *o*-coumaric, acid and *m*-coumaric acid [1]. Chlorogenic acid is the major phenolic present in the fruit [3]. It has been shown that chlorogenic acid can scavenge reactive oxygen species (ROS) and alkylperoxyl radicals, protecting macromolecules like deoxyribonucleic acid (DNA), proteins and membranes [4]. These properties convert chlorogenic acid into a good candidate to be used in the food industry, pharmacology and cosmetology. Then, the industrial production of antioxidant compounds from blueberries seems to be desirable. However, the supply of secondary metabolites has many limitations from fruits or leaves grown under field conditions such as biomass availability and vary in their chemical composition which is strongest affected by the environment [5]. There has been considerable interest in plant tissue culture as a potential alternative for the production of secondary metabolites [6]. The major advantages of a plant tissue culture system over the conventional cultivation of whole plants are as follows: the synthesis of bioactive secondary metabolites runs in a controlled environment, independent of climatic and soil conditions; negative biological influences that affect secondary metabolites production in the nature are eliminated (microorganisms and insects); and automated control of cell growth and rational regulation of metabolite processes would reduce labor costs and improve productivity [7].

In this work, we evaluated the use of *in vitro* culture of highbush blueberry cultivars of commercial importance in Chile as a continuous source of antioxidant compounds. In addition, we characterize the chemical composition of extracts by using HPLC-DAD and LC-MS/MS in order to produce standardized extracts.

2. Experimental Section

2.1. Chemicals

All solvents used were HPLC grade. Ethanol, methanol and acetonitrile were purchased from J.T Baker Chemical Co. (Phillipsburg, NJ, USA), phosphoric acid and hydrochloridric acid were purchased from Riedel-de-Haëhn (Seelze, Germany). Folin-Ciocalteu reagent, sodium carbonate anhydrous and formic acid were purchased from Merck Chemical Co. (Darmstadt, Germany), 1,1-diphenyl-2-picrylhidrazyl (DPPH•), 2,4,6-tris(2-pyridyl)-*s*-triazine (TPTZ), $FeCl_3 \cdot 6H_2O$, sodium acetate and HPLC standards (gallic acid, quercetin, rutin, chlorogenic acid, caffeic acid, syringic acid, myricetin, *p*-coumaric acid, ellagic acid, kaempferol, naringenin, isoquercitrin, morin, genistein and luteolin) were purchased from Sigma-Aldrich Chemical Co. (St. Louis, MO, USA), Lloyd-McCown medium base and 6-γ,γ-dimethylallylaminopurine (2-iP) were purchased from Phytotechnology labs. (Kansas City, MO, USA), and finally acetic acid and agar-agar were purchased from Winkler Ltda. (Santiago, Chile).

2.2. Plant Material

In vitro cultures of *V. corymbosum cv.* Duke, Legacy, Brigitta, Elliott, Misty and Bluegold were started from shoot apices of pathogen free certified plants. Explants were cultured in a Lloyd-McCown medium base [8] supplemented with 2% of sucrose, 2.5 mg/L of 2-iP (cytokinin) and 7.5 g/L of agar-agar. Cultures were maintained during 30 days at 23 ± 2 °C with 16/8 light/darkness photoperiod.

2.3. Extracts Preparation

A total of 100 mg of fresh *in vitro* culture medium free shoots were mixed with 1 mL of ethanol (85% v/v) and sonicated at 50–60 Hz of frequency for two hours at 25 °C according the method previously described [9]. Extracts were filtered in a 0.45 μm pore filter (Millipore, Billerica, MA, USA).

2.4. DPPH• Free Radical-Scavenger Spectrophotometric Assay

The radical scavenging activity of the blueberry extracts was determined by using the 2,2-diphenyl-1-picrylhydrazyl radical (DPPH•) according the procedure previously described [10]. In its radical form, DPPH• has an absorption band at 517 nm, which disappears upon reduction by an antiradical compound. Briefly, 20 μL of each extract were added to 980 μL of daily-prepared DPPH• ethanolic solution (0.78 absorbance units). Absorbance at 517 nm was measured with an Agilent 8453 UV-Vis spectrophotometer (Palo Alto, CA, USA), 4 min after starting the reaction. Results were expressed as % of DPPH• consumed.

To determinate the 50% of inhibitory concentration (IC_{50}), serial dilutions of extracts and/or phenolic standards were used to measure the scavenging of DPPH• radical as a function of serial dilution. Each determination was performed in triplicate and repeated at least three times and using a linear regression to calculate the concentration to scavenge the 50% of DPPH•.

2.5. Ferric Reducing/Antioxidant Power (FRAP)

The FRAP activity of the extracts was measured according to the method previously described [11]. Briefly, 5 μL of ethanolic extract was mixed with 900 μL of FRAP reagent and 95 μL of water (FRAP reagent was prepared to mix acetate buffer (300 mM, pH 3.6): TPTZ solution (10 mM in chloride acid) and ferric chloride solution (20 mM) (10:1:1 v/v). Absorbance at 593 nm was measured with an Agilent 8453 UV-Vis spectrophotometer, 4 min after starting the reaction. Results were expressed as ascorbic acid equivalents (AAE) per gram of dry weight (DW). Each determination was performed in triplicate and repeated at least three times.

2.6. Total Phenolic Content (TPC)

The total phenolic content was measured according to the method previously described [12]. 40 μL of the extract to be tested were added with 100 μL of Folin-Ciocalteu's reagent, 560 μL of deionized water; reaction was stopped after 5 min at room temperature with 300 μL of 7% aqueous sodium carbonate. The absorbance was measured at 660 nm [13] on an Agilent 8453 UV-Vis spectrophotometer.

The results were expressed in gallic acid equivalents (GAE) per gram of DW. Each determination was performed in triplicate and repeated at least three times.

2.7. Analysis of Extracts by HPLC-DAD

The determination of phenolic compounds was carried out using an Agilent high performance liquid chromatography system equipped with a UV-Vis photodiode-array detector (HPLC-DAD, 1100 series, Palo Alto, CA, USA). The chromatographic separation was obtained by a RP-C18 column (Zorbax, Eclipse XDB-C18 4.6 × 150 mm, 5 μm, Agilent Technologies, Inc., Santa Clara, CA, USA) with solvent A (acetonitrile) and B (1% phosphoric acid) under gradient conditions: 0 min, 10% of A, 5 min, 25% of A, 8 min, 35% of A, 15 min, 60% of A, 17 min, 35% of A and final 20 min, 10% of A. The flow rate was 1 mL min^{-1} and the column were thermostatically controlled at 25 °C. UV-detection was performed at 254, 280, 314 and 340 nm, the results were expressed as mg per gram of DW. The standards used were gallic acid, quercetin, rutin, chlorogenic acid, caffeic acid, syringic acid, myricetin, *p*-coumaric acid, ellagic acid, kaempferol, naringenin, isoquercitrin, morin, genistein and luteolin.

2.8. Analysis of Extracts by LC-MS/MS

The compounds in the extracts were analyzed by LC-MS/MS. LC-MS/MS was performed using an Agilent triple quadrupole mass spectrometer (MS/MS, 6400) equipped with an Agilent LC 1200 series. A RP-C18 column (Zorbax, Eclipse XDB-C18 4.6 × 150 mm, 5 μm) was used at flow rates of 1 mL min^{-1} at room temperature. Conditions for MS analysis include a capillary voltage of 4000 V, a nebulizing pressure of 40 psi, and the drying gas temperature of 330 °C. HPLC gradient was in acetonitrile (A) and 0.1% formic acid (B), as follows: 0 min, 10% of A, 5 min, 25% of A, 8 min, 35% of A, 15 min, 60% of A, 17 min, 35% of A and 20 min, 10% of A. Compounds were analyzed by both negative and positive ion mode.

2.9. Statistical Analysis

Statistical differences were determined using analysis of variance (ANOVA) with Tukey's post-test for all samples. Significant differences were determined using 95% of confidence ($p < 0.05$). We use $n \geq 3$ replicas in all experiments and measurements.

3. Results and Discussion

3.1. In Vitro Culture Establishment

The six cultivars were successfully established using the McCown-Lloyd medium base [8] supplemented with 2-iP. The replication index of all cultivars was 8 after 30 days of culture. Figure 1 shows the appearance of blueberry *cv.* Misty. The other cultivars presented a very similar appearance. Shoots of one month were used to obtain the extracts used.

Figure 1. *In vitro* blueberry plants *cv.* Misty in supplemented Lloyd-McCown medium (see Experimental Section).

3.2. Antioxidant Activity of Extracts

Blueberry fruits are known to have a high content of polyphenols, e.g., anthocyanins, flavonols, isoflavonols, *etc.*, that are potent antioxidants and that may have protective effects against diseases related to free radical production, such as cancer and cardiovascular diseases.

This work reports differences in antioxidant activity and composition between cultivars of blueberry cultured *in vitro*. Ethanolic extracts from *in vitro* blueberry cultivars displayed antioxidant activity (Figure 2A). The results show two groups of plants: those with a high antioxidant capacity (up to 80% DPPH• consumption) such as Duke, Legacy, Elliott and Bluegold, and those with lower antioxidant capacity (lower to 80% DPPH• consumption) such as Brigitta and Misty. The IC_{50} (Figure 2B), shows that the cultivars with better antioxidant activity were Elliott and Bluegold (IC_{50} of 13.4 ± 0.4 and 11.5 ± 0.7 μg, respectively) and with less antioxidant capacity were Duke, Brigitta and Misty (IC_{50}, 27.4 ± 0.5, 32.9 ± 3.5 and 27.6 ± 2.7 μg, respectively).

The FRAP assay (Figure 3) show a similar pattern to DPPH• scavenge assay. The FRAP values shows that cultivars Duke, Legacy, Elliott and Bluegold, have a higher reducing power, while Brigitta and Misty cultivars have a lower reducing power ($p < 0.05$). However, a clear difference was observed between the cultivars Brigitta and Misty ($p < 0.05$), which could not be observed in the DPPH• assay. The difference between these two assays may be explained by the methodology. The DPPH• assay considers both synergic and antagonist process and reaches a saturation point, while the FRAP assay does not reach a saturation point and is linear in a wide range. For these reasons both methods are considered complementary [14–16].

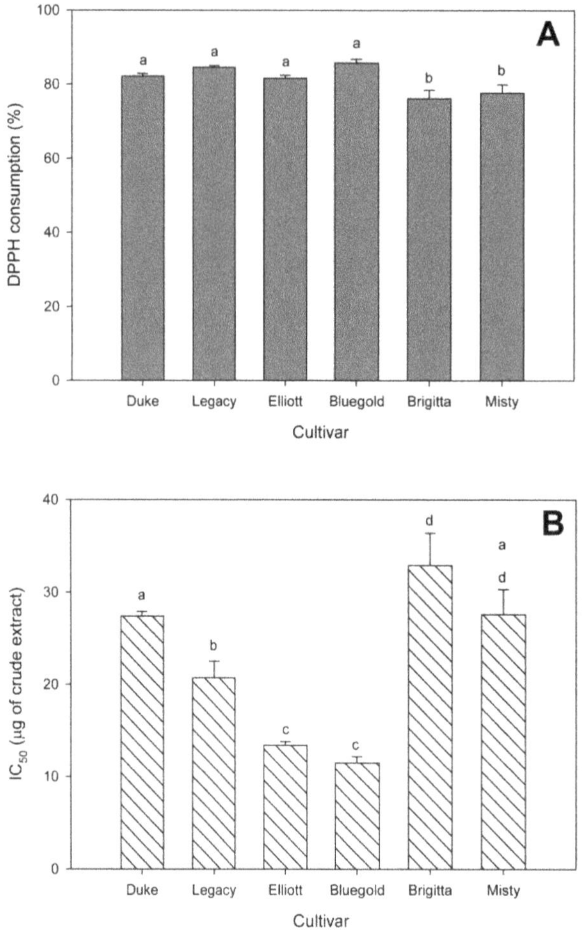

Figure 2. 1,1-diphenyl-2-picrylhydrazin (DPPH•) free radical-scavenging activity of six *in vitro* highbush blueberry cultivars. (**A**) Total scavenging activity (as a percentage) from 20 μL of mass normalized hydroethanolic extract; (**B**) IC_{50} index of DPPH• scavenging activity of hydroethanolic extracts. Each bar corresponds at the mean of nine independent measurements ± standard error; letters represent significant differences ($p < 0.05$).

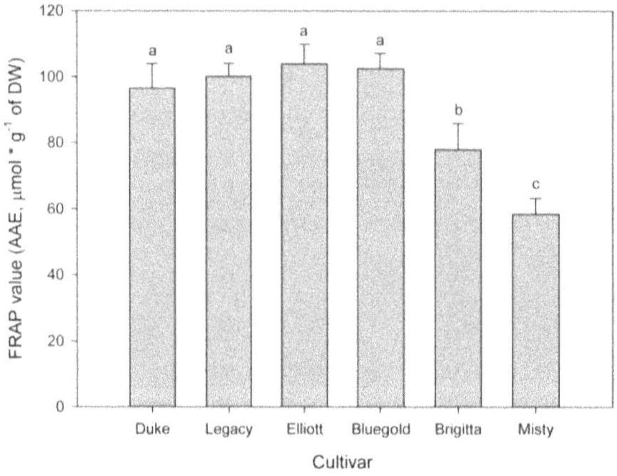

Figure 3. Antioxidant power measured by the Ferric reducing antioxidant power (FRAP) method for six *in vitro* highbush blueberry cultivars. Each bar corresponds at the mean of nine independent measurements ± standard error; letters represent significant differences ($p < 0.05$).

3.3. Total Phenolic Compound Content Determined by Folin-Ciocalteu's Assay

The total content of phenolic compounds present in each cultivar is reported in Figure 4. Cultivar Bluegold contains higher amount of phenolic compounds than the other cultivars (about twice, $p < 0.05$).

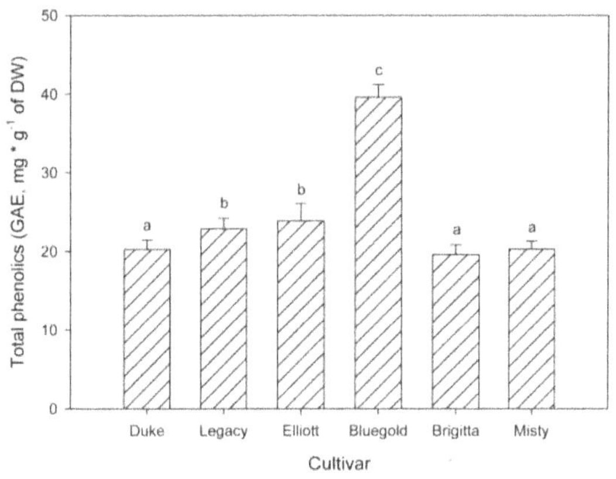

Figure 4. Total phenolic content of six *in vitro* highbush blueberry cultivars. Each bar corresponds at the mean of nine independent measurements ± standard error; letters represent significant differences ($p < 0.05$).

3.4. Phenolic Profile Determined by HPLC-DAD and LC-MS/MS

The phenolic profile of the extracts obtained by HPLC-DAD is shown in Figure 5 and Table 1. By comparing their retention time and UV-Vis spectral properties with those of pure standards, we detected 9 peaks, seven of which were present in all cultivars. The main compounds present in all cultivars were chlorogenic acid, syringic acid and rutin (Figure 5, Table 1). In addition, myricetin (peak #6) and peak #8, was not detected in cultivars Brigitta and Legacy (Figure 6). The ultraviolet spectra and LC-MS/MS analysis corroborates the presence of chlorogenic acid, syringic acid, rutin and myricetin. In addition, peak #1 was identified as caffeic acid hexoside, peak #5 was identified as apigenin-7-*O*-glucoside [17], peaks #7 and #8 were identified luteolin 3-*O*-glucuronide and myricitrin, respectively [18] and finally peak #9 was identified as kaempferol 3-*O*-glycosyde [19].

The IC_{50} of identified compounds compared with IC_{50} of crude extract (Figure 2) and determined by DPPH assay, suggest that the antioxidant activity is mainly due to chlorogenic acid and myricetin (Table 1). Chlorogenic acid is widely distributed in berry crops used as natural antioxidants [20]. Their antioxidant activity is associated to some extent with the number of hydroxyl groups in their molecular structure [21]. Flavonols, such as myricetin, showed high antioxidant activity that correlates with their structure [22].

Table 1. Main phenolic compounds identified in blueberry extracts by HPLC and LC-MS/MS. The numbers correspond to peaks in Figure 5 and structures of Figure 7.

Peak Number	Retention Time	UV Spectra	Compound	IC_{50} (µg) *
1	3.5	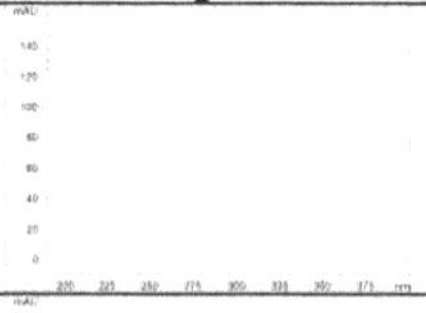	Caffeic acid hexoside	nd
2	4.2	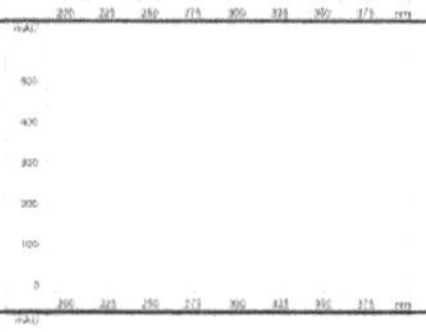	Chlorogenic acid	4.18
3	4.7	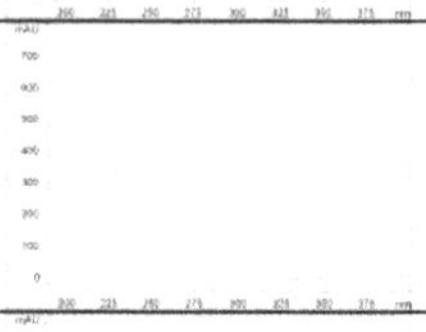	Syringic acid	7.28
4	6.4	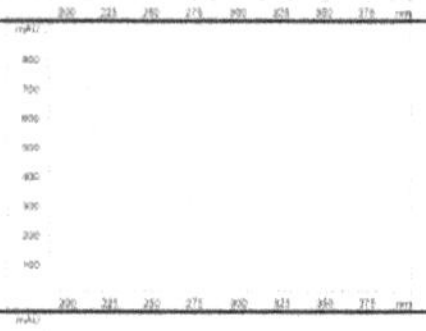	Rutin	9.25
5	6.9	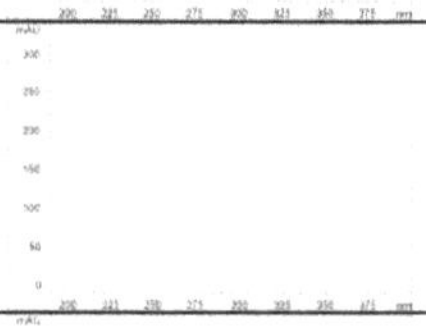	Apigenin-7-*O*-glucoside	nd
6	8.1	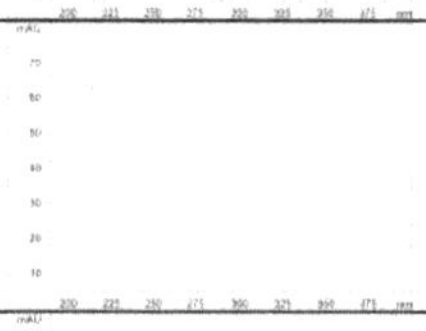	Myricetin	3.20
7	15.8	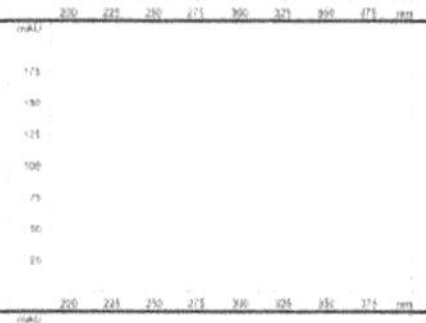	Luteolin 3-*O*-glucuronide	nd
8	16.2	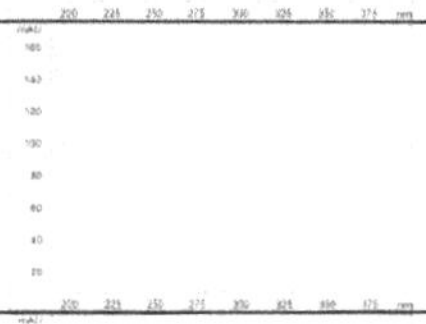	Myricitrin	nd
9	17.4	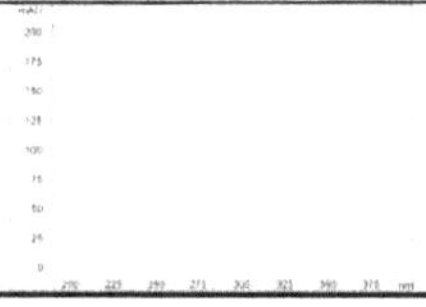	Kempferol 3-*O*-glucosyde	nd

* IC_{50} as calculated using DPPH assay and commercial standards with serial dilutions and lineal regression (n = 6), nd = non-determined.

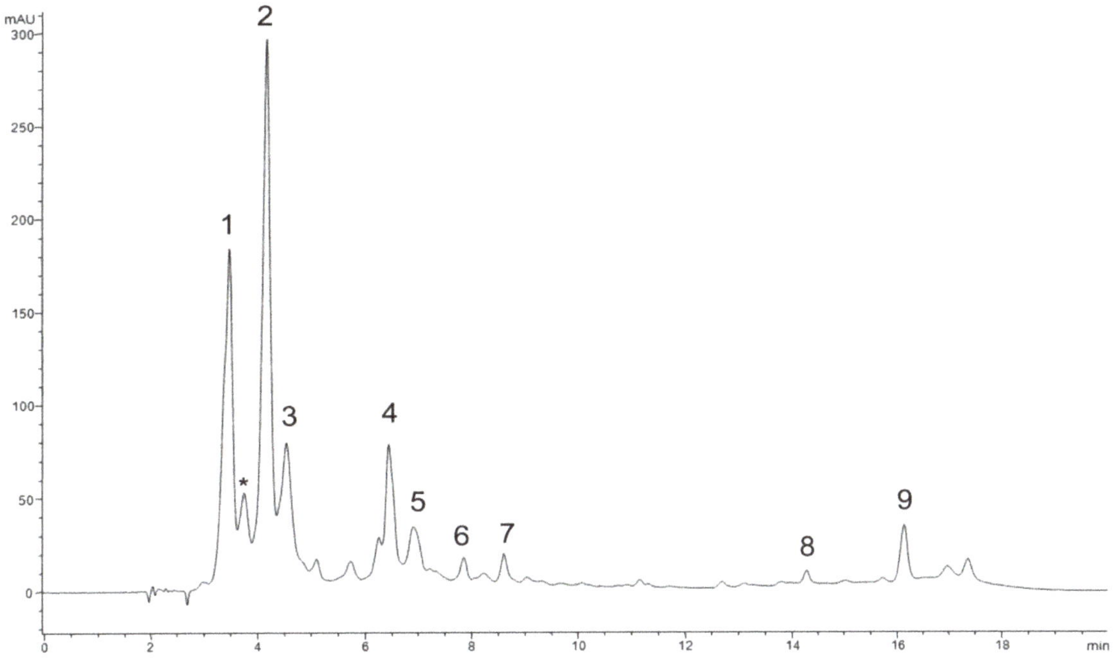

Figure 5. High performance liquid chromatography with diode array detector (HPLC-DAD) chromatogram of hydroethanolic extract of highbush blueberry *cv.* Bluegold. Chromatogram was registered at 280 nm; * background of detector, does not present a UV spectra.

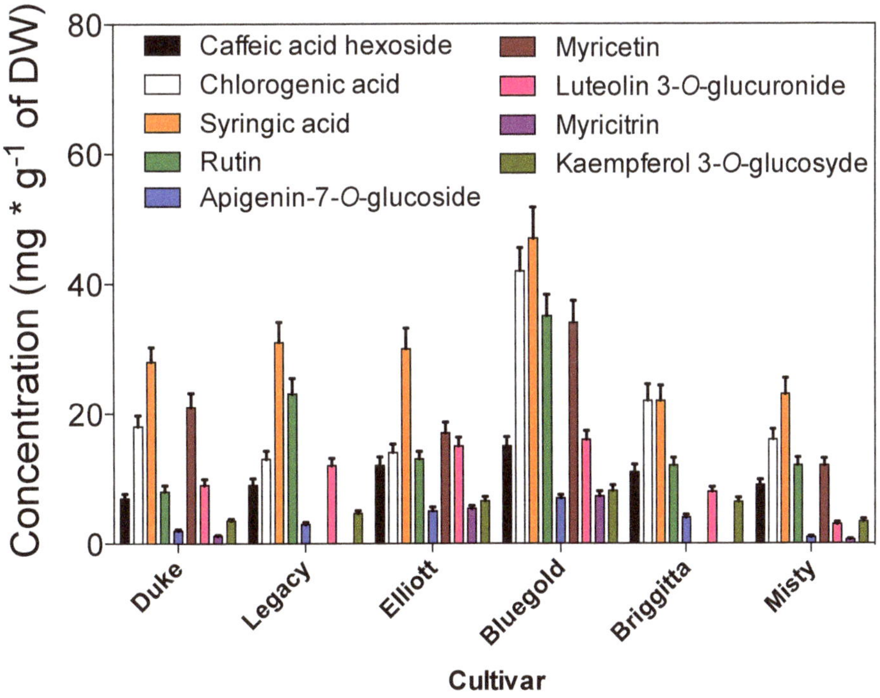

Figure 6. Total concentration of identified compounds in the six *in vitro* highbush blueberry cultivars. Each bar corresponds at the mean of three independent measurements ± standard error ($p < 0.05$).

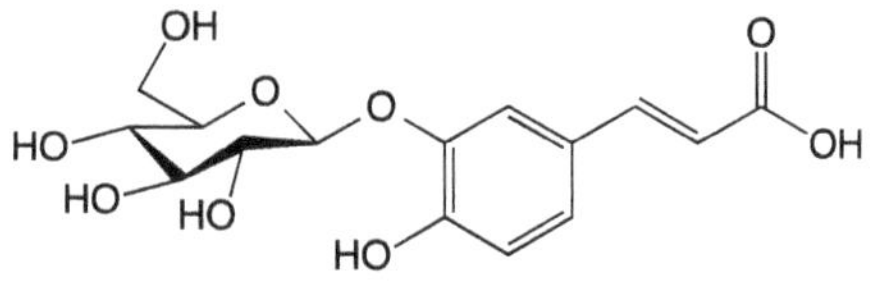

(1) Caffeic acid hexoside

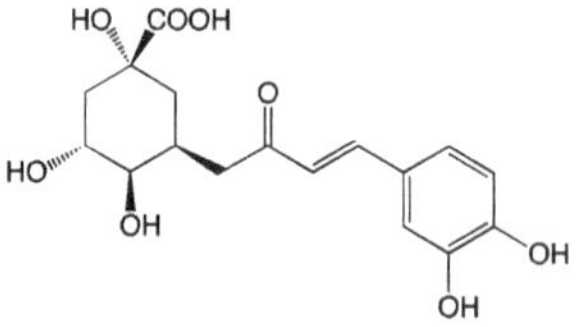

(2) Chlorogenic acid

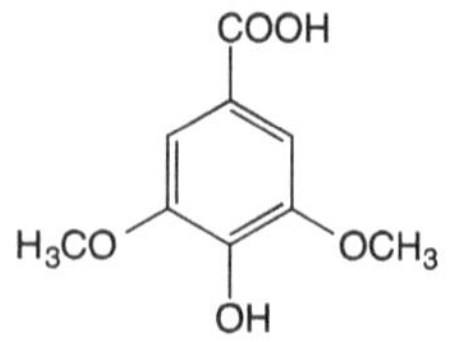

(3) Syringic acid

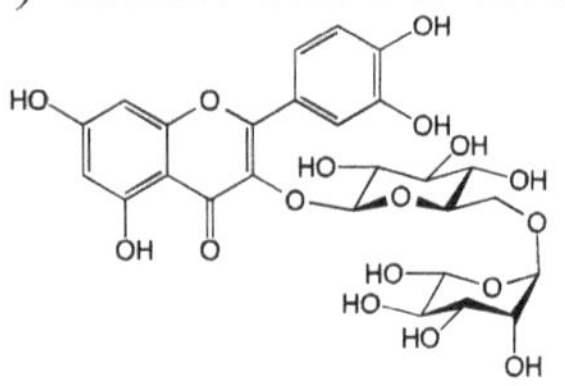

(4) Rutin

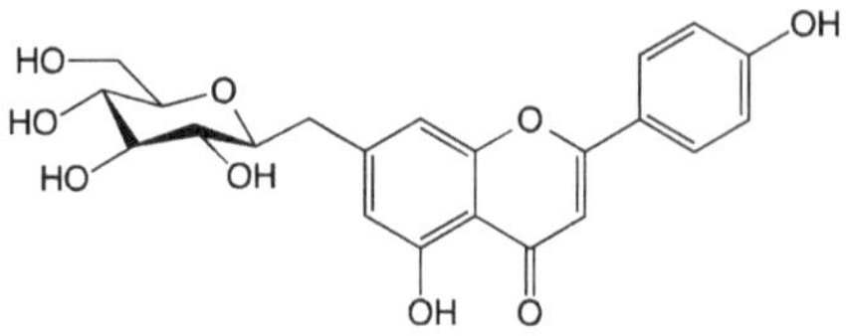

(5) Apigenin-7-*O*-glucoside

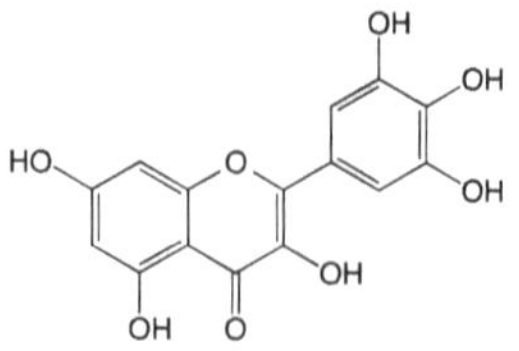

(6) Myricetin

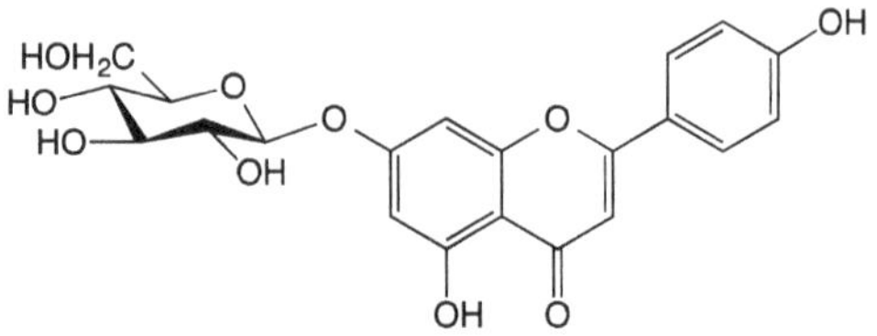

(7) Luteolin 3-*O*-glucuronide

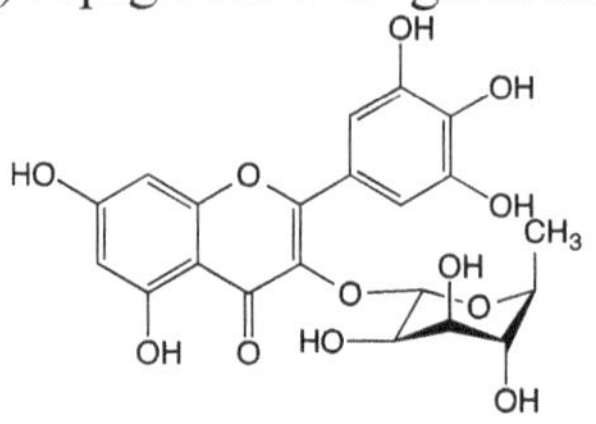

(8) Myricitrin

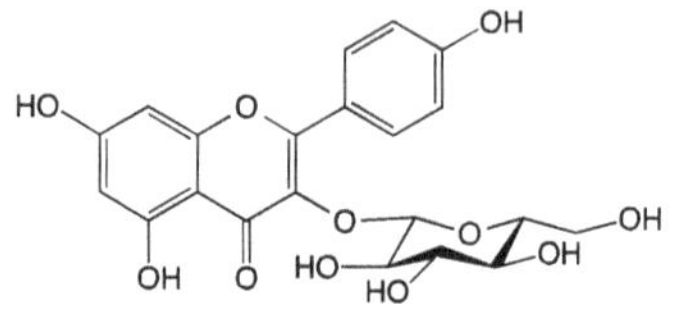

(9) Kaempferol 3-*O*-glucoside

Figure 7. Structures determined for highbush blueberry using UV and MS/MS spectra.

The antioxidant activity of the extracts analyzed was comparable to blueberry fruits (e.g., IC_{50} of Misty fruits is 21.95 μg v/s *in vitro* plants that are 27.6 μg). The very high content of these phytochemicals, especially of chlorogenic acid, syringic acid, rutin and myricetin in extracts of blueberry plants growing *in vitro* conditions, suggest that *in vitro* culture of blueberry is a valuable alternative to produce phenolic compounds with the high demand in the market, especially chlorogenic acid [20].

Plant tissue culture represents an alternative for the production of secondary metabolites with biological activity that are of interest to humans. The obtaining of phytochemicals from the leaves or fruits is affected by environmental factors, such as UV-B radiation, water availability, temperature and diseases [23]. Then, plant organ culture is of greatzhe interest as an alternative for obtaining chemicals from blueberry shoots by reducing the time interval to harvest. Moreover, it allows continuous biomass production. This paper was shown that *in vitro* culture of blueberry is a good alternative to obtain antioxidant molecules such as chlorogenic acid.

The amount (expressed in mg g^{-1} of DW) of each identified compound is reported in Figure 6. In the analyzed extracts from *in vitro* plants, chlorogenic acid was the most abundant compound followed by syringic acid and rutin and myricetin.

4. Conclusions

In conclusion, our analyses demonstrate that *in vitro* cultures of blueberries are a good source of antioxidant compounds, especially Bluegold cultivar. Through tissue culture, plant material can be produced continuously to make extracts, whose chemical composition is highly reproducible. Our analytical measurement suggests that the antioxidant activity of the extracts is mainly related to the content of chlorogenic acid, myricetin, syringic acid and rutin. Further analyses are needed to evaluate the cellular toxicity of the *in vitro* shoots extract.

Acknowledgments

To Conicyt doctoral fellowships to RAC, HK and MP. Financed by Project AT24120963 of Conicyt (RAC) and VRIDEI-USACH (GEZ).

Author Contributions

RAC worked in the experimental design, experimental development and wrote the paper. HK, MP contributed to experimental development. GEZ is the PI of our laboratory, correct the paper and approve the experiments. All authors read and approved the final manuscript.

Conflicts of Interest

The authors declare no conflict of interest.

References

1. Sellappan, S.; Akoh, C.C.; Krewer, G. Phenolic compounds and antioxidant capacity of Georgia-grown blueberries and blackberries. *J. Agric. Food Chem.* **2002**, *50*, 2432–2438.
2. Amakura, Y.; Umino, Y.; Tsuji, S.; Tonogai, Y. Influence of jam processing on the radical scavenging activity and phenolic content in berries. *J. Agric. Food Chem.* **2009**, *48*, 6294–6297.
3. Zheng, W.; Wang, S.Y. Oxygen radical absorbing capacity of phenolics in blueberries, cranberries, chokeberries, and lingonberries. *J. Agric. Food Chem.* **2003**, *51*, 502–509.
4. Sawa, T.; Nakao, M.; Akaike, T.; Ono, K.; Maeda, H. Alkylperoxyl radical-scavenging activity of various flavonoids and other phenolic compounds: Implications for the anti-tumor-promoter effect of vegetables. *J. Agric. Food Chem.* **1999**, *47*, 397–402.
5. Rodney, C.; Toni, M.; Kutchan, N.; Lewis, G. Natural products. In *Biochemistry and Molecular Biology of Plants*; Buchanan, B., Gruissem, W., Jones, R., Eds.; Wiley: Rockville, MD., USA, 2000; pp. 1253–1348.
6. Karuppusamy, S. A review on trends in production of secondary metabolites from higher plants by *in vitro* tissue, organ and cell cultures. *J. Med. Plant Res.* **2009**, *3*, 1222–1239.
7. Hussain, S.; Fareed, S.; Ansari, S.; Rahman, A.; Zareen, A.; Saeed, M. Current approaches toward production of secondary plant metabolites. *J. Pharm. Bioallied Sci.* **2012**, *4*, 10–20.
8. Lloyd, G.; McCown, B. Commercially feasible micropropagation of mountain laurel, *Kalmia latifolia*, by use of shoot tip culture. *Comb. Proc. Int. Plant Propag. Soc.* **1980**, *30*, 421–427.
9. Adam, M.; Dobiáš, P.; Eisner, A.; Ventura, K. Extraction of antioxidants from plants using ultrasonic methods and their antioxidant capacity. *J. Sep. Sci.* **2009**, *32*, 288–294.
10. Shyu, Y.-S.; Lin, J.-T.; Chang, Y.-T.; Chiang, Ch.-J.; Yang, D.-J. Evaluation of antioxidant ability of ethanolic extract from dill (*Anethum graveolens* L.) flower. *Food Chem.* **2009**, *115*, 515–521.
11. Benzie, I.F.F.; Strain, J.J. The ferric reducing ability of plasma (FRAP) as a measure of "antioxidant power": The FRAP assay. *Anal. Biochem.* **1996**, *239*, 70–76.
12. Singleton, V.L.; Rossi, J.A. Colorimetry of total phenolics with phosphomolybdic-phosphotungstic acid reagents. *Am. J. Enol. Vitic.* **1965**, *16*, 144–188.

13. Alhakmani, F.; Kumar, S.; Khan, S.A. Estimation of total phenolic content, *in vitro* antioxidant and anti-inflammatory activity of flowers of *Moringa oleifera*. *Asian Pac. J. Trop. Biomed.* **2013**, *3*, 623–627.
14. Thaipong, K.; Boonprakob, U.; Crosby, K.; Cisneros-Zeballos, L.; Hawkins, D. Comparision of ABTS, DPPH, FRAP, and ORAC assays for estimating antioxidant activity from guava fruit extracts. *J. Food Compos. Anal.* **2006**, *19*, 669–675.
15. Antolovich, M.; Prenzler, P.D.; Patsalides, E.; McDonald, S.; Robards, K. Methods for testing antioxidant activity. *Analyst* **2002**, *127*, 183–198.
16. Huang, D.; Ou, B.; Prior, R.L. The chemistry behind the antioxidant capacity assays. *J. Agric. Food Chem.* **2005**, *53*, 1841–1856.
17. Barreca, D.; Bellocco, E.; Caristi, C.; Leuzzi, U.; Gattuso, G. Distribution onf *C*- and *O*-glycosyl flavonoids, (3-hydroxy-3-methylglutaryl)glycosyl flavanones and furanocoumarins in *Citrus aurantium* L. juice. *Food Chem.* **2011**, *124*, 576–582.
18. Gournelis, D. Flavonoids of *Erica verticillata*. *J. Nat. Prod.* **1995**, *58*, 1065–1069.
19. Bohm, B.A.; Saleh, A.M. The flavonoids of *Cladothamnus pyrolaeflorus*. *Can. J. Bot.* **1972**, *50*, 2081–2083.
20. Shibata, H.; Sakamoto, Y.; Oka, M.; Kono, Y. Natural antioxidant, chlorogenic acid, protects against DNA breakage caused by monochloramine. *Biosci. Biotecnhol. Biochem.* **1999**, *63*, 1295–1297.
21. Pietta, P.G. Flavonoids as antioxidants. *J. Nat. Prod.* **2000**, *63*, 1035–1042.
22. Rice-Evans, C.A.; Miller, N.J.; Paganga, G. Structure-antioxidant activity relationships of flavonoids and phenolic acids. *Free Radic. Biol. Med.* **1996**, *20*, 933–956.
23. Ramakrishna, A.; Ravishankar, G.A. Influence of abiotic stress signals on secondary metabolites in plants. *Plant Signal. Behav.* **2011**, *6*, 1720–1731.

13

Systematic Study of the Content of Phytochemicals in Fresh and Fresh-Cut Vegetables

María Isabel Alarcón-Flores, Roberto Romero-González, José Luis Martínez Vidal and Antonia Garrido Frenich *

Department of Chemistry and Physics (Analytical Chemistry Area), Research Centre for Agricultural and Food Biotechnology (BITAL), University of Almería, Agrifood Campus of International Excellence, ceiA3, E-04120 Almería, Spain; E-Mails: maf400@ual.es (M.I.A.F.); rromero@ual.es (R.R.G.); jlmartin@ual.es (J.L.M.V.)

* Author to whom correspondence should be addressed; E-Mail: agarrido@ual.es

Academic Editors: Antonio Segura-Carretero and David Arráez-Román

---

Abstract: Vegetables and fruits have beneficial properties for human health, because of the presence of phytochemicals, but their concentration can fluctuate throughout the year. A systematic study of the phytochemical content in tomato, eggplant, carrot, broccoli and grape (fresh and fresh-cut) has been performed at different seasons, using liquid chromatography coupled to triple quadrupole mass spectrometry. It was observed that phenolic acids (the predominant group in carrot, eggplant and tomato) were found at higher concentrations in fresh carrot than in fresh-cut carrot. However, in the case of eggplant, they were detected at a higher content in fresh-cut than in fresh samples. Regarding tomato, the differences in the content of phenolic acids between fresh and fresh-cut were lower than in other matrices, except in winter sampling, where this family was detected at the highest concentration in fresh tomato. In grape, the flavonols content (predominant group) was higher in fresh grape than in fresh-cut during all samplings. The content of glucosinolates was lower in fresh-cut broccoli than in fresh samples in winter and spring sampling, although this trend changes in summer and autumn. In summary, phytochemical concentration did show significant differences during one-year monitoring, and the families of phytochemicals presented different behaviors depending on the matrix studied.

Keywords: one-year monitoring; phytochemicals; vegetables; fruits; fresh; fresh-cut

---

1. Introduction

Vegetables and fruits are considered particularly protective for human health [1]. This characteristic is linked to the chemical composition of these foods, particularly to phytochemicals (known as bioactive compounds) [2], which include flavones, flavonols, isoflavones, phenolic acids and glucosinolates [3]. These phytochemicals seem to play a role against the development of different types of cancer and cardiovascular diseases, because these compounds could provide antioxidant capacity (AOC) [4], anti-inflammation properties [5,6], lipid profile modification [7,8] and antitumor effects [9,10]. In addition to these beneficial properties of phytochemicals in human health, they are responsible for the color, flavor and smell of fruits and vegetables [11].

Currently, all kinds of vegetables and fruits can be bought in all seasons, but their beneficial properties are not the same at every time of year. There are some studies that indicated a variation of phytochemical content in vegetables with the season, *i.e.*, glucosinolates in brassica vegetables [12,13], carotenoids in spinach, parsley and green onion [14], flavonols in tomato [15,16], onion and lettuce [16] and anthocyanidins in grape [17]. In general, most of these studies indicate that spring/summer season plants, which are grown at intermediate temperatures, high light intensity, longer days and dry conditions, have the highest content of phytochemicals. Conversely, the conditions in autumn/winter seasons are different (lower temperatures, lower light intensity, shorter days and higher water availability), and plants tend to have the lowest levels of these types of compounds. However, variation has also been observed throughout the harvest. For instance, Rosa and Rodrígues reported that some broccoli cultivars had higher glucosinolate concentrations in late season compared to the early season crop [18]. Moreover, tomato's lycopene is inhibited at high temperatures (>30 °C) due to overheating [19]. Therefore, it can be observed that the seasons directly affected the concentration of phytochemicals in these matrices. Several factors, such as temperature, precipitation and radiation, can influence the accumulation of phytochemicals in vegetables and fruits. For instance, light exposure, especially ultraviolet-B rays, can provoke an increment in the content of these types of compounds [16]. Consequently, growing plants in greenhouses, which blocks ultraviolet light, reduces the flavonol content in plants. Vegetables grown outdoors contained four- to five-fold more flavonols than those cultivated in greenhouses [20]. Moreover, there is a trend indicating that higher flavonol levels were observed in warmer, sunnier climates than in cooler regions. Thus, Stewart *et al.* commented that growing conditions in Spain induce the accumulation of relatively high flavonols content in tomato fruits throughout most of the year [15].

These papers analyzed phytochemicals in vegetables and fruits from two [12,14] to four seasons [15,16], but few phytochemicals were determined. Therefore, the purpose of this work has been to provide a comprehensive view of the evolution of the content of more than 30 phytochemicals belonging to several families (phenolic acids, flavonols, flavones, glucosinolates and isoflavones) in fresh and fresh-cut products stored under modified atmosphere packaging (MAP), such as tomato, eggplant, grape, carrot and broccoli, during one year.

2. Experimental Section

2.1. Chemicals and Reagents

Commercial glucosinolate standards, such as progoitrin, gluconasturtiin and glucoraphanin, were supplied by PhytoLab GmbH & Co (Vestenbergsgreuth, Germany). Glucotropaeolin, glucoerucin and glucoiberin were purchased from Scharlab (Barcelona, Spain). Genistein, apigenin, quercetin, quercetin-3-*O*-glucoside, gallic acid, sulforaphane, ferulic acid, baicalein, gallic acid and caffeic acid were purchased from Sigma-Aldrich (Steinheim, Germany). Other standards, such as daidzein, glycitein, luteolin-4-*O*-glucoside, luteolin-7-*O*-glucoside, apigenin-7-*O*-neohesperoside, kaempferol, kaempferol-3-*O*-glucoside, kaempferol-3-*O*-rutinoside, luteolin, luteolin-6-*C*-glucoside, luteolin-8-*C*-glucoside, apigenin-7-*O*-glucoside, apigenin-6-*C*-glucoside, apigenin-8-*C*-glucoside, quercetin-3-*O*-ramnoside, quercetin-3-*O*-galactoside, quercetin-3-*O*-rutinoside, quercetin-3-*O*-ramnoside, quercetin-3-*O*-galactoside, isorhamnetin, isorhamnetin-3-*O*-rutinoside, isorhamnetin-3-*O*-glucoside, apigenin-7-*O*-rutinoside and tamarixetin, were purchased from Extrasynthese (Genay, France). Stock standard solutions of individual compounds (with concentrations between 200 and 300 mg/L) were prepared by exact weighing of the powder and dissolved in 10 mL of HPLC-grade methanol or in a mixture of methanol:water (50:50, v/v). Then, they were stored at −20 °C in dark bottles. A multi-compound working standard solution at a concentration of 5 mg/L of each compound was prepared by appropriate dilution of the stock solutions with methanol and stored in screw-capped glass tubes at −20 °C. The solutions were prepared every six months. Ultrapure water was obtained from a Milli-Q Gradient water system (Millipore, Bedford, MA, USA). Ammonium acetate was purchased from Panreac (Barcelona, Spain). Formic acid (purity > 98%), HPLC-grade methanol and dimethylsulfoxide were provided by Sigma (Madrid, Spain). Millex-GN nylon filters of 0.20 μm were provided by Millipore (Millipore, Carrightwohill, Ireland).

2.2. Apparatus and Software

Chromatographic analyses were carried out using an Agilent series 1290 RRLC instrument (Agilent, Santa Clara, CA, USA) equipped with a high-performance autosampler (G4226A), a binary pump (G4220A), a column compartment thermostat (G1316C) and an autosampler thermostat (G1330B). The system was coupled to an Agilent triple quadrupole mass spectrometer (6460A) with a Jet Stream ESI ion source (G1958-65138). For the chromatographic separation of the extracts, a Zorbax Eclipse Plus C18 column (100 mm × 2.1 mm, 1.8-μm particle size) from Agilent was used. Chromatographic conditions optimized in a previous work were used in this study [21].

An Agilent Mass Hunter Quantitative analyzer (Agilent Technologies, Inc.) was used for data acquisition and quantification of samples.

Statistical analysis (Analysis of Variance, ANOVA) was carried out with SPSS v21 (Armonk, New York, NY, USA).

Lyophilizer Alpha from Martin Christ (Osterode, Germany) was also used; an analytical balance AB204-S from Mettler Toledo (Greifensee, Switzerland), a Reax-2 rotary agitator from Heidolph (Schwabach, Germany) and vacuum pump from Vacuubrand (Wertheim, Germany) were also utilized.

2.3. Analysis of Samples

Samples were homogenized and were transferred to a Petri dish, and they were weighed and cooled to −18 °C. Then, all samples were processed according to the extraction method optimized in a previous work [21]. Briefly, 150 mg of lyophilized sample were weighed in a 15-mL polypropylene centrifuge tube and 3 mL of a mixture methanol:water (80:20, v/v) were added. The mixture was agitated for 30 min with a rotary shaker. After that, the extract was filtered and transferred into a vial containing mobile phase (50:50 v/v of Eluents A and B). A chromatographic method based on ultra-high performance liquid chromatography coupled to triple quadrupole mass spectrometry (UHPLC-QqQ-MS/MS) was used for the determination of the target compounds [21].

2.4. Samples

The sampling was performed at different seasons: February (winter), April (spring), July (summer) and October (autumn). For each sampling, fresh and fresh-cut vegetables and fruits were analyzed, and the same variety of each fruit and vegetable was evaluated.

Fresh tomato, broccoli, carrot, grape and eggplant samples was obtained from different supermarkets located in the province of Almería (southeast of Spain). On the other hand, fresh-cut products were obtained from Guzmán Gastronomía (Barcelona, Spain). In both cases, six samples of each matrix were analyzed. However, for fresh-cut carrot and fresh-cut eggplant, twelve samples per matrix were analyzed, bearing in mind that two types of cuts were evaluated for each matrix (grated and sliced for carrots and diced and sliced for eggplant).

Samples were chopped and homogenized at room temperature. After that, they were kept frozen at −18 °C until lyophilization.

3. Results and Discussion

3.1. Phytochemical Content in Tomato

The concentrations of phytochemicals in fresh and fresh-cut tomato are shown in Table 1. The effects of processing and sampling season were evaluated in tomato by two-way analysis of variance (ANOVA) study. When ANOVA was performed, it was observed that there is no statistical difference between fresh and fresh-cut samples during a year in the total content of phytochemicals, neither in phenolic acids nor flavones. However, there is a significant difference in the content of flavonols ($p < 0.02$), the content being lower in fresh-cut than in fresh tomato. In Figure 1a, it can be observed that the total content of phytochemicals in fresh-cut samples is lower in winter and autumn than in fresh samples, whereas in spring and summer, the total content is similar.

Table 1. Phytochemical content (expressed as mg/kg, dry weight (DW)) in fresh and fresh-cut in tomato, broccoli and grape in different seasons.

Matrix	Phytochemicals	Winter Sampling		Spring Sampling		Summer Sampling		Autumn Sampling	
		Fresh *	Fresh-Cut *	Fresh *	Fresh-Cut *	Fresh *	Fresh-Cut *	Fresh *	Fresh-Cut *
Tomato	Phenolic acids	102 (30) ^{a}	54 (19) ^{1}	225 (31) ^{b}	294 (26) ^{2}	97 (46)	107 (43)	148 (46)	127 (54)
	Flavones	29 (7) ^{a}	21 (5) ^{1}	37 (8) ^{a}	24 (5) ^{1}	39 (9) ^{a}	62 (12) ^{1}	104 (36) ^{b}	113 (55) ^{2}
	Flavonols	44 (20) ^{a}	21 (11) ^{1}	66 (16) ^{a}	34 (17) ^{1}	92 (17) ^{a}	59 (12) ^{1}	165 (29) ^{b}	130 (94) ^{2}
Broccoli	Phenolic acids	38 (11) ^{a,c,d}	30 (8) ^{1,3,4}	61 (16) ^{b}	71 (26) ^{2}	21 (9) ^{a,c,d}	50 (12) ^{1,3,4}	29 (12) ^{a,c,d}	32 (15) ^{1,3,4}
	Flavones	26 (4) ^{a}	22 (10) ^{1,2}	35 (5) ^{b,c}	26 (10) ^{1,2}	43 (3) ^{c,d}	61 (27) ^{3,4}	53 (6) ^{d}	74 (19) ^{3,4}
	Flavonols	13 (4) ^{a,d}	5 (2) ^{1,2}	35 (10) ^{b,d}	15 (4) ^{2,1}	58 (18) ^{c}	41 (16) ^{3,4}	32 (5) ^{a,b,d}	43 (15) ^{3,4}
	Glucosinolates	4449 (105) ^{a}	3308 (198)	2929 (986) ^{b}	2432 (945)	884 (237) ^{c,d}	1517 (856)	919 (331) ^{c,d}	1526 (877)
Grape	Phenolic acids	1 (1)	Not detected ^{1,2}	3 (2)	2 (1)	1 (1)	1 (1)^{12}	2 (1)	9 (8) ^{3}
	Flavones	35 (7) ^{a,b}	49 (22) ^{1,2}	48 (8) ^{a,b}	47 (7) ^{1,2}	74 (4)	77 (31)	123 (66) ^{c}	90 (21) ^{3}
	Flavonols	230 (155)	149 (92)	213 (52)	112 (27)	207 (138)	114 (28)	220 (109)	139 (48)

* Mean value. The standard deviation is given in parenthesis ($n = 6$); mean values in italics indicate significant differences between fresh and fresh-cut matrices; in fresh samples, the mean values in a row with different superscript letters are significantly different ($p < 0.05$) using the Bonferroni test; in fresh-cut samples, the mean values in a row with different superscript numbers are significantly different ($p < 0.05$) using the Bonferroni test.

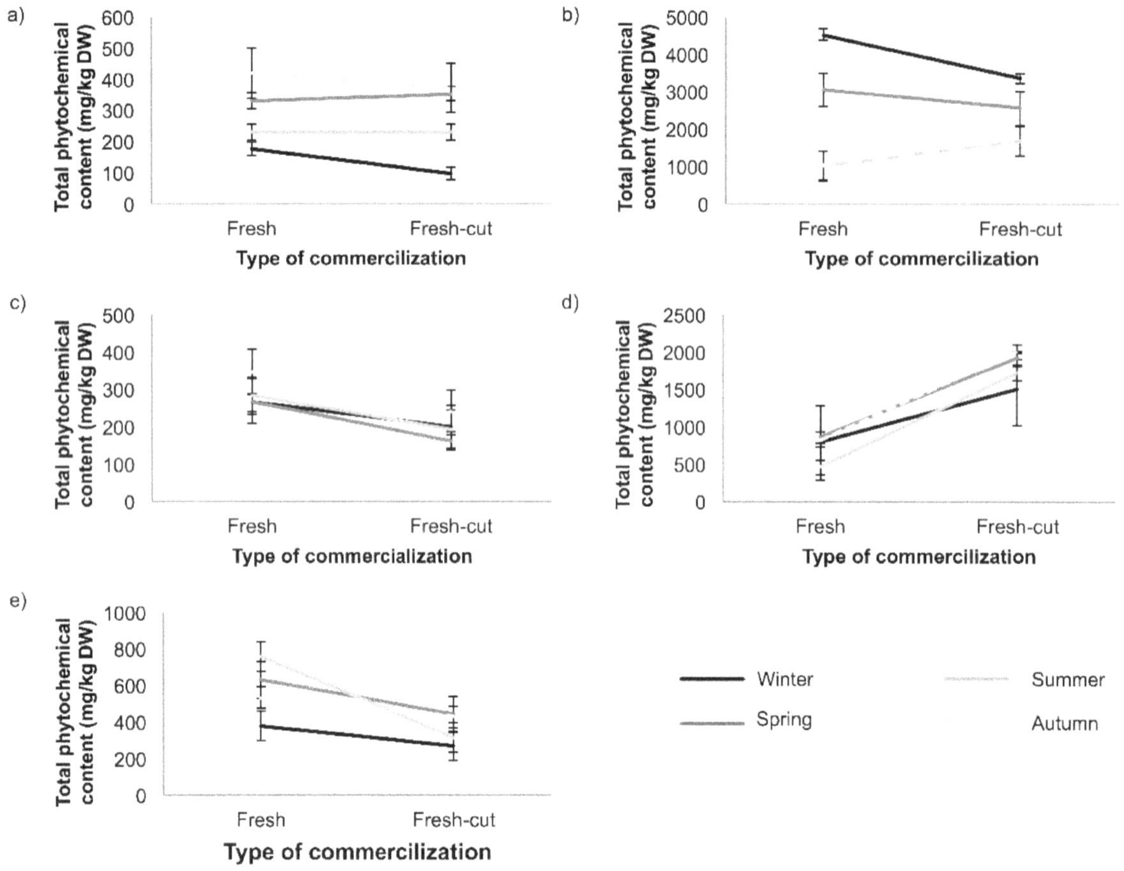

Figure 1. Total phytochemical content (mg/kg, dry weight (DW)) during the different seasons of the year in fresh and fresh-cut: (**a**) tomato; (**b**) broccoli; (**c**) grape; (**d**) eggplant; and (**e**) carrot.

On the other hand, the total content of phytochemicals in each sampling was evaluated, being the difference only significant ($p = 0.01$) in the winter sampling (175 and 96 mg/kg dry weight (DW) for fresh and fresh-cut samples, respectively). Concerning the different families of phytochemicals evaluated, phenolic acids are the most abundant compounds in tomato; the level of this family fluctuated between fresh and fresh-cut in all samplings, but the difference was only significant ($p = 0.02$) in winter sampling.

Total phytochemical content, as well as the different families studied were affected by season. Phenolic acids presented the highest concentration in spring sampling (225 and 294 mg/kg fresh and fresh-cut, respectively), although this difference was only significant when it was compared with winter sampling (102 and 54 mg/kg fresh and fresh-cut, respectively). Our data could not confirm that the amount of phenolic acids increases with increasing light intensity due to lower concentrations of phenolic acids (not significant) obtained in summer than in autumn and spring. This can be explained considering that high temperatures (>30 °C) can induce an overheating in tomatoes, which inhibits the biosynthesis of some compounds, as happens with lycopene in tomato [19]. On the other hand, growing conditions in Spain can lead to the accumulation of a relatively high phenolic acids content in tomato throughout most of the year. In relation to the levels of flavonols (165 and 130 mg/kg in fresh and fresh-cut samples, respectively) and flavones (104 and 113 mg/kg in fresh and fresh-cut samples, respectively) in autumn, they were significantly higher than in other samplings. However, Stewart *et al.* [15] and Raffo *et al.* [22] did not find any difference between tomatoes harvested in different seasons, observing that chlorogenic acid was the phenolic acid detected at the highest concentrations (from 26.7 to 54.4 mg/kg). In this study, chlorogenic acid was also the predominant phenolic acid in tomato in all samplings, as can be observed in Figure 2a, although the highest concentration was obtained in spring, which is in accordance with previous studies [22]. These results could indicate that chlorogenic acid content is affected by typical spring environmental conditions, although further studies should be performed in order to get definitive conclusions, because other factors, such as biotic and abiotic stress, can influence the final concentration.

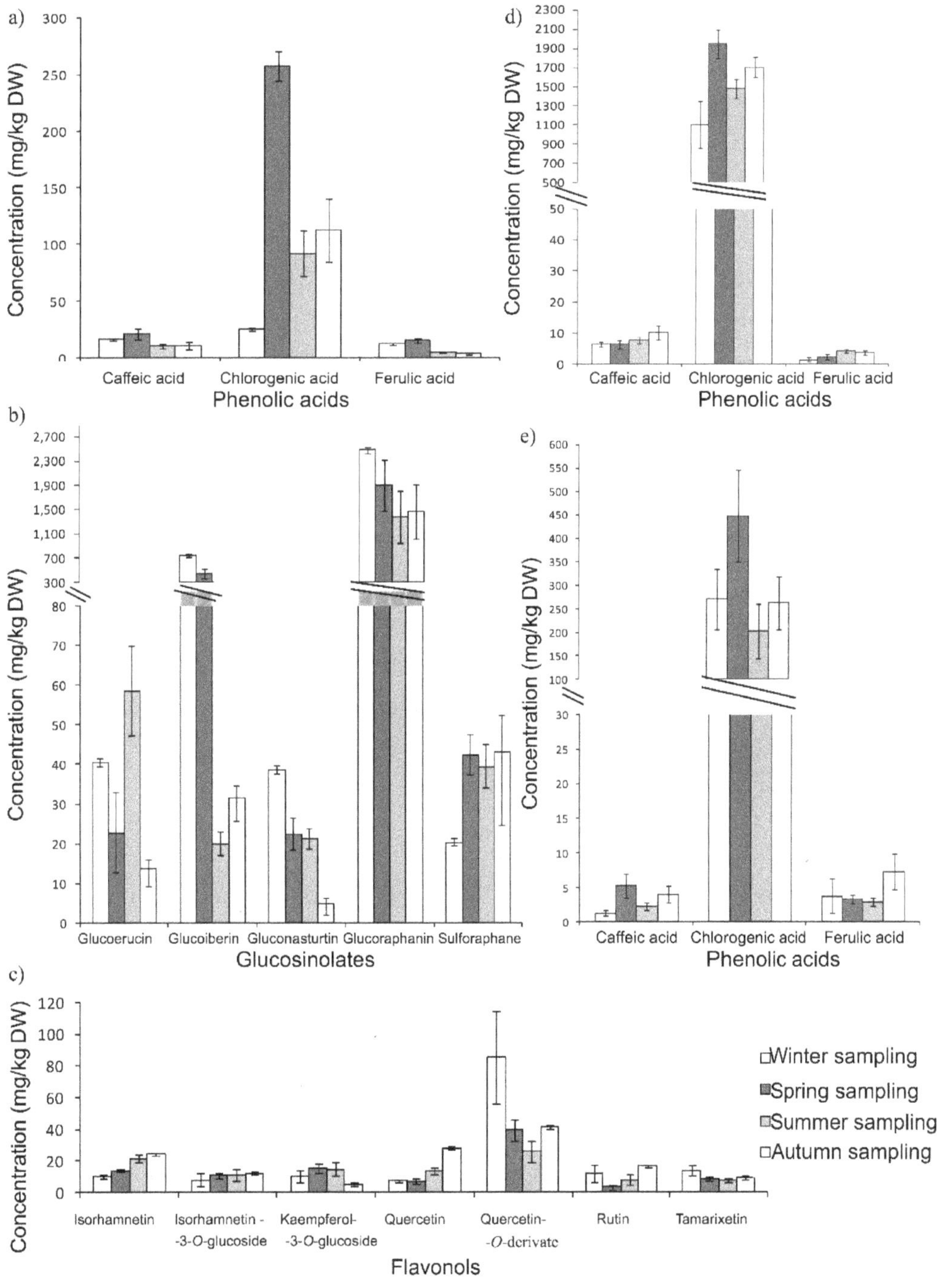

Figure 2. Concentration (mg/kg, dry weight (DW)) of the main family of phytochemicals in: (**a**) fresh-cut tomato; (**b**) fresh-cut broccoli; (**c**) fresh-cut grape; (**d**) fresh-cut eggplant; and (**e**) fresh-cut carrot during the different seasons of the year.

3.2. Phytochemical Content in Broccoli

The concentration of the phytochemicals in different seasons in fresh and fresh-cut broccoli samples is shown in Table 1. In general, there was no significant difference in the total content of phytochemicals according to the type of commercialization (*p*-value = 0.83) throughout the year (Figure 1b). It can be observed that the total content of phytochemicals was very similar. When the content of the different families was evaluated, only flavonols showed significant differences between both types of products ($p < 0.01$), and their content is higher in fresh broccoli throughout the year.

The total content of phytochemicals was significantly affected by season, specifically between winter-spring sampling and summer-autumn sampling. This difference is attributed to the diminution of glucosinolates in summer and autumn, and these levels are higher in winter (4449 and 3308 mg/kg DW in fresh and fresh-cut samples, respectively) and spring (2929 and 2432 mg/kg DW in fresh and fresh-cut samples, respectively) than in the other seasons. This is in accordance with the results obtained from Henning *et al.*, who found higher levels of glucosinolates concentration in the spring season than in autumn [23]. However, Aires *et al.* [12] evaluated the content of glucosinolates in two different years, obtaining the highest concentration of this family in summer-winter in 2005–2006, whilst in 2006–2007, the highest concentration was obtained in spring-summer (3942 μmol/100 g DW). It was observed in Figure 2b that glucoraphanin was the predominant glucosinolate in broccoli in all samplings, as can be observed in previous studies [12], obtaining the highest concentration in the winter sampling (3737 mg/kg DW). In relation to the other families, significant differences were obtained, as can be observed in Table 1. The highest concentration of phenolic acids was obtained in spring (61 and 71 mg/kg DW in fresh and fresh-cut samples, respectively), which was statistically different than in the other seasons. The lowest concentration of flavonols was obtained in winter, which was statistically different than the content obtained during the rest of the samplings. The content of flavones was also affected by seasons, especially between the winter-spring sampling and the summer-autumn sampling, and flavones presented the highest concentration in the summer-autumn seasons.

3.3. Phytochemical Content in Grape

The concentration of phytochemicals in fresh and fresh-cut grape obtained during the four seasons evaluated is shown in Table 1. The results indicated that the total content of phytochemicals is not influenced by the type of commercialization. However, in all samplings, it can be noted that fresh grape has the highest concentration of phytochemicals (Figure 1c) ranging from 264 to 345 mg/kg DW, while in fresh-cut grape, the content ranged from 161 to 238 mg/kg DW. In relation to the families of compounds studied, no significant differences were observed among fresh and fresh-cut grapes.

Regarding the influence of the seasons, neither the total content of phenolic compounds nor the content of flavonols show significant differences. However, the content of flavones is significantly different depending on the season, because of the high concentration found in autumn (53 and 74 mg/kg DW for fresh and fresh-cut product, respectively).

In order to explain these results, the conclusions obtained by Ferrer-Gallego *et al.* must be considered [24]. They reported that there was a direct relationship between solar radiation exposure and concentration of phenolic compounds in grape, whereas there is an inverse relationship between accumulated rainfall and contents of phenolic compounds. Moreover, temperature does not have an influence on the content of flavonols in grape [25]. According to these results, it could be indicated that the grapes evaluated in this study have been cultivated in a place where the level of precipitation is low and solar radiation is high during the four seasons, because there were no significant differences in the content of flavonols (predominant group) and, consequently, in the total content of phenolic compounds. It can be observed in Figure 2c that quercetin 3-*O*-derivate was the most abundant flavonol in all samplings, although the highest concentration was obtained in the winter sampling (85.4 mg/kg DW in fresh-cut grape). In relation to other matrices, it can be noted that there is not a common phytochemical trend between different sampling times, indicating that each compound could be affected in a different way by the same environmental conditions.

3.4. Phytochemical Content in Eggplant

The obtained results showed a variation of the total content of phytochemicals between fresh and fresh-cut eggplant in all samplings (Figure 1d). This variation is mainly due to the fluctuations of phenolic acids among both types of products (fresh and fresh-cut), due to this family is the predominant group in eggplant. The highest concentration of phenolic acids was detected in fresh-cut eggplant, ranging from 1474 to 1875 mg/kg DW, whereas the content in fresh eggplant ranged from 385–799 mg/kg DW (Table 2). The higher levels in cut samples can be explained considering that during the preparation of the diced or sliced samples, the synthesis of the phenylpropanoid pathway is induced, and therefore, the content of phenolic compounds could increase [26].

It can be observed in Figure 2d that chlorogenic acid was the predominant phenolic acid in eggplant in all samplings, as can be observed in previous studies [27]. Flavones and flavonols also showed significant differences among fresh and fresh-cut samples throughout the year ($p < 0.01$ in both cases). Different types of cut in fresh-cut eggplant, diced and sliced, were analyzed, but in general, significant differences were not found. However, in a previous study [28], it was reported that there were significant differences in the content of phenolic acids between diced and sliced fresh-cut eggplant. This difference between both studies can be explained considering that in the current study, the effect of the type of presentation was evaluated throughout a year, whereas in the previous one, a single season only was evaluated.

When the ANOVA study was carried out to check if the sampling season affects the total content of phytochemicals in the assayed products, it was observed that this factor has no significant influence on the total content ($p = 0.82$). Bearing in mind the different families included in this study, it can be noted that the season significantly influenced the content of flavones and flavonols ($p < 0.01$), obtaining for both families the highest concentrations in autumn (62 and 87 mg/kg DW of flavones in fresh and fresh-cut samples, respectively). The content of flavonols was 53 and 48 mg/kg DW in fresh and fresh-cut samples, respectively, whereas the lowest level of flavones was obtained in spring (10 and 25 mg/kg in fresh and fresh-cut samples, respectively) and in winter for flavonols (24 and 10 mg/kg fresh and fresh-cut samples, respectively). However, no significant differences were observed for the content of phenolic acids ($p = 0.86$).

Table 2. Phytochemical content (expressed as mg/kg, dry weight (DW)) in fresh and fresh-cut in carrot and eggplant in different seasons.

Matrix	Phytochemicals	Winter Sampling		Spring Sampling		Summer Sampling		Autumn Sampling	
		Fresh *	Fresh-cut *	Fresh *	Fresh-cut *	Fresh *	Fresh-cut *	Fresh *	Fresh-cut *
Eggplant	Phenolic acids	733 (846)	1474 (649)	799 (137)	1867 (202)	385 (67)	1632 (197)	703 (241)	1875 (205)
	Flavones	30 (6) ^{a,c}	30 (7) ^{1,2}	10 (7) ^{b}	25 (8) ^{1,2}	39 (8) ^{a,c}	56 (10) ^{3}	62 (15) ^{d}	87 (30) ^{4}
	Flavonols	24 (14) ^{a}	10 (6) ^{1}	48 (22)	24 (12) ^{2,3}	30 (8)	31 (6) ^{2,3}	53 (9) ^{b}	48 (16) ^{4}
Carrot	Phenolic acids	340 (155)	236 (117)	564 (208)	378 (175)	652 (224)	239 (123)	407 (118)	249 (94)
	Flavones	26 (4) ^{a}	24 (5) ^{1,2}	46 (8) ^{b}	45 (14) $^{1-3}$	69 (8) ^{c,d}	54 (6) ^{2,3}	87 (17) ^{c,d}	114 (42) ^{4}
	Flavonols	10 (5) ^{a,b}	8 (4) ^{1,2}	18 (3) ^{a,b}	19 (5) $^{1-3}$	35 (9) ^{c,d}	21 (3) ^{2,3}	39 (7) ^{c,d}	64 (20) ^{4}

* Mean value. The standard deviation is given in parenthesis (n = 6 for fresh samples and n = 12 for fresh-cut samples); mean values in italics indicate significant differences between fresh and fresh-cut matrices; in fresh samples, the mean values in a row with different superscript letters are significantly different ($p < 0.05$) using the Bonferroni test; in fresh-cut samples, the mean values in a row with different superscript numbers are significantly different ($p < 0.05$) using the Bonferroni test.

3.5. Phytochemical Content in Carrot

The concentration of the total content of phytochemicals in fresh and fresh-cut carrots in different seasons is shown in Table 2. In relation to the content of total phytochemicals, it can be observed (Figure 1e) that there is a significant variation between fresh and fresh-cut carrot (p-value < 0.01), and phenolic acids are the predominant phytochemical family; the results for this family are shown in Figure 2e. Regarding the type of commercialization, significant effects (p-values < 0.01) were observed for the total content of phenolic acids, obtaining the highest values in fresh carrots, ranging from 340 to 652 mg/kg DW, whereas the content in fresh-cut carrots ranged from 236 to 378 mg/kg DW. These results are different from those obtained in a previous study [28], which reported that although differences between total phytochemicals and phenolic acids contents in fresh and fresh-cut carrots were observed, they were not significant. The discrepancy between both studies can be explained considering that in the current study, the effect of the type of commercialization was evaluated throughout a year, whereas in the previous study, a single season was only tested.

Season is a significant factor in the average content of total phenolic acids ($p < 0.01$) and, consequently, in the total content of phenolic compounds, the content in spring (564 mg/kg DW in fresh matrix) and summer (652 mg/kg DW in fresh matrix) samplings are significantly higher than in winter sampling (340 mg/kg DW in fresh carrot). In relation to the content of flavones and flavonols, it was high in autumn and low in winter, being these differences statistically significant ($p < 0.01$ in both cases), whereas in spring and summer, the detected concentrations were in the middle. These results are in agreement with those obtained by Leja *et al.* [29], who published that in the season of lower rainfall, higher amounts of phenolic compounds were accumulated in carrots. The found concentrations depended on the type of cultivar evaluated, and they ranged from 242 to 3115 mg/kg fresh weight.

Different types of cut in fresh-cut carrot, grated and sliced, were analyzed in all samplings, but no significant differences were observed in the total content of phenolic compounds (p-value = 0.16), as well as in the content of the different families in relation to the type of presentation of carrot throughout the year.

4. Conclusions

According to the data obtained in this work, it can be indicated that in relation to the matrix, the different families of phytochemicals have different behaviors when they are analyzed throughout a year. The several matrices harvested at different seasons of the year showed differences in the total content of phytochemicals, although these were not always significant. Autumn is the season with the highest concentration of phytochemicals in tomato, grape and eggplant, while the best season for broccoli was winter and spring, and summer for carrot. Although variations in total content or different families of phytochemicals according to the season have been observed, neither a clear seasonal trend, nor a correlation between phytochemicals content and mean solar radiation, nor average temperature was found. The total content of phytochemicals obtained was higher in fresh matrix than in fresh-cut matrix, as in carrot or grape. However, for eggplant, the highest concentration of phytochemicals was obtained in fresh-cut eggplant. In relation to broccoli, lower concentrations of phytochemicals in fresh-cut broccoli than in fresh broccoli were obtained in winter and spring.

Acknowledgments

The authors are grateful to Andalusian Regional Government (Regional Ministry of Innovation, Science and Enterprise), the Centre for Industrial Technological Development (CDTI) and FEDER for financial support Project Ref. P11-AGR-7034 and IDI-20110017 respectively. The authors also are grateful to Guzmán Gastronomía for supplying the fresh-cut samples. María Isabel Alarcón-Flores acknowledges her grant from the University of Almería (Spain). Roberto Romero-González is also grateful for personal funding through the Ramon y Cajal Program (Spanish Ministry of Economy and Competitiveness-European Social Fund).

Author Contributions

María Isabel Alarcón Flores and Roberto Romero-González have performed the measurements and analyzed the data. Antonia Garrido Frenich and José Luis Martínez Vidal have designed and supervised the research. All authors have contributed to the writing of the manuscript and have approved the final paper.

Conflicts of Interest

The authors declare no conflict of interest.

References

1. Hasler, C.M. Functional foods: Benefits, concerns and challenges—A position paper from the American Council on Science and Health. *J. Nutr.* **2002**, *132*, 3772–3781.
2. Mudgal, V.; Madaan, N.; Mudgal, A.; Mishras, A. Dietary polyphenols and human health. *Asian J. Biochem.* **2010**, *5*, 154–162.
3. Luthria, D.L.; Natarajan, S.S. Influence of sample preparation on the assay of isoflavones. *Planta Med.* **2009**, *75*, 704–710.
4. Kim, D.O.; Padilla-Zakour, O.I.; Griffiths, P.D. Flavonoids and antioxidant capacity of various cabbage genotypes at juvenile stage. *J. Food Sci.* **2004**, *69*, 685–689.
5. Vincent, H.K.; Bourguignon, C.M.; Taylor, A.G. Relationship of the dietary phytochemical index to weight gain, oxidative stress and inflammation in overweight young adults. *J. Hum. Nutr. Diet.* **2010**, *23*, 20–29.
6. González-Gallego, J.; García-Mediavilla, M.V.; Sánchez-Campos, S.; Tuñón, M.J. Fruit polyphenols, immunity and inflammation. *Br. J. Nutr.* **2010**, *104*, S15–S27.
7. Wang, S.; Melnyk, J.P.; Tsao, R.; Marcone, M.F. How natural dietary antioxidants in fruits, vegetables and legumes promote vascular health. *Food Res. Int.* **2011**, *44*, 14–22.
8. Perez-Vizcaino, F.; Duarte, J. Flavonols and cardiovascular disease. *Mol. Asp. Med.* **2010**, *31*, 478–494.
9. Stan, S.D.; Kar, S.; Stoner, G.D.; Singh, S.V. Bioactive food components and cancer risk reduction. *J. Cell. Biochem.* **2008**, *104*, 339–356.
10. Prasain, J.K.; Barnes, S. Metabolism and bioavailability of flavonoids in chemoprevention: Current analytical strategies and future prospectus. *Mol. Pharm.* **2007**, *4*, 846–864.

11. Miglio, C.; Chiavaro, E.; Visconti, A.; Flogliano, V.; Pellegrini, N. Effects of different cooking methods on nutritional and physicochemical characteristics of selected vegetables. *J. Agric. Food Chem.* **2008**, *56*, 139–147.
12. Aires, A.; Fernandes, C.; Carvalho, R.; Bennett, R.N.; Saavedra, M.J.; Rosa, E.A.S. Seasonal effects on bioactive compounds and antioxidant capacity of six economically important Brassica vegetables. *Molecules* **2011**, *16*, 6816–6832.
13. Björkman, M.; Klingen, I.; Birch, A.N.E.; Bones, A.M.; Johansen, T.J.; Meadow, R.; Mølmann, J.; Seljåsen, R.; Bruce, T.J.A.; Smart, L.E.; *et al.* Phytochemicals of Brassicaceae in plant protection and human health—Influences of climate, environment and agronomic practice. *Phytochemistry* **2011**, *72*, 538–556.
14. Azevedo-Meleiro, C.H.D.; Rodriguez-Amaya, D.B. Carotenoids of endive and New Zealand spinach as affected by maturity, season and minimal processing. *J. Food Compos. Anal.* **2005**, *18*, 845–855.
15. Stewart, A.J.; Bozonnet, S.; Mullen, W.; Jenkins, G.I.; Lean, M.E.J.; Crozier, A. Occurrence of flavonols in tomatoes and tomato-based products. *J. Agric. Food Chem.* **2000**, *48*, 2663–2669.
16. Crozier, A.; Lean, M.E.J.; McDonald, M.S.; Black. C. Quantitative analysis of the flavonoid content of commercial tomatoes, onions, lettuce, and celery. *J. Agric. Food Chem.* **1997**, *45*, 590–595.
17. Xu, C.; Zhang, Y.; Zhu, L.; Huang, Y.; Lu, J. Influence of growing season on phenolic compounds and antioxidant properties of grape berries from vines grown in subtropical climate. *J. Agric. Food Chem.* **2011**, *59*, 1078–1086.
18. Rosa, E.A.S.; Rodrigues, A.S. Total and individual glucosinolate content in 11 broccoli cultivars grown in early and late seasons. *Hortscience* **2001**, *36*, 56–59.
19. Toor, R.K.; Savagea, G.P.; Lister, C.E. Seasonal variations in the antioxidant composition of greenhouse grown tomatoes. *J. Food Compos. Anal.* **2006**, *19*, 1–10.
20. Aisling Aherne, S.; O'Brien, N.M. Dietary flavonols: Chemistry, food content, and metabolism. *Nutrition* **2002**, *18*, 75–81.
21. Alarcón-Flores, M.I.; Romero-González, R.; Martinez Vidal, J.L.; Garrido Frenich, A. Multiclass determination of phytochemicals in vegetables and fruits by ultra high performance liquid chromatography coupled to tandem mass spectrometry. *Food Chem.* **2013**, *141*, 1120–1129.
22. Raffo, A.; La Malfa, G.; Fogliano, V.; Maiani, G.; Quagli, G. Seasonal variations in antioxidant components of cherry tomatoes (*Lycopersicon esculentum* cv. Naomi F1). *J. Food Compos. Anal.* **2006**, *19*, 11–19.
23. Hennig, K.; Verkerk, R.; Bonnema, G.; Dekker, M. Rapid estimation of glucosinolate thermal ddegradation rate constants in leaves of chinese kale and broccoli (*Brassica oleracea*) in two seasons. *J. Agric. Food Chem.* **2012**, *60*, 7859−7865.
24. Ferrer-Gallego, R.; Hernández-Hierro, J.M.; Rivas-Gonzalo, J.C.; Escribano-Bailón, M.T. Influence of climatic conditions on the phenolic composition of *Vitis vinifer*a L. cv. Graciano. *Anal. Chim. Acta* **2012**, *732*, 73–77.
25. Spayd, S.E.; Tarara, J.M.; Mee, D.L.; Ferguson, J.C. Separation of sunlight and temperature effects on the composition of *Vitis vinifera* cv. Merlot berries. *Am. J. Enol. Vitic.* **2002**, *53*, 171–182.

26. Kang, H.-M.; Saltveit, M.E. Wound-induced increases in phenolic content of fresh-cut lettuce is reduced by a short immersion in aqueous hypertonic solutions. *Postharvest Biol. Technol.* **2003**, *29*, 271–277.
27. Concellón, A.; Zaro, M.J.; Chaves, A.R.; Vicente, A.R. Changes in quality and phenolic antioxidants in dark purple American eggplant (*Solanum melongena* L. cv. Lucía) as affected by storage at 0 °C and 10 °C. *Postharvest Biol. Technol.* **2012**, *66*, 35–41.
28. Alarcón-Flores, M.I.; Romero-González, R.; Garrido Frenich, A.; Martinez Vidal, J.L.; Egea González, F.J. Monitoring of phytochemicals in fresh and fresh-cut vegetables. *Food Chem.* **2014**, *142*, 392–399.
29. Leja, M.; Kamińska, I.; Kramer, M.; Maksylewicz-Kaul, A.; Kammerer, D.; Carle, R.; Baranski, R. The content of phenolic compounds and radical scavenging activity varies with carrot origin and root color. *Plant Food Hum. Nutr.* **2013**, *68*, 163–170.

14

Food Inhibits the Oral Bioavailability of the Major Green Tea Antioxidant Epigallocatechin Gallate in Humans

Nenad Naumovski [1,2,*], Barbara L. Blades [2] and Paul D. Roach [2]

[1] School of Public Health and Nutrition, University of Canberra, Canberra 2601, ACT, Australia

[2] School of Environmental & Life Sciences, University of Newcastle, Ourimbah 2258, NSW, Australia; E-Mails: Barbara.blades@newcastle.edu.au (B.L.B.); Paul.Roach@newcastle.edu.au (P.D.R.)

* Author to whom correspondence should be addressed; E-Mail: nenad.naumovski@canberra.edu.au

Academic Editors: Antonio Segura-Carretero and David Arráez-Román

Abstract: The bioavailability of the most abundant and most active green tea antioxidant, epigallocatechin gallate (EGCG) remains uncertain. Therefore, the systemic absorption of EGCG was tested in healthy fasted humans. It was administered as capsules with water or with a light breakfast, or when incorporated within a strawberry sorbet. The results for plasma EGCG clearly revealed that taking EGCG capsules without food was better; the AUC was 2.7 and 3.9 times higher than when EGCG capsules were taken with a light breakfast (p = 0.044) or with EGCG imbedded in the strawberry sorbet (p = 0.019), respectively. This pattern was also observed for C_{max} and C_{av}. Therefore, ingesting food at the same time as EGCG, whether it was imbedded or not in food, substantially inhibited the absorption of the catechin. As with some types of medications that are affected by food, it appears that EGCG should be taken without food in order to maximise its systemic absorption. Therefore, based on these findings, ingesting EGCG with water on an empty stomach is the most appropriate method for the oral delivery of EGCG in clinical trials where EGCG is to be investigated as a potential bioactive nutraceutical in humans.

Keywords: EGCG; systemic absorption; green tea catechins; functional foods

1. Introduction

Green tea (*Camellia sinensis*) is one of the oldest beverages known to man in that it has been consumed for thousands of years. In recent times, an increased consumption of green tea has been linked with a wide range of different health benefits such as an increase in antioxidative potential [1–4], a reduction in the mortality rate from cardiovascular disease [5–9], a reduced development of coronary artery disease [10] and a lowering of plasma cholesterol [11]. Furthermore, there has been a rise in research on the individual green tea components, particularly its major component epigallocatechin gallate (EGCG), as attractive targets for the development of nutraceuticals [12,13] and functional foods [14–16].

As the most abundant green tea (GT) constituent, EGCG has been the focus of research in relation to the reduction of morbidity and mortality as a consequence of cardiovascular disease. Several mechanisms of action have been proposed as to why EGCG may benefit cardiovascular health in humans, such as the lowering of plasma cholesterol through a reduction of cholesterol absorption [17], a decrease in cholesterol synthesis [18] and/or an increase in the cholesterol clearance rate through an upregulation of the LDL receptor [19–22]. However, many of the studies of this type in humans have used hot/cold GT preparations or whole GT extracts in powder form that considerably vary in their content of EGCG, other catechins and other components naturally found in GT such as caffeine [18,21,23–25]. Therefore, despite the fact that EGCG is the predominant compound found in these preparations it is still uncertain whether pure EGCG would exhibit the same health properties in humans as EGCG in a polyphenolic matrix such as GT or GT extracts.

The relatively recent availability of pure, crystalline, stable and affordable EGCG has enabled the design of studies with controlled amounts of EGCG [26–28]. Consequently, capsules containing precisely measured amounts of EGCG can be utilised in controlled clinical trials and measured amounts of the crystalline form of EGCG can be incorporated into food products for use as nutraceuticals or functional foods. However, incorporating EGCG into food products requires dissolving the crystalline form in various liquids which could affect the stability of the EGCG and potentially lead to its degradation [29–32].

The processing temperature and pH of EGCG solutions can influence the catechin's structural stability and cause degradation and epimerization [31]. Although these processes have been observed to mainly occur at high temperatures and high pH, significant degradation of resolubilised EGCG has also been reported at temperatures below 25 °C [33,34]. Therefore, to minimise the loss of EGCG, solutions need to be made at pH below 4 [34] and kept at low temperatures prior to being used. The use of reducing agents such as ascorbic acid [35] can also be useful.

The modification and use of foods in order to deliver beneficial health outcomes, commonly referred to as functional foods, is a fast growing area of research [36]. Apart from the health benefits that these functional foods can bring, the delivery of a functional component within a food can increase the convenience of the component's consumption and may also reduce the fasting time. A recent study by Hirun and Roach [37] found that EGCG imbedded in a strawberry sorbet had very good stability during frozen storage for at least 4 months, a benefit ascribed to the low temperature and pH. In addition, this study [37] and a study by Green *et al.* [38] showed that the stability of EGCG under simulated digestion conditions increased when GT was in the strawberry sorbet or mixed with fruit juices, respectively. However, there are no reports on the effectiveness of EGCG once it is incorporated into a food product relative to any health outcomes in humans.

Irrespective of the format in which it is ingested, EGCG needs to be bioavailable in order to have health outcomes [39]. Studies on the bioavailability of pure EGCG in humans are limited [27,40,41] as such studies have primarily focused on the pharmacokinetic properties of preparations where EGCG was imbedded in a polyphenolic complex, as part of GT extracts [24,40–42]. In one study, the oral administration of pure EGCG at a dose of 1.6 g in healthy human volunteers [27] produced physiologically-relevant plasma EGCG concentrations (greater than 1 μmol/L) capable of having beneficial health effects. Although there were variations between individuals, the peak EGCG concentrations were reached between 1.3 and 2.2 h after ingestion and the mean elimination half-life ranged from 1.9 to 4.6 h. However, in this study, only encapsulated EGCG was studied as the method of delivery and only after a 10 h fast [27].

To date, only one study in humans has examined the oral bioavailability of EGCG, provided with and without food (a light breakfast) and the EGCG was imbedded in a GT extract [42]. The findings of this study indicated that EGCG (400 and 800 mg), when taken in the form of a polyphenolic complex, showed a greater systemic absorption when the extract was taken on an empty stomach after an overnight fast and it was very well tolerated with only mild symptoms of discomfort reported [42].

Based on the findings of the previous studies [24,27,40–42] the present study aimed to determine whether the systemic absorption of pure EGCG would be similarly decreased by the presence of food in the form of a light breakfast, as it was for EGCG in the complex polyphenolic GT extract [42]. Another aim was to test whether EGCG imbedded in a low pH food product such as a strawberry sorbet, could enhance the absorption of EGCG. Therefore, the systemic absorption of EGCG was tested in healthy human volunteers after an overnight fast with the EGCG administered by itself within capsules without breakfast, in capsules taken with a light breakfast or incorporated within a strawberry sorbet, to determine which of the three methods of delivery was the most effective.

2. Experimental Section

2.1. Materials and Reagents

The EGCG (Teavigo, DSM Nutritionals, Heerlen, The Netherlands) was purchased from RejuvaCare International Pty Ltd. (Sydney, NSW, Australia). All chemicals, (+)-catechin, (−)-epigallocatechin gallate (EGCG), ascorbic acid, disodium ethylene tetra acetate (EDTA), potassium dihydrogen phosphate (KH_2PO_4), formic acid, acetonitrile, ethyl acetate, 4-aminosalycilic acid and carboxyl methyl cellulose were purchased from Sigma-Aldrich (Castle Hill, NSW, Australia). Ultrapure deionized water was prepared on the day using a Millipore Milli-Q water purification system (Millipore Australia, North Ryde, NSW, Australia). Gelatin capsule casings were purchased from Melbourne Food Ingredient Depot (Melbourne, VIC, Australia), whey protein isolate was purchased from Vital Strength Nutraceuticals (Marrickville, NSW, Australia) while all food ingredients were purchased from local supermarkets.

Preparation and Quality Control of EGCG Products for Oral Delivery

The concentration (purity) of the EGCG in the Teavigo was stated by the manufacturer (DSM Nutritionals, Heerlen, The Netherlands) to be at least 92% (w/w) EGCG. However, in our hands, the powder was measured by HPLC to be 100% ± 2% (w/w) EGCG using EGCG purchased from Sigma-Aldrich (Castle Hill, NSW, Australia) as the external standard, taking into account that the EGCG from Sigma is 97% pure. Therefore, the Teavigo was taken to be 100% EGCG in this study.

The clear gelatin capsule casings were filled with EGCG using a portable capsule filler (Cap-M-Quick, Murietta, USA) to 250 ± 5 mg and capsules showing a variation greater than 2.5% from the target weight were discarded and not used in the study.

The strawberry sorbet was prepared based on a previously published method [37] with the ingredients shown in Table 1. However, as the preparation of the strawberry sorbet required solubilisation in an aqueous medium, heating and exposure to light, conditions that are not favorable for EGCG [31,33,34], the amount of EGCG left in the sorbet after preparation was determined. Briefly, 100 mL of methanol, containing 100 mmol/L 4-aminosalycilic acid solution as an internal standard, was added to 50 g of strawberry sorbet thawed at 4 °C and was incubated on a shaking water bath (Ratek Instruments, Boronia, VIC, Australia) at 60 °C for 1 h and samples were filtered under vacuum through a Whatman No. 1 filter paper (Sigma-Aldrich, Castle Hill, NSW, Australia), passed through a 0.45 µm syringe nylon filter (Phenomenex, Pennant Hills, NSW, Australia) and injected onto a high pressure liquid chromatography (HPLC) system for analysis as described previously [43,44].

Table 1. Ingredients used in the preparation of the strawberry sorbet.

Ingredient	Weight (g)	% (w/w)
Strawberries (Creative Gourmet Pty Ltd., Silverwater, NSW, Australia)	250	76.63
Caster Sugar (CSR, Yarraville, VIC, Australia)	45	13.79
Strawberry Flavoured WPI (Vital Strength Nutraceuticals, Marrickville, NSW, Australia)	30	9.20
Carboxyl methyl cellulose (Sigma-Aldrich, Castle Hill, NSW, Australia)	0.40	0.12
EGCG (RejuvaCare International, Sydney, NSW, Australia)	0.84	0.25
Total Weight	326.24	100

The extraction was done in duplicate on five different storage containers of strawberry sorbet with EGCG and on one storage container of strawberry sorbet prepared without EGCG (control). To test for the presence of peaks in the sorbet, which could interfere with the 4-aminosalycilic acid internal standard peak as well as with the EGCG peak, the control strawberry sorbet was also extracted without the addition of the internal standard.

2.2. Treatments

2.2.1. Ethics

Ethics approval for this study was granted by the Human Ethics Committee of the University of Newcastle, NSW, Australia (H2008-0089) and informed written consent was obtained from all the participants prior to commencement of the study.

2.2.2. Participants and Selection Criteria

Four participants (3 males and 1 female) completed the study. They were healthy and aged between 18–64 years of age. Participants were excluded from the study if they were on any medication, dietary supplement or functional food to lower cholesterol or triglycerides. Exclusion criteria also included the following: baseline triglyceride levels ≥4.0 mmol/L, a history of coronary heart disease, a body mass index of ≥35 kg/m^2, uncontrolled resting hypertension (≥160/95 mmHg), any known active pulmonary, hematologic, hepatic, gastrointestinal, renal, pre-malignant or malignant disease, diabetic, thyroid dysfunction or any pathology values known to be abnormal based on internationally accepted guidelines [45] and previous similar research conducted in humans using EGCG supplementation [27,42,46].

2.2.3. Study Clinics

The participants attended three clinics in total, one for each of the three EGCG delivery methods. On each clinic visit, the standard anthropometric measurements of height and weight were taken and their BMI was calculated. Resting blood pressure was determined using a mercury sphygmomanometer (Livingstone International, Roseberry, NSW, Australia) with participants in the sitting position; the first and fifth Korotkoff sounds were taken to represent the systolic (SBP) and diastolic blood pressure (DBP), respectively. Measurements were done in triplicate to the nearest 2 mmHg and the average value was determined to be the participants' blood pressure [47].

On each of the three clinics, after an overnight fast of at least 10 h, the participants ingested 500 mg of EGCG given either as two capsules (2 × 250 mg) with 100 mL of water only (without any breakfast), two capsules (2 × 250 mg) provided with 50 g of Special K breakfast cereal (Kellogg's, Pagewood, NSW, Australia) served with 200 mL of full cream milk or EGCG (500 mg) incorporated in 200 g of strawberry sorbet (Table 1). All treatments were ingested within 5 min and no additional food was taken for a further 4 h. All participants were provided with a lunch and a drink three hours after ingesting the EGCG, which consisted of a sandwich (bread, ham, cheese, tomato, lettuce) and 200 mL of orange juice.

2.3. Blood Collection and Handling

An intravenous catheter (BD, North Ryde, NSW, Australia) connected to a SmartSite® needle-free valve (Altaris Cardinal Health, Sydney, NSW, Australia) was inserted in the median cubital vein of the antecubital fossa of one of the participants arms and it was left in for 8 h. Approximately 5 mL of blood was collected into lithium heparin tubes (BD, North Ryde, NSW, Australia) before the ingestion of the EGCG (500 mg) as capsules or strawberry sorbet followed by an additional six collections of blood 30 min, 1 h, 2 h, 3 h, 5 h and 8 h after the ingestion of EGCG was completed. Immediately after collection, the blood samples were stored on ice in the dark and within 60 min of blood collection, plasma was separated by centrifugation, aliquoted under red light and samples were stored at −84 °C until assayed.

2.4. Determination of Free EGCG in Plasma Samples

2.4.1. Preparation and Extraction of EGCG from Plasma Samples

Thawed plasma samples (200 μL) were diluted with 200 μL of pH 2 VcEDTA solution containing 5% (w/v) KH_2PO_4; 20% (w/v) ascorbic acid and 0.1% (w/v) EDTA) to enhance the stability of the EGCG and 200 μL of 0.05% (w/v) (+)-catechin to serve as an internal standard. The extraction of the catechins was performed by adding 1 mL of ethyl acetate and vortexing for 5 min and the organic layer was then transferred into a new tube and placed in a heating block (Ratec Instruments; Boronia; VIC; Australia) set at 50 °C and evaporated to dryness under a stream of nitrogen. The dried samples were re-dissolved in 20% (w/v) ascorbic acid pH 2 in 15% (*v*/*v*) acetonitrile and sequentially injected onto the HPLC. An external standard curve was prepared with pooled plasma samples spiked with known (+)-catechin and EGCG (15.6–1000 ng/mL) concentrations.

2.4.2. Equipment and Chromatographic Conditions

The HPLC analysis was performed using a Finnigan Surveyor System (Thermo Fisher Electron Corporation, Sydney, NSW, Australia) equipped with a quaternary pump (Surveyor LC pump 1.4) and autosampler (Surveyor AS1.4) set at 4 °C and fitted with a 100 μL sample injection loop. Separation was performed using an analytical Prodigy ODS(3) 250 × 4.6 mm 5 μ column, protected by a guard column (Phenomenex, Pennant Hills, NSW, Australia) with the oven set at 26 °C. The elution times for (+)-catechin and EGCG were monitored using a Surveyor Photo-Diode array detector (Surveyor PDA 1.4) and UV-VIS absorption spectra was acquired over the range of 200–500 nm. The outlet of the PDA detector was directly connected to a mass spectrometer (Surveyor MSQ Plus 1.4 SUR 1). The full portion of the effluent was delivered into the ion source of the electrospray ionisation mass spectrometer (ESI-MS).

The mobile phases consisted of (A) 0.2% (*v*/*v*) formic acid (pH 2) and (B) acetonitrile. The auto-injector and needle wash solution consisted of acetonitrile:DI water (50:50 *v*/*v*). All solvents were filtered through a 0.45 μm Millipore cellulose filter (Millipore Australia, North Ryde, NSW, Australia) and degassed just prior to use.

The system was run at a flow rate of 1 mL/min with 87.5% mobile phase A and 12.5% of mobile phase B for the first 10 min after each injection. Then, for the following 30 min, the system was switched to a linear gradient concentration increase of mobile phase B to 25%. Mobile phase B was held for an additional 10 min at this concentration (25%) and then gradually decreased to 12.5% over the next 10 min. For the following 20 min, mobile phase A and B were run as described for the starting conditions (87.5% A and 12.5% B) to allow the column to re-equilibrate before the injection of the next sample.

The signals from the detectors (PDA and MS) were recorded and analyzed using the ExcaliburTM (v1.4 SR1) software (Thermo Fisher Electron Corporation, Sydney, Australia) installed on the computer assigned as a remote control operating system.

2.4.3. Tuning and Setting Parameters of the ESI-MS

The tuning of the ESI-MS was optimised after the injection of 1 µg/mL pure EGCG or (+)-catechin in 0.2% formic acid (pH 2): acetonitrile (87.5:12.5 v/v) at a flow rate of 1 mL/min representing the chromatographic conditions prior to injection into the MS. The ESI probe temperature was set to 629 °C, the cone voltage to 75 V and the needle source voltage to 4 kV. Nitrogen was used as the nebulising gas at a flow rate of 50 L/h in order to create a fine spray of sample/solvent droplets.

From the optimisation procedure it was determined that the (+)-catechin would elute in the time range between 9 and 13 min and EGCG between 19 and 26 min. For subsequent analyses, the MS was operated in the selective ion monitoring mode (SIM) with negative polarity to detect the IS and EGCG as they eluted from the HPLC column.

Samples from pooled plasma spiked with 500 ng/mL EGCG were also used as a quality control for the plasma storage, extraction and assay system. This showed that the EGCG was stable during, storage, extraction and handling and that the extraction/analysis system was very robust; 98% ± 2% of the EGCG was recovered from the spiked plasma and the intra- and inter-assay variations were 2.2% and 3.7%, respectively.

2.5. Statistical Analysis

Pharmacokinetic analysis was performed in accordance with current industry guidance for orally administered pharmaceutical products [45]. The maximum concentration of EGCG from time 0 to 8 h was defined as C_{max}, with T_{max} being the time required to reach the C_{max}. The concentration of plasma EGCG at the end of the dosing interval was defined as C_{min} and the mean concentration during the dosing interval was defined as C_{av}. The degree of fluctuation (DF) was also determined based on the formula $(C_{max} - C_{min})/C_{av}$ while the swing of plasma EGCG was determined using $(C_{max} - C_{min})/C_{min}$. The plasma EGCG elimination half-life ($T_{1/2}$) was calculated based on the formula $T_{1/2} = 0.693/Ke$ where Ke is the slope of the logarithmically transformed (ln) linear regression of the plasma EGCG concentrations [24,27]. The area under the curve ($AUC_{0–8}$) analysis was determined using the linear trapezoidal rule from 0–8 h.

The statistical analysis of the pharmacokinetic variables and the anthropometric measurements was performed using the Statistical Package for Social Sciences (PASW Statistics 17) (SPSS Inc., Chicago, IL, USA). The one-way ANOVA and the Bonferonni *post-hoc* test were used to determine differences between the mean values of the anthropometric and pharmacokinetic variables obtained for the three EGCG delivery methods. The threshold for all statistical significances was set at $p < 0.05$ level. All pharmacokinetic data was calculated and presented in accordance with internationally accepted and standardized methods [45,48]. The correlations were also analyzed using linear regression analysis of the PASW Statistics 17.

3. Results

3.1. Recovery of EGCG in Strawberry Sorbet

The preparation of the strawberry sorbet exposed EGCG to conditions, such as solubilisation in an aqueous media, heating and exposure to light, which could have affected its stability and possibly have led to its degradation in the food matrix [29,31,33,34], Therefore, in order to ascertain that the amount of EGCG added in the sorbet was the amount that the participants would be receiving, EGCG was extracted from the strawberry sorbet after it had been prepared and frozen for at least one week at −20 °C.

Importantly, there was no trace of EGCG or the internal standard detected on the HPLC chromatogram of the extract from the control sorbet sample and there were no other interfering peaks (Figure 1); on the chromatogram obtained for the control sample (Figure 1a) there was no trace of any peaks at the respective retention times of the peaks for the internal standard and EGCG, as indicated in Figure 1b for a sample of the strawberry sorbet prepared with added EGCG (Table 1) and extracted after the addition of the internal standard, 4-aminosalycilic acid.

The amount of EGCG extracted was measured and the recovery of EGCG from the strawberry sorbet was expressed as a percentage of the EGCG originally added during the preparation of the strawberry sorbet. The results (Table 2) demonstrated that more than 97% of the EGCG originally added to the strawberry sorbet was extracted and therefore, it was ascertained that the amount of EGCG that the participants would be receiving was very close to the amount of EGCG added in the sorbet (Table 1).

Table 2. Recovery of EGCG from the strawberry sorbet.

n	EGCG Added to Strawberry Sorbet (mg/g)	EGCG Extracted from Strawberry Sorbet (mg/g) †	EGCG Recovery (%) †
5	2.50	2.44 ± 0.03	97.4 ± 1.3

† Values represent the Mean ± SD for five different sorbet containers.

3.2. Measurements of EGCG in Human Plasma

The methods used for the HPLC analysis of individual catechins in tea samples commonly use UV detection. However, the amounts observable in plasma are well below the detection limits of UV detectors. The use of a MS as a detector for HPLC offers greater detection sensitivity and also greater specificity for the measurement of EGCG in biological fluids such as plasma. Therefore, the HPLC-ESI-MS analysis system was used to measure EGCG in human plasma but it first needed to be validated for specificity and sensitivity. Furthermore, (+)-catechin was used as the internal standard instead of 4-aminosalycilic acid; because its structure is closer to EGCG than 4-aminosalycilic acid, (+)-catechin was more detectable at the conditions for which the ESI probe was optimally tuned to detect EGCG: probe temperature at 629 °C, cone voltage at 75 V and needle source voltage to 4 kV.

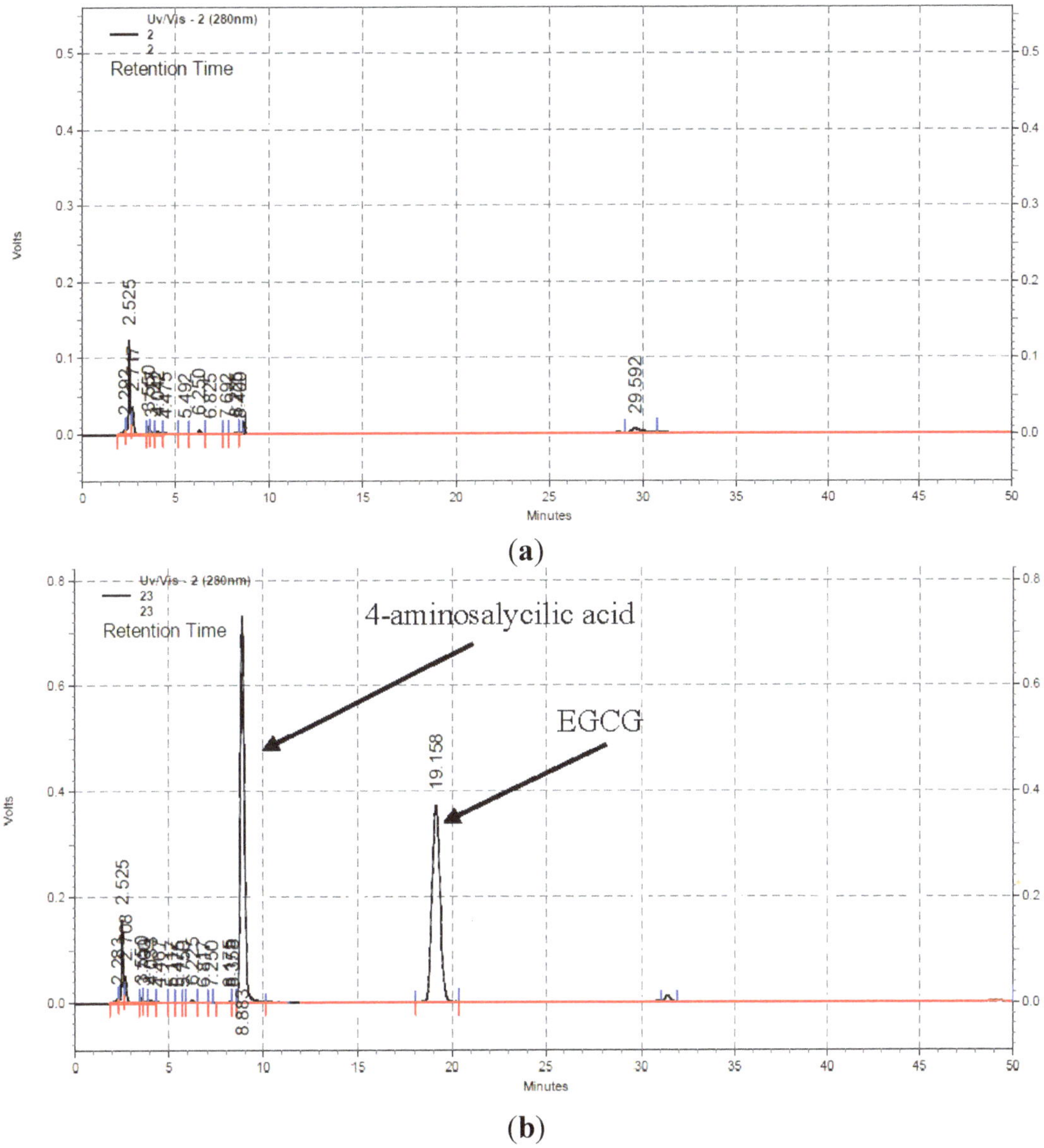

Figure 1. Typical HPLC-UV chromatograms (280 nm) of extracts from strawberry sorbet without (**a**) and with (**b**) the addition of EGCG (2.5 mg/g) and 4-aminosalycilic acid acting as internal standard (100 mmol/L).

As seen in Figure 2, excellent chromatographic separation of the (+)-catechin (11.28 min), acting as the internal standard, and EGCG (22.25 min), was achieved without any interference from other plasma constituents. Therefore, using the ESI-MS, set in SIM mode, as the detector gave a clean HPLC chromatogram, which allowed clear identification and quantification of EGCG in the human plasma samples.

To determine the limit of quantification (LOQ), the limit of detection (LOD) and the linearity of the response, EGCG was added to plasma at seven different concentrations in quintuplicates (Table 3) and measured using the HPLC and ESI-MS system. The LOQ was evaluated as the EGCG concentration that resulted in a coefficient of variation (CV) of 5% or less; the results indicated that, using this system of extraction and detection, the lowest measureable concentration of EGCG in terms of precision was 31.25 ng/mL, with a CV of 4.78%. The LOD was identified to be lower than the LOQ at

15.63 ng/mL EGCG. Furthermore, the response, shown in Figure 3, was clearly very linear over the range of 15.6 to 1000 ng/mL for EGCG.

Table 3. Mean, standard deviation and coefficient of variation for seven concentrations of EGCG spiked into human plasma and detected using the optimised HPLC electrospray ionisation mass spectrometer (ESI-MS) system.

EGCG Concentration (ng/mL)	*n*	Mean Peak Area Ratio (EGCG/Catechin)	SD Peak Area Ratio	CV (%)
1000	5	3.18	0.004	1.21
500	5	1.79	0.041	2.32
250	5	0.88	0.023	2.67
125	5	0.35	0.014	4.01
62.5	5	0.11	0.003	2.82
31.25 †	5	0.07	0.004	4.78
15.63 *	5	0.03	0.003	8.77

† LOQ—the lowest concentration of EGCG with a CV < 5%; * LOD—the lowest detectable concentration of EGCG; *n*—the number of replicates.

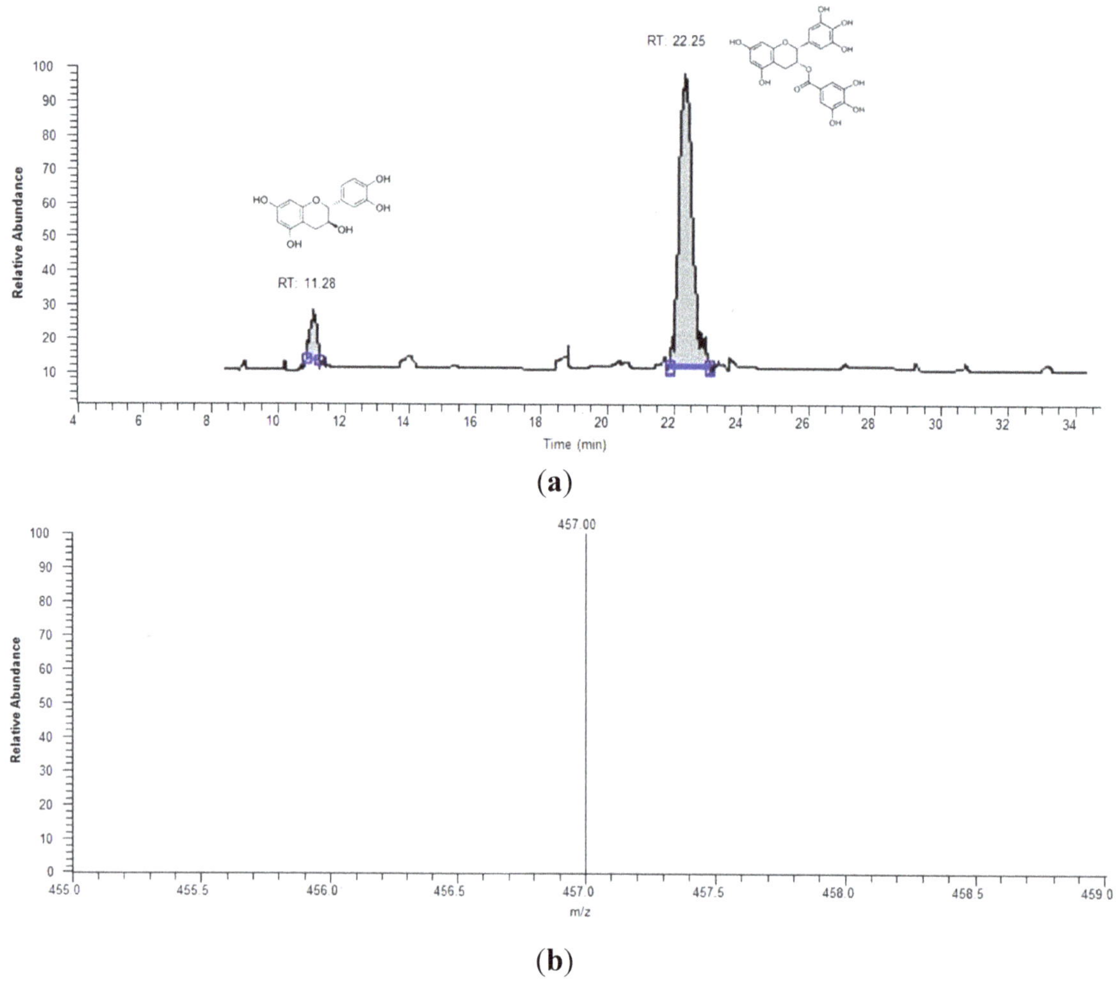

Figure 2. Typical chromatogram of the internal standard, (+)-catechin (100 ng/mL) and EGCG (500 ng/mL) analyzed by HPLC and ESI-MS set in negative polarity SIM mode. When the EGCG peak in (**a**) was selected the molecular ion for EGCG (molecular mass-1) was clearly seen at 457 m/z (**b**).

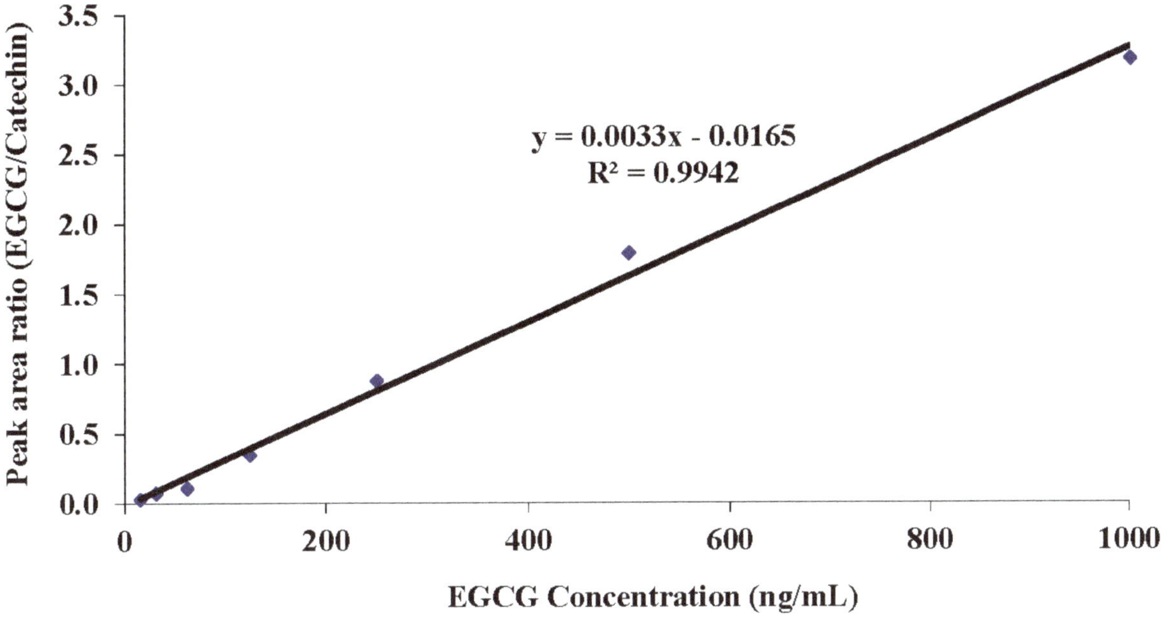

Figure 3. Calibration curve for EGCG spiked into human plasma at increasing concentrations and detected using the optimised HPLC ESI-MS system.

3.3. Participants' Anthropometric and Blood Pressure Data

Four participants, three males and 1 female with an average age of 31.25 ± 9.54 years, completed the study. Due to the limited number of participants, the results (Table 4) are presented indiscriminate of gender. The participants' anthropometric data, standing height and weight, and their systolic and diastolic blood pressures were measured at the start of each clinic prior the ingestion of product containing EGCG.

The subjects' measurements remained stable over the period of the study. Using the one-way ANOVA and the Bonferonni *post-hoc* test, it was found that there was no significant difference in BMI (p = 0.999), systolic blood pressure (p = 0.998), or diastolic blood pressure (p = 0.999) between the three clinic visits (Table 4).

Table 4. The anthropometric, BMI and BP results for the four participants at each clinic visit.

	Clinic 1	**Clinic 2**	**Clinic 3**
Weight (kg)	70.5 ± 10.7	70.7 ± 10.8	70.5 ± 10.8
Height (m)	1.75 ± 0.08	1.75 ± 0.08	1.75 ± 0.08
BMI (kg/m^2)	23.1 ± 3.5	23.2 ± 3.6	23.1 ± 3.5
SBP (mmHg)	112.5 ± 5.0	115.3 ± 6.1	112.5 ± 5.0
DBP (mmHg)	85.0 ± 4.1	85.5 ± 3.3	82.5 ± 5

BMI—Body Mass Index; SBP—Systolic Blood Pressure; DBP—Diastolic Blood Pressure.

3.4. Plasma EGCG Concentration-Time Results

The plasma EGCG concentration was measured just before and over 8 h after the administration of 500 mg of EGCG delivered as two 250 mg capsules without breakfast, two 250 mg capsules with breakfast or 500 mg in 200 g of strawberry sorbet. The EGCG concentration curves over this period of time for each delivery method and for each participant (Figure 4) as well as the arithmetic mean EGCG concentration curves for each delivery method in all four participants (Figure 5) are presented.

Taking the two capsules of EGCG without breakfast resulted in a noticeably higher response than the other two treatments of the capsules of EGCG with breakfast or the strawberry sorbet containing EGCG (Figure 4). Moreover, the arithmetic mean concentration for plasma EGCG (Figure 5) reached a maximum value at 1 h (T_{max} = 60 min) for capsules taken without the breakfast, while the maximum concentrations for EGCG taken in capsules with breakfast and imbedded into the strawberry sorbet were reached at 2 h (T_{max} = 120 min) after the subjects ingested the EGCG.

3.5. Pharmacokinetic Parameters of Plasma EGCG

After the one-way ANOVA analysis and the Bonferonni *post-hoc* test it was noted that the AUC_{0-8} for EGCG taken as capsules without breakfast (Table 5) was significantly higher than the AUC_{0-8} for EGCG taken as capsules with breakfast (p = 0.044) and the AUC_{0-8} for EGCG taken incorporated in strawberry sorbet (p = 0.019) t . However, the difference between the AUC_{0-8} for EGCG taken as capsules with breakfast and the AUC_{0-8} for EGCG taken in the strawberry sorbet was not statistically significant (p = 1.000) (Table 5).

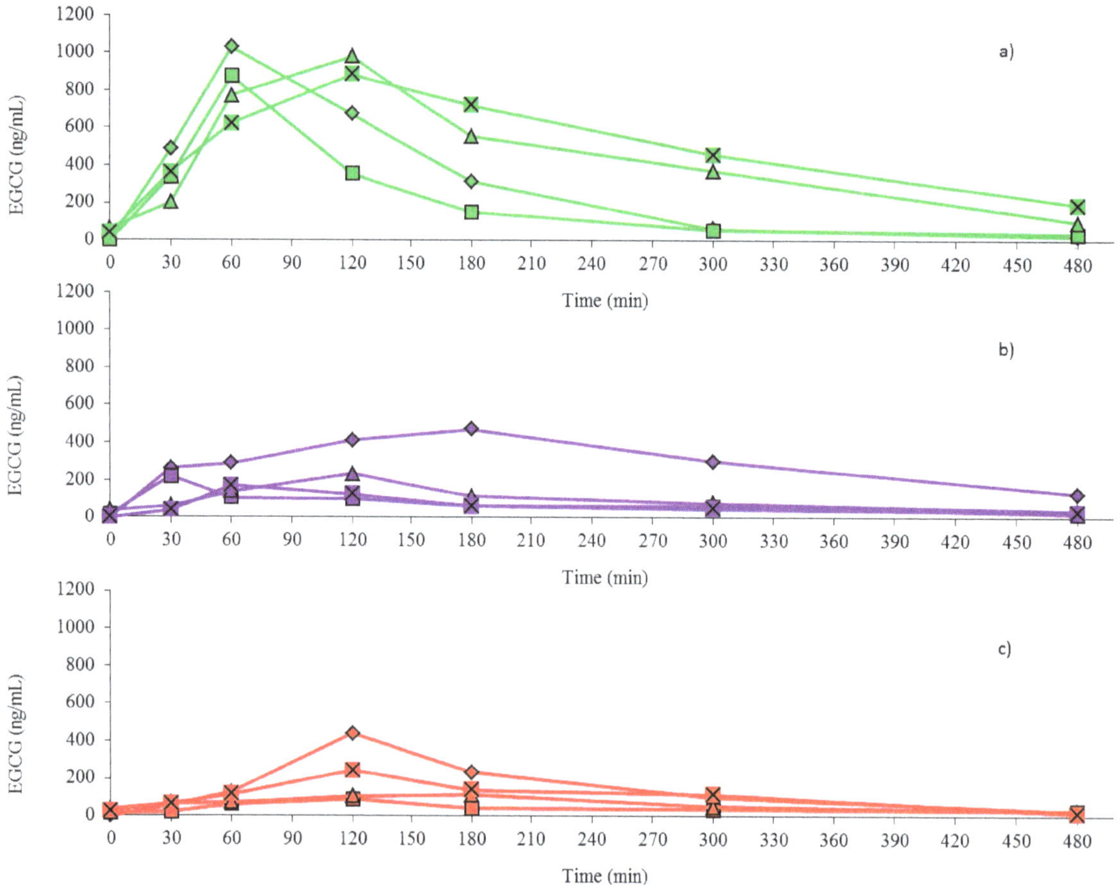

Figure 4. Plasma EGCG concentration-time curves for each of the four individual participants for the three different methods of EGCG oral delivery: (**a**) capsules without breakfast; (**b**) capsules with breakfast; and (**c**) strawberry sorbet.

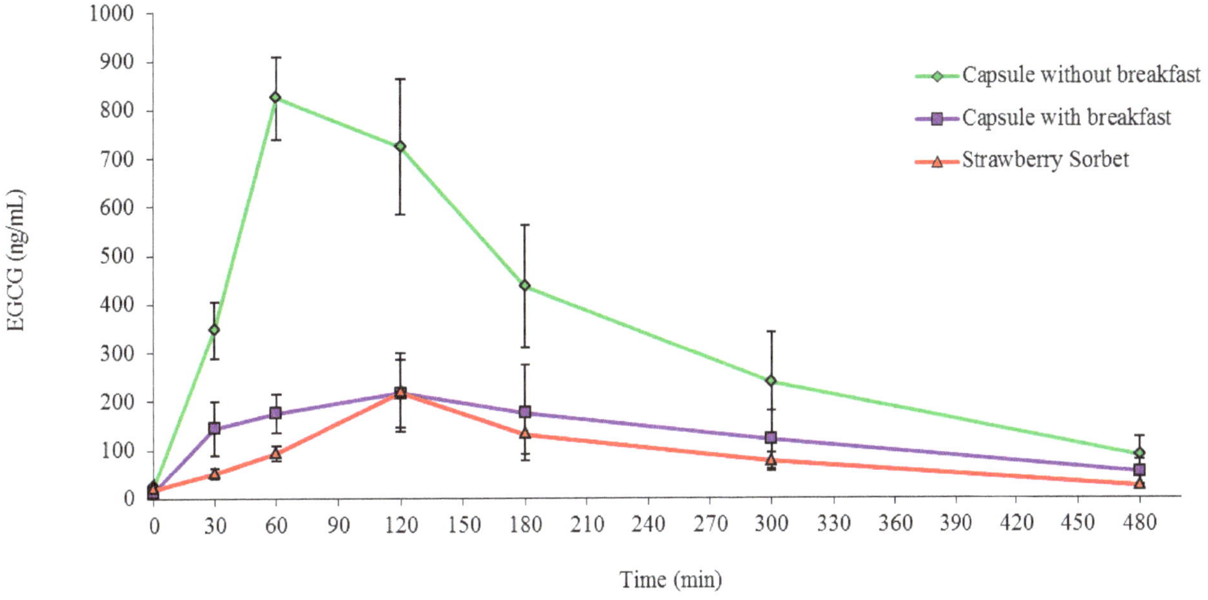

Figure 5. The mean plasma EGCG concentration-time curves for the three different methods of oral delivery: capsules without breakfast (green), capsules with breakfast (purple) and strawberry sorbet (red). The values are means ± the standard error of the means for the four participants ($n = 4$).

Table 5. Plasma kinetic parameters for EGCG after the three different methods of ingestion.

	EGCG Capsules Without Breakfast	EGCG Capsules With Breakfast	EGCG in Strawberry Sorbet
$AUC_{0\text{–}8}$ (μg/mL/8 h)	173.8 ± 67.6 *	64.1 ± 53.7	44.5 ± 22.9
C_{max} (ng/mL)	824.2 ± 75.1 **	231.8 ± 134.3	218.0 ± 160.0
C_{av} (ng/mL)	382.6 ± 92.5 *	92.0 ± 46.0	87.6 ± 43.0
C_{min} (ng/mL)	86.9 ± 75.2	49.3 ± 24.6	25.0 ± 5.8
$T_{1/2}$ (min)	78.6 ± 94.2	136.2 ± 85.2	228.6 ± 97.2
T_{max} (min)	60 ± 34.6 **	120 ± 34.6	120 ± 34.6
DF	1.93 ± 0.77	1.39 ± 0.46	2.20 ± 0.76
Swing	8.49 ± 17.19	3.36 ± 3.04	7.70 ± 6.61

Values are means ± standard deviations for the four participants ($n = 4$). * Value is significantly different from the other values in the row at the level of $p < 0.05$. ** Value is significantly different from the other values in the row at the level of $p < 0.001$. **$AUC_{0\text{–}8}$**—area under the curve from 0 to 8 h; **C_{max}**—maximum concentration; **C_{av}**—Average concentration; **C_{min}**—minimum concentration (at the end of the treatment); **DF**—degree of fluctuation; **T_{max}**—time required to reach the maximal concentration; **$T_{1/2}$**—half-life.

The C_{max} of EGCG in plasma (Table 5) after the ingestion of EGCG as capsules without breakfast was also significantly higher than when EGCG was taken as capsules with breakfast ($p < 0.001$) and higher than when it was taken in strawberry sorbet ($p < 0.001$). However, there was no significant difference in the C_{max} between the EGCG taken as capsules with breakfast and EGCG taken in strawberry sorbet ($p = 1.000$) (Table 5).

The C_{av} of EGCG in plasma for the eight hours after ingestion was also significantly higher for EGCG taken as capsules without breakfast (Table 5) than when taken as capsules with breakfast

($p = 0.004$) or taken in strawberry sorbet ($p = 0.002$). However, there was no significant difference between EGCG capsules taken with breakfast and taken in strawberry sorbet ($p = 1.000$) (Table 5).

From Table 5, it can also be noted that the C_{min}, achieved at the end of the 8 h of monitoring time, the T_{max}, the $T_{1/2}$, the DF and swing for EGCG were not significantly different between the three treatment groups ($p > 0.05$).

4. Discussion

The results for the concentration of EGCG in plasma over the 8 h period after ingestion (Figures 4 and 5, Table 5) clearly revealed that EGCG taken in capsule form by itself without food gave higher plasma values; the AUC was 2.7 times higher than when EGCG was taken in capsules with a light breakfast ($p = 0.044$) and 3.9 times higher than when EGCG was imbedded in the strawberry sorbet ($p = 0.019$). Furthermore, there was no significant difference in the AUC values (Table 5) between the EGCG taken as capsules with a light breakfast or taken in the strawberry sorbet ($p = 1.000$). This pattern was also observed when the values for C_{max} and C_{av} were compared between the three ingestion conditions (Table 5). Therefore, ingesting food at the same time as EGCG whether it was imbedded or not in food, substantially inhibited the absorption of the catechin. As with some types of medications that are affected by food, it appears that EGCG should be taken without food in order to maximise its intestinal absorption.

The concentrations of EGCG measured in the plasma after its ingestion in the present study are consistent with those of the only previous study that used similar doses of administered pure EGCG [27]. When 500 mg of EGCG was given in capsule form without food in the present study, the plasma EGCG concentration curve (Figures 4 and 5) was very similar to the plasma EGCG values obtained when a similar dose of 400 mg EGCG was given without food to humans by Ullmann *et al.* [27]. Furthermore, the C_{max} and C_{av} values in the present study, 824 ng/mL and 383 ng/mL, respectively (Table 5), were very comparable to the C_{max} and C_{av} values, 862 ng/mL, and 568 ng/mL, respectively, observed in the Ullmann *et al.* [27] study.

The results in the present study are also very consistent with the previous findings of Chow *et al.* [42], who showed that food intake also clearly interfered with the systemic absorption of EGCG when it was given as part of a GT extract. Similar to the present study (Table 5), Chow *et al.* [42] also reported markedly higher plasma EGCG AUC (3.5 times higher) and C_{max} (5.7 times higher) values when the catechin extract was taken without food compared to when it was taken with a standardised breakfast consisting of muffins.

With varied food types, such as muffins [49], a breakfast cereal plus full cream milk and a strawberry sorbet (Figures 4 and 5, Table 5), all decreasing the plasma EGCG concentration measured after ingestion, it is evident that food is an important factor, which can affect the intestinal absorption and systemic levels of orally administered EGCG, whether it is in a GT extract or in pure form. However, it is not known how food gives rise to this effect or which particular food component plays a role in decreasing the absorption of EGCG.

This is also not entirely surprising because it is well known that the absorption of some pharmaceutical medications is decreased when they are taken with various food products. From what is known about the interactions of food with medications and plant bioactives [50,51], several factors may influence the bioavailability of EGCG. These factors can be divided into three broad categories: (1) the effect of the vehicle in which the EGCG was administered; (2) the effect of the biological fluids on EGCG prior to it reaching its absorption site and; (3) effects on EGCG due to physiological responses to the ingested food products.

Relative to the first category—the effect of the vehicle in which the EGCG was administered—incorporating EGCG in a food product like a strawberry sorbet could have been expected to improve the absorption of EGCG compared to taking the catechin in capsule form with a breakfast. It is well known that EGCG is relatively unstable and susceptible to degradation [29] at high temperatures [31] and at pH values above four [34]. Therefore, imbedding the EGCG in a food product like strawberry sorbet, which is stored at −20 °C and has a pH below 4, may have increased the stability of the EGCG. This was supported by the finding that the EGCG was very stable in this food product, as evidenced by the very high percentage (over 97%) observed for its recovery from the strawberry sorbet after storage at −20 °C (Table 1). This finding indicated that the method used to prepare the strawberry sorbet and its acidic environment preserved the EGCG extremely well [37] and therefore, the amount of chemically intact EGCG (500 mg) ingested in the 200 g of sorbet given to the human volunteers was the intended amount.

Evidently however, the fact that the EGCG was chemically intact and stable in the strawberry sorbet did not improve its bioavailability (Figures 4 and 5, Table 5). Obviously, similar to the breakfast cereal and the full cream milk in the present study and the muffins in the study by Chow *et al.* [42], the presence of food components in the strawberry sorbet must have played a role in reducing the bioavailability of EGCG and resulted in a significantly lower AUC for EGCG than when it was taken on its own without food.

Relative to the second category—the effect of the biological fluids on EGCG prior to it reaching its absorption site—the principal biological fluids EGCG would come into contact with, are the gastric juice and the pancreatic/biliary juices. Saliva was unlikely to be a factor, as the EGCG capsules taken with the light breakfast were very quickly swallowed with water and therefore, the EGCG was unlikely to have been exposed to much saliva.

In the stomach, little degradation of EGCG is expected to occur because of the acidic nature of the gastric secretions. In effect, as shown by Record and Lane [52], EGCG was stable in acidic solutions (pH < 3) made up to mimic those found in the fasting stomach environment (pH 1.5–2) [53]. However, any increase in pH caused by protein or any other component of food could have led to an accelerated degradation of EGCG and reduced its bioavailability [27,42,54].

In the small intestine, the acidic chyme that is pushed down from the stomach is quickly neutralised by the bicarbonate solution secreted by the pancreas into the duodenum [50,55]. This is where most of the EGCG is expected to be lost. As shown by Record and Lane [52], EGCG is particularly unstable under conditions which mimic digestion fluids in the small intestines, with only 1% of the EGCG still measurable after an hour incubation.

Hirun and Roach [37] and Green *et al.* [38] showed that the stability of EGCG under these simulated digestion conditions increased when EGCG was in a strawberry sorbet or mixed with fruit juices, respectively. However, the current studies suggest that this is unlikely to happen under the true small intestine conditions in humans, as imbedding the EGCG in a strawberry sorbet in the present study did not lead to any improvement in absorption compared to taking the EGCG in capsules with the light breakfast. Most likely, the amount of bicarbonate solution secreted by the pancreas into the duodenum was enough to fully neutralise any acidity brought down to the small bowel by the strawberry sorbet.

Evidently, the possibility that the acidic nature of the strawberry sorbet could keep the pH of the small intestine less basic either did not occur or was not a factor in the bioavailability of the EGCG it contained. Clearly, other ways of protecting the EGCG from the basic pH in the small intestines are needed. Several studies have reported different methods of preserving the EGCG such as encapsulation using oil-in-water sub-micrometer emulsions [56], lyposomes [57] and protein/polyphenols microparticles [58], but whether they can preserve the EGCG from degradation in the small intestine and increase its bioavailability remains to be determined.

It is possible that taking the capsules without food may have allowed the EGCG to survive longer in the small intestines and therefore, enhance its chances of being absorbed, because it did not elicit strong responses from the stomach and the pancreas. The EGCG capsules by themselves could be expected to have caused the stomach to produce much less acidic chyme than when the EGCG was taken along with food, either in the form of the strawberry sorbet or the breakfast cereal with full cream milk. Consequently, this could have elicited a less strong response from the pancreas to secrete bicarbonate solution to neutralise the chyme coming down from the stomach.

Relative to the third category—effects on EGCG due to physiological responses to the ingested food products—the small intestine is the primary absorption site for EGCG and the rate at which EGCG is presented into the upper portion of the small intestine and travels down to its absorption site can determine the bioavailability of the catechin. It is known that the ingestion of food can delay the rate of gastric emptying [51,53] and that the rate of gastric emptying is one of the most important factors known to influence the absorption rate of orally administered pharmaceuticals from the gastrointestinal tract [51].

In concurrence with this, EGCG taken with the breakfast and in the strawberry sorbet showed a delay in time it took to reach its maximum concentration in plasma (T_{max} in Table 5) compared to EGCG taken without food. Therefore, a slower gastric emptying in the presence of food most likely prolonged the time needed for EGCG to travel into the upper portion of small intestine. However, given that the bioavailability of the EGCG was much lower when it was taken with food, some of the extra time is likely to have been spent transiting through the small intestine where exposure to a high pH for longer could have contributed to a greater degradation.

Another possibility, which could have explained the higher plasma EGCG concentrations observed when the catechin was taken without food, was an increased clearance rate of EGCG from the plasma when it was taken with food. However, the mean elimination half-life ($T_{1/2}$ in Table 5) was not significantly different between the three ingestion methods ($p > 0.05$). Therefore, the ingestion of EGCG with food or incorporated in a food did not appear to significantly influence the clearance rate of free EGCG from the systemic circulation once it was absorbed.

The main limitation of this study is the low number of participants ($n = 4$) and it would be useful to repeat the study with a higher number of subjects. Nevertheless, despite the low numbers, the inhibitory effect of food on the systemic absorption of the pure EGCG was strong (3–4 times lower with food than without food) and unequivocal (statistically significant). The strength of the cross-over study design may have helped; with each participant acting as their own control in this design, the impact of inter-individual variation is minimised. Therefore, the results were unequivocal in revealing that the EGCG was best absorbed when consumed in capsule form without any food after the overnight fast, a finding which supports the results of Chow *et al.*, the only previous study to study the effect of food on the systemic absorption of EGCG, although it was given as part of green tea extract and not as a pure catechin.

5. Conclusions

In conclusion, the systemic absorption was significantly higher for EGCG taken in capsules without food after an overnight fast than it was when it was taken in capsules with a light breakfast or imbedded in a strawberry sorbet. Therefore, based on these findings, ingesting EGCG with water on an empty stomach is the most appropriate method for the oral delivery of EGCG in future clinical trials where EGCG is to be investigated as a potential bioactive nutraceutical in humans.

Acknowledgments

We acknowledge funding support from the University of Newcastle in the form of a Scholarship for NN. The authors would also like to acknowledge Dr. James Krahe and Leonie Holmesby for their assistance with the HPLC-DAD-MS.

Author Contributions

Nenad Naumovski participated in the experimental design, collection and interpretation of the data, and in manuscript design and preparation. Barbara Blades and Paul Roach participated in the experimental design, data interpretation and in the manuscript design and preparation. All authors read and approved the final manuscript.

Conflicts of Interest

The authors declare no conflict of interest.

References

1. Gundimeda, U.; McNeill, T.H.; Barseghian, B.A.; Tzeng, W.; Rayudu, D.; Cadenas, E.; Gopalakrishna, R. Polyphenols from green tea prevent antineuritogenic action of Nogo-A via 67-kDa laminin receptor and hydrogen peroxide. *J. Neurochem.* **2015**, *132*, 70–84.
2. Narotzki, B.; Reznick, A.Z.; Mitki, T.; Aizenbud, D.; Levy, Y. Green tea drinking improves erythrocytes and saliva oxidative status in the elderly. *Adv. Exp. Med. Biol.* **2015**, *832*, 25–33.
3. Zhao, C.; Li, C.; Liu, S.; Yang, L. The galloyl catechins contributing to main antioxidant capacity of tea made from Camellia sinensis in China. *Sci. World J.* **2014**, *2014*, 1–11.
4. Erba, D.; Riso, P.; Bordoni, A.; Foti, P.; Biagi, P.L.; Testolin, G. Effectiveness of moderate green tea consumption on antioxidative status and plasma lipid profile in humans. *J. Nutr. Biochem.* **2005**, *16*, 144–149.

5. Kuriyama, S.; Shimazu, T.; Ohmori, K.; Kikuchi, N.; Nakaya, N.; Nishino, Y.; Tsubono, Y.; Tsuji, I. Green tea consumption and mortality due to cardiovascular disease, cancer, and all causes in Japan: The Ohsaki study. *JAMA* **2006**, *296*, 1255–1265.
6. Bhatti, S.K.; O'Keefe, J.H.; Lavie, C.J. Coffee and tea: Perks for health and longevity? *Curr. Opin. Clin. Nutr. Metab. Care* **2013**, *16*, 688–697.
7. Mineharu, Y.; Koizumi, A.; Wada, Y.; Iso, H.; Watanabe, Y.; Date, C.; Yamamoto, A.; Kikuchi, S.; Inaba, Y.; Toyoshima, H.; *et al.* Coffee, green tea, black tea and oolong tea consumption and risk of mortality from cardiovascular disease in Japanese men and women. *J. Epidemiol. Community Health* **2011**, *65*, 230–240.
8. Nabavi, S.M.; Daglia, M.; Moghaddam, A.H.; Nabavi, S.F.; Curti, V. Tea consumption and risk of ischemic stroke: A brief review of the literature. *Curr. Pharm. Biotechnol.* **2014**, *15*, 298–303.
9. Suzuki, E.; Yorifuji, T.; Takao, S.; Komatsu, H.; Sugiyama, M.; Ohta, T.; Ishikawa-Takata, K.; Doi, H. Green tea consumption and mortality among Japanese elderly people: The prospective Shizuoka elderly cohort. *Ann. Epidemiol.* **2009**, *19*, 732–739.
10. Sano, J.; Inami, S.; Seimiya, K.; Ohba, T.; Sakai, S.; Takano, T.; Mizuno, K. Effects of green tea intake on the development of coronary artery disease. *Circ. J.* **2004**, *68*, 665–670.
11. Tokunaga, S.; White, I.R.; Frost, C.; Tanaka, K.; Kono, S.; Tokudome, S.; Akamatsu, T.; Moriyama, T.; Zakouji, H. Green tea consumption and serum lipids and lipoproteins in a population of healthy workers in Japan. *Ann. Epidemiol.* **2002**, *12*, 157–165.
12. Mielgo-Ayuso, J.; Barrenechea, L.; Alcorta, P.; Larrarte, E.; Margareto, J.; Labayen, I. Effects of dietary supplementation with epigallocatechin-3-gallate on weight loss, energy homeostasis, cardiometabolic risk factors and liver function in obese women: Randomised, double-blind, placebo-controlled clinical trial. *Br. J. Nutr.* **2014**, *111*, 1263–1271.
13. Uchiyama, Y.; Suzuki, T.; Mochizuki, K.; Goda, T. Dietary supplementation with (−)-epigallocatechin-3-gallate reduces inflammatory response in adipose tissue of non-obese type 2 diabetic Goto-Kakizaki (Gk) rats. *J. Agric. Food Chem.* **2013**, *61*, 11410–11417.
14. Nagle, D.G.; Ferreira, D.; Zhou, Y.D. Epigallocatechin-3-gallate (EGCG): Chemical and biomedical perspectives. *Phytochemistry* **2006**, *67*, 1849–1855.
15. Ahmad, R.S.; Butt, M.S.; Huma, N.; Sultan, M.T. Green tea catechins based functional drink (Green cool) improves the antioxidant status of SD rats fed on high cholesterol and sucrose diets. *Pak. J. Pharm. Sci.* **2013**, *26*, 721–726.
16. Munir, K.M.; Chandrasekaran, S.; Gao, F.; Quon, M.J. Mechanisms for food polyphenols to ameliorate insulin resistance and endothelial dysfunction: Therapeutic implications for diabetes and its cardiovascular complications. *Am. J. Physiol. Endocrinol. Metab.* **2013**, *305*, E679–E686.
17. Raederstorff, D.G.; Schlachter, M.F.; Elste, V.; Weber, P. Effect of EGCG on lipid absorption and plasma lipid levels in rats. *J. Nutr. Biochem.* **2003**, *14*, 326–332.
18. Koo, S.I.; Noh, S.K. Green tea as inhibitor of the intestinal absorption of lipids: Potential mechanism for its lipid-lowering effect. *J. Nutr. Biochem.* **2007**, *18*, 179–183.
19. Bursill, C.A.; Roach, P.D. A green tea catechin extract upregulates the hepatic low-density lipoprotein receptor in rats. *Lipids* **2007**, *42*, 621–627.

20. Bursill, C.A.; Roach, P.D. Modulation of cholesterol metabolism by the green tea polyphenol (−)-epigallocatechin gallate in cultured human liver (HepG2) cells. *J. Agric. Food. Chem.* **2006**, *54*, 1621–1626.
21. Bursill, C.A.; Abbey, M.; Roach, P.D. A green tea extract lowers plasma cholesterol by inhibiting cholesterol synthesis and upregulating the LDL receptor in the cholesterol-fed rabbit. *Atherosclerosis* **2007**, *193*, 86–93.
22. Bursill, C.; Roach, P.D.; Bottema, C.D.; Pal, S. Green tea upregulates the low-density lipoprotein receptor through the sterol-regulated element binding Protein in HepG2 liver cells. *J. Agric. Food Chem.* **2001**, *49*, 5639–5645.
23. Nantz, M.P.; Rowe, C.A.; Bukowski, J.F.; Percival, S.S. Standardized capsule of Camellia sinensis lowers cardiovascular risk factors in a randomized, double-blind, placebo-controlled study. *Nutrition* **2009**, *25*, 147–154.
24. Stalmach, A.; Troufflard, S.; Serafini, M.; Crozier, A. Absorption, metabolism and excretion of Choladi green tea flavan-3-ols by humans. *Mol. Nutr. Food Res.* **2009**, *53* (Suppl. S1), S44–S53.
25. Fu, T.; Liang, J.; Han, G.; Lv, L.; Li, N. Simultaneous determination o fthe major active components of tea polyphenols in rat plasma by a simple and specific HPLC assay. *J. Chromatogr. B* **2008**, *875*, 363–367.
26. Ullmann, U.; Haller, J.; Bakker, G.C.; Brink, E.J.; Weber, P. Epigallocatechin gallate (EGCG) (TEAVIGO) does not impair nonhaem-iron absorption in man. *Phytomedicine* **2005**, *12*, 410–415.
27. Ullmann, U.; Haller, J.; Decourt, J.P.; Girault, N.; Girault, J.; Richard-Caudron, A.S.; Pineau, B.; Weber, P. A single ascending dose study of epigallocatechin gallate in healthy volunteers. *J. Int. Med. Res.* **2003**, *31*, 88–101.
28. Wolfram, S.; Raederstorff, D.; Wang, Y.; Teixeira, S.R.; Elste, V.; Weber, P. TEAVIGO (epigallocatechin gallate) supplementation prevents obesity in rodents by reducing adipose tissue mass. *Ann. Nutr. Metab.* **2005**, *49*, 54–63.
29. Roach, P.D.; Le, V.H.; Naumovski, N.; Blades, B. Despite its instability, epigallocatechin gallate lowers serum cholesterol in the hypercholesterolaemic rabbit. *Atheroscler. Suppl.* **2006**, *7*, 441.
30. Wang, R.; Zhou, W. Stability of tea catechins in the breadmaking process. *J. Agric. Food Chem.* **2004**, *52*, 8224–8229.
31. Wang, R.; Zhou, W.; Jiang, X. Reaction kinetics of degradation and epimerization of epigallocatechin gallate (EGCG) in aqueous system over a wide temperature range. *J. Agric. Food Chem.* **2008**, *56*, 2694–2701.
32. Wang, R.; Zhou, W.; Wen, R.A. Kinetic study of the thermal stability of tea catechins in aqueous systems using a microwave reactor. *J. Agric. Food Chem.* **2006**, *54*, 5924–5932.
33. Chen, Z.; Zhu, Q.Y.; Tsang, D.; Huang, Y. Degradation of green tea catechins in tea drinks. *J. Agric. Food Chem.* **2001**, *49*, 477–482.
34. Proniuk, S.; Liederer, B.M.; Blanchard, J. Preformulation study of epigallocatechin gallate, a promising antioxidant for topical skin cancer prevention. *J. Pharm. Sci.* **2002**, *91*, 111–116.
35. Dube, A.; Ng, K.; Nicolazzo, J.A.; Larson, I. Effective use of reducing agents and nanoparticle encapsulation in stabilizing catechins in alkaline solution. *Food Chem.* **2010**, *122*, 662–667.
36. Laparra, J.M.; Sanz, Y. Interactions of gut microbiota with functional food components and nutraceuticals. *Pharmacol. Res.* **2010**, *61*, 219–225.

37. Hirun, S.; Roach, P. A study of stability of (−)-epigallocatechin gallate (EGCG) from green tea in a frozen product. *Int. Food Res. J.* **2011**, *18*, 1261–1264.
38. Green, R.J.; Murphy, A.S.; Schulz, B.; Watkins, B.A.; Ferruzzi, M.G. Common tea formulations modulate *in vitro* digestive recovery of green tea catechins. *Mol. Nutr. Food Res.* **2007**, *51*, 1152–1162.
39. Mereles, D.; Hunstein, W. Epigallocatechin-3-gallate (EGCG) for clinical trials: More pitfalls than promises? *Int. J.Mol. Sci.* **2011**, *12*, 5592–5603.
40. Lee, M.J.; Maliakal, P.; Chen, L.; Meng, X.; Bondoc, F.Y.; Prabhu, S.; Lambert, G.; Mohr, S.; Yang, C.S. Pharmacokinetics of tea catechins after ingestion of green tea and (−)-epigallocatechin-3-gallate by humans: Formation of different metabolites and individual variability. *Cancer Epidemiol. Biomark. Prev.* **2002**, *11*, 1025–1032.
41. Chow, H.H.; Cai, Y.; Hakim, I.A.; Crowell, J.A.; Shahi, F.; Brooks, C.A.; Dorr, R.T.; Hara, Y.; Alberts, D.S. Pharmacokinetics and safety of green tea polyphenols after multiple-dose administration of epigallocatechin gallate and polyphenon E in healthy individuals. *Clin. Cancer Res.* **2003**, *9*, 3312–3319.
42. Chow, H.H.; Hakim, I.A.; Vining, D.R.; Crowell, J.A.; Ranger-Moore, J.; Chew, W.M.; Celaya, C.A.; Rodney, S.R.; Hara, Y.; Alberts, D.S. Effects of dosing condition on the oral bioavailability of green tea catechins after single-dose administration of polyphenon E in healthy individuals. *Clin. Cancer Res.* **2005**, *11*, 4627–4633.
43. Vuong, Q.V.; Golding, J.B.; Nguyen, M.; Roach, P.D. Extraction and isolation of catechins from tea. *J. Sep. Sci.* **2010**, *33*, 3415–3428.
44. Vuong, Q.V.; Golding, J.B.; Stathopoulos, C.E.; Nguyen, M.H.; Roach, P.D. Optimizing conditions for the extraction of catechins from green tea using hot water. *J. Sep. Sci.* **2011**, *34*, 3099–3106.
45. FDA. Bioavailability and bioequivalence studies for orally administered drug products–General considerations. In *Guidance for Industry*, 1st ed.; Food and Drug Administration, Ed.; U.S. Department of Health and Human Services: Rockville, MD, USA, 2003; p. 23.
46. Maron, D.J.; Lu, G.P.; Cai, N.S.; Wu, Z.G.; Li, Y.H.; Chen, H.; Zhu, J.Q.; Jin, X.J.; Woulters, B.C.; Zhao, J. Cholesterol-lowering effect of a theaflin-enriched green tea extract. *Arch. Intern. Med.* **2003**, *163*, 1448–1453.
47. Adiyaman, A.; Verhoeff, R.; Lenders, J.W.; Deinum, J.; Thien, T. The position of the arm during blood pressure measurement in sitting position. *Blood Pressure Monit.* **2006**, *11*, 309–313.
48. Sauter, R.; Steinijans, V.W.; Diletti, E.; Bohm, A.; Schulz, H.U. Presentation of results from bioequivalence studies. *Int. J. Clin. Pharm.* **1992**, *30*, 233–256.
49. Welling, P.G. Effects of food on drug absorption. *Pharmacol. Ther.* **1989**, *43*, 425–441.
50. Spencer, J.P.; Schroeter, H.; Rechner, A.R.; Rice-Evans, C. Bioavailability of flavan-3-ols and procyanidins: Gastrointestinal tract influences and their relevance to bioactive forms *in vivo*. *Antioxid. Redox Signal.* **2001**, *3*, 1023–1039.
51. Holladay, J.W. Biofarmaceutics of orally ingested products. In *Handbook of Food-Drug Interactions*, McCabe, B.J., Wolfe, J.J., Frankel, E.H., Eds.; CRC Press LLC: Boca Raton, FL, USA, 2003.

52. Record, I.R.; Lane, J.M. Simulated intestinal digestion of green and black teas. *Food Chem.* **2001**, *73*, 481–486.
53. Fleisher, D.; Li, C.; Zhou, Y.; Pao, L.H.; Karim, A. Drug, meal and formulation interactions influencing drug absorption after oral administration. Clinical implications. *Clin. Pharmacokinet.* **1999**, *36*, 233–254.
54. Chen, L.; Lee, M.J.; Li, H.; Yang, C.S. Absorption, distribution, elimination of tea polyphenols in rats. *Drug Metab. Dispos.* **1997**, *25*, 1045–1050.
55. Spencer, J.P. Metabolism of tea flavonoids in the gastrointestinal tract. *J. Nutr.* **2003**, *133*, 3255S–3261S.
56. Ru, Q.; Yu, H.; Huang, Q. Encapsulation of epigallocatechin-3-gallate (EGCG) using oil-in-water (O/W) submicrometer emulsions stabilized by ι-carrageenan and β-lactoglobulin. *J. Agric. Food Chem.* **2010**, *58*, 10373–10381.
57. Fang, J.Y.; Hwang, T.L.; Huang, Y.L.; Fang, C.L. Enhancement of the transdermal delivery of catechins by liposomes incorporating anionic surfactants and ethanol. *Int. J. Pharm.* **2006**, *310*, 131–138.
58. Shutava, T.G.; Balkundi, S.S.; Lvov, Y.M. (−)-Epigallocatechin gallate/gelatin layer-by-layer assembled films and microcapsules. *J. Colloid Interface Sci.* **2009**, *330*, 276–283.

Permissions

All chapters in this book were first published in Antioxidants, by MDPI; hereby published with permission under the Creative Commons Attribution License or equivalent. Every chapter published in this book has been scrutinized by our experts. Their significance has been extensively debated. The topics covered herein carry significant findings which will fuel the growth of the discipline. They may even be implemented as practical applications or may be referred to as a beginning point for another development.

The contributors of this book come from diverse backgrounds, making this book a truly international effort. This book will bring forth new frontiers with its revolutionizing research information and detailed analysis of the nascent developments around the world.

We would like to thank all the contributing authors for lending their expertise to make the book truly unique. They have played a crucial role in the development of this book. Without their invaluable contributions this book wouldn't have been possible. They have made vital efforts to compile up to date information on the varied aspects of this subject to make this book a valuable addition to the collection of many professionals and students.

This book was conceptualized with the vision of imparting up-to-date information and advanced data in this field. To ensure the same, a matchless editorial board was set up. Every individual on the board went through rigorous rounds of assessment to prove their worth. After which they invested a large part of their time researching and compiling the most relevant data for our readers.

The editorial board has been involved in producing this book since its inception. They have spent rigorous hours researching and exploring the diverse topics which have resulted in the successful publishing of this book. They have passed on their knowledge of decades through this book. To expedite this challenging task, the publisher supported the team at every step. A small team of assistant editors was also appointed to further simplify the editing procedure and attain best results for the readers.

Apart from the editorial board, the designing team has also invested a significant amount of their time in understanding the subject and creating the most relevant covers. They scrutinized every image to scout for the most suitable representation of the subject and create an appropriate cover for the book.

The publishing team has been an ardent support to the editorial, designing and production team. Their endless efforts to recruit the best for this project, has resulted in the accomplishment of this book. They are a veteran in the field of academics and their pool of knowledge is as vast as their experience in printing. Their expertise and guidance has proved useful at every step. Their uncompromising quality standards have made this book an exceptional effort. Their encouragement from time to time has been an inspiration for everyone.

The publisher and the editorial board hope that this book will prove to be a valuable piece of knowledge for researchers, students, practitioners and scholars across the globe.

List of Contributors

Francisco Segovia Gómez
Chemical Engineering Department, Technical University of Catalonia, Avda. Diagonal 647, 08028 Barcelona, Spain

Sara Peiró Sánchez
Chemical Engineering Department, Technical University of Catalonia, Avda. Diagonal 647, 08028 Barcelona, Spain

Maria Gabriela Gallego Iradi
Chemical Engineering Department, Technical University of Catalonia, Avda. Diagonal 647, 08028 Barcelona, Spain

Nurul Aini Mohd Azman
Chemical Engineering Department, Technical University of Catalonia, Avda. Diagonal 647, 08028 Barcelona, Spain

María Pilar Almajano
Chemical Engineering Department, Antonio José de Sucre National Experimental Polytechnic University, Avenida Corpahuaico, 3001 Barquisimeto, Venezuela

Nurul Aini Mohd Azman
Chemical Engineering Department, Technical University of Catalonia, Avda. Diagonal 647, 08028 Barcelona, Spain
Chemical and Natural Resources Engineering Faculty, University Malaysia Pahang, LebuhrayaTunRazak, Pahang 26300, Malaysia

Francisco Segovia
Chemical Engineering Department, Technical University of Catalonia, Avda. Diagonal 647, 08028 Barcelona, Spain

Xavier Martínez-Farré
Agro-Food Engineering and Biotechnology Department, EsteveTerradas, 8, 08860 Castelldefels, Spain

Emilio Gil
Agro-Food Engineering and Biotechnology Department, EsteveTerradas, 8, 08860 Castelldefels, Spain

María Pilar Almajano
Chemical Engineering Department, Technical University of Catalonia, Avda. Diagonal 647, 08028 Barcelona, Spain

Azizah Abdul Hamid
Department of Food Science, Faculty of Food Science and Technology, Universiti Putra Malaysia, UPM Serdang, Selangor 43400, Malaysia

Mohd Sabri Pak Dek
Department of Food Science, Faculty of Food Science and Technology, Universiti Putra Malaysia, UPM Serdang, Selangor 43400, Malaysia

Chin Ping Tan
Department of Food Technology, Faculty of Food Science and Technology, Universiti Putra Malaysia, UPM Serdang, Selangor 43400, Malaysia

Mohd Asraf Mohd Zainudin
Department of Food Science, Faculty of Food Science and Technology, Universiti Putra Malaysia, UPM Serdang, Selangor 43400, Malaysia

Evelyn Koh Wee Fang
Department of Food Science, Faculty of Food Science and Technology, Universiti Putra Malaysia, UPM Serdang, Selangor 43400, Malaysia

Shyamchand Mayengbam
Department of Human Nutritional Sciences, University of Manitoba, Winnipeg, MB R3T 2N2, Canada

Ayyappan Aachary
Department of Human Nutritional Sciences, University of Manitoba, Winnipeg, MB R3T 2N2, Canada
Richardson Centre for Functional Foods and Nutraceuticals & Department of Human Nutritional Sciences, University of Manitoba, Winnipeg, MB R3T 2N2, Canada

Usha Thiyam-Holländer
Department of Human Nutritional Sciences, University of Manitoba, Winnipeg, MB R3T 2N2, Canada
Richardson Centre for Functional Foods and Nutraceuticals & Department of Human Nutritional Sciences, University of Manitoba, Winnipeg, MB R3T 2N2, Canada

Sara Peiró
Department of Health Microbiology and Parasitology, Faculty of Pharmacy, Barcelona University, Avenue Joan XXIII s/n, 08028 Barcelona, Spain
IRIS-Innovació I Recerca Industrial i Sostenible, Avda. Carl Friedrich Gauss nº11, 08860 Barcelona, Spain
Chemical Engineering Department, Technical University of Catalonia, Avda Diagonal 647, 08028 Barcelona, Spain

Michael H. Gordon
Department of Food and Nutritional Science, University of Reading, Whiteknights, P.O. Box 226, Reading RG6 6AP, UK

Mónica Blanco
Department of Applied Mathematics III, Technical University of Catalonia, ESAB, Campus del Baix Llobregat, Esteve Terradas 8, 08860 Barcelona, Spain

Francisca Pérez-Llamas
Department of Physiology and Pharmacology, School of Biology, University of Murcia, Campus de Espinardo, 30100 Murcia, Spain

Francisco Segovia
Chemical Engineering Department, Technical University of Catalonia, Avda Diagonal 647, 08028 Barcelona, Spain

María Pilar Almajano
Chemical Engineering Department, Technical University of Catalonia, Avda Diagonal 647, 08028 Barcelona, Spain

Chloe D. Goldsmith
School of Environmental & Life Sciences, University of Newcastle, Ourimbah, NSW 2258, Australia

Quan V. Vuong
School of Environmental & Life Sciences, University of Newcastle, Ourimbah, NSW 2258, Australia

Costas E. Stathopoulos
Faculty of Bioscience Engineering, Ghent University Global Campus, Incheon 406-840, South Korea

Paul D. Roach
School of Environmental & Life Sciences, University of Newcastle, Ourimbah, NSW 2258, Australia

Christopher J. Scarlett
School of Environmental & Life Sciences, University of Newcastle, Ourimbah, NSW 2258, Australia

Mirian Pateiro
Centro Tecnológico de la Carne de Galicia, Rúa Galicia No. 4, Parque Tecnológico de Galicia, San Cibrao das Viñas, 32900 Ourense, Spain

Roberto Bermúdez
Centro Tecnológico de la Carne de Galicia, Rúa Galicia No. 4, Parque Tecnológico de Galicia, San Cibrao das Viñas, 32900 Ourense, Spain

José Manuel Lorenzo
Centro Tecnológico de la Carne de Galicia, Rúa Galicia No. 4, Parque Tecnológico de Galicia, San Cibrao das Viñas, 32900 Ourense, Spain

Daniel Franco
Centro Tecnológico de la Carne de Galicia, Rúa Galicia No. 4, Parque Tecnológico de Galicia, San Cibrao das Viñas, 32900 Ourense, Spain

Bertrand Matthäus
Max Rubner-Institut, Federal Research Institute for Nutrition and Food, Schützenberg 12, 32756 Detmold, Germany

Mehmet Musa Özcan
Department of Food Engineering, Faculty of Agriculture, University of Selcuk, 42079 Konya, Turkey

Nurul Aini Mohd Azman
Chemical Engineering Department, Technical University of Catalonia, Av. Diagonal 647, 08028 Barcelona, Spain
Chemical and Natural Resources Engineering Faculty, University Malaysia Pahang, Lebuhraya Tun Razak, 26300 Pahang, Malaysia

Maria Gabriela Gallego
Chemical Engineering Department, Technical University of Catalonia, Av. Diagonal 647, 08028 Barcelona, Spain

Luis Juliá
Química Biològica i Modelització Molecular, Institut de Química Avançada de Catalunya (CSIC), Jordi Girona 18-26, 08034 Barcelona, Spain

Lluis Fajari
Química Biològica i Modelització Molecular, Institut de Química Avançada de Catalunya (CSIC), Jordi Girona 18-26, 08034 Barcelona, Spain

MaríaPilar Almajano
Chemical Engineering Department, Technical University of Catalonia, Av. Diagonal 647, 08028 Barcelona, Spain

Clara Albano
Institute of Sciences of Food Production (ISPA), CNR, Lecce Unit, 73100 Lecce, Italy

Carmine Negro
Department of Biological and Environmental Sciences and Technologies (DISTeBA), Salento University, 73100 Lecce, Italy

Noemi Tommasi
Institute of Sciences of Food Production (ISPA), CNR, Lecce Unit, 73100 Lecce, Italy

Carmela Gerardi
Institute of Sciences of Food Production (ISPA), CNR, Lecce Unit, 73100 Lecce, Italy

Giovanni Mita
Institute of Sciences of Food Production (ISPA), CNR, Lecce Unit, 73100 Lecce, Italy

Antonio Miceli
Department of Biological and Environmental Sciences and Technologies (DISTeBA), Salento University, 73100 Lecce, Italy

Luigi De Bellis
Department of Biological and Environmental Sciences and Technologies (DISTeBA), Salento University, 73100 Lecce, Italy

Federica Blando
Institute of Sciences of Food Production (ISPA), CNR, Lecce Unit, 73100 Lecce, Italy

Rodrigo A. Contreras
Laboratorio de Fisiología y Biotecnología Vegetal, Departamento de Biología, Facultad de Química y Biología, Universidad de Santiago de Chile. L. B. O'Higgins Ave. #3363, Estación Central, Santiago of Chile, 9170022, Chile

Hans Köhler
Laboratorio de Fisiología y Biotecnología Vegetal, Departamento de Biología, Facultad de Química y Biología, Universidad de Santiago de Chile. L. B. O'Higgins Ave. #3363, Estación Central, Santiago of Chile, 9170022, Chile

Marisol Pizarro
Laboratorio de Fisiología y Biotecnología Vegetal, Departamento de Biología, Facultad de Química y Biología, Universidad de Santiago de Chile. L. B. O'Higgins Ave. #3363, Estación Central, Santiago of Chile, 9170022, Chile

Gustavo E. Zúñiga
Laboratorio de Fisiología y Biotecnología Vegetal, Departamento de Biología, Facultad de Química y Biología, Universidad de Santiago de Chile. L. B. O'Higgins Ave. #3363, Estación Central, Santiago of Chile, 9170022, Chile

María Isabel Alarcón-Flores
Department of Chemistry and Physics (Analytical Chemistry Area), Research Centre for Agricultural and Food Biotechnology (BITAL), University of Almería, Agrifood Campus of International Excellence, ceiA3, E-04120 Almería, Spain

Roberto Romero-González
Department of Chemistry and Physics (Analytical Chemistry Area), Research Centre for Agricultural and Food Biotechnology (BITAL), University of Almería, Agrifood Campus of International Excellence, ceiA3, E-04120 Almería, Spain

José Luis Martínez Vidal
Department of Chemistry and Physics (Analytical Chemistry Area), Research Centre for Agricultural and Food Biotechnology (BITAL), University of Almería, Agrifood Campus of International Excellence, ceiA3, E-04120 Almería, Spain

Antonia Garrido Frenich
Department of Chemistry and Physics (Analytical Chemistry Area), Research Centre for Agricultural and Food Biotechnology (BITAL), University of Almería, Agrifood Campus of International Excellence, ceiA3, E-04120 Almería, Spain

Nenad Naumovski
School of Public Health and Nutrition, University of Canberra, Canberra 2601, ACT, Australia
School of Environmental & Life Sciences, University of Newcastle, Ourimbah 2258, NSW, Australia

Barbara L. Blades
School of Environmental & Life Sciences, University of Newcastle, Ourimbah 2258, NSW, Australia

Paul D. Roach
School of Environmental & Life Sciences, University of Newcastle, Ourimbah 2258, NSW, Australia

www.ingramcontent.com/pod-product-compliance
Lightning Source LLC
Chambersburg PA
CBHW070152240326
41458CB00126B/4415
9781682862001